VOLUME THREE HUNDRED AND FIFTY ONE

International Review of
CELL AND MOLECULAR BIOLOGY

Cell Death Regulation in Health and Disease - Part A

INTERNATIONAL REVIEW OF CELL AND MOLECULAR BIOLOGY

VOLUME THREE HUNDRED AND FIFTY ONE

INTERNATIONAL REVIEW OF CELL AND MOLECULAR BIOLOGY

Cell Death Regulation in Health and Disease - Part A

Edited by

JOHAN K.E. SPETZ
Harvard T.H. Chan School of Public Health, Boston, MA, USA

LORENZO GALLUZZI
Weill Cornell Medical College, New York, NY, United States

ELSEVIER

ACADEMIC PRESS
An imprint of Elsevier

Academic Press is an imprint of Elsevier
50 Hampshire Street, 5th Floor, Cambridge, MA 02139, United States
525 B Street, Suite 1650, San Diego, CA 92101, United States
The Boulevard, Langford Lane, Kidlington, Oxford OX5 1GB, United Kingdom
125 London Wall, London, EC2Y 5AS, United Kingdom

First edition 2020

ISBN: 978-0-12-820133-6
ISSN: 1937-6448

For information on all Academic Press publications
visit our website at https://www.elsevier.com/books-and-journals

Publisher: Zoe Kruze
Acquisitions Editor: Ashlie M. Jackman
Editorial Project Manager: Leticia M. Lima
Production Project Manager: Denny Mansingh
Cover Designer: Matthew Limbert

Typeset by SPi Global, India

Contents

Contributors

Geert Bultynck
KU Leuven, Laboratory of Molecular and Cellular Signaling, Department of Cellular and Molecular Medicine and Leuven Kanker Instituut (LKI), Leuven, Belgium

Zhipeng Cao
Department of Biochemistry and Genetics, La Trobe University, Bundoora, VIC, Australia

Hector Flores-Romero
Interfaculty Institute of Biochemistry, Eberhard-Karls-Universität Tübingen, Tübingen, Germany

Ana J. García-Sáez
Interfaculty Institute of Biochemistry, Eberhard-Karls-Universität Tübingen, Tübingen, Germany

Tina Garofalo
Department of Experimental Medicine, Sapienza University, Rome, Italy

Jason Andrew Glab
Department of Biochemistry and Genetics, La Trobe University, Bundoora, VIC, Australia

Elisabetta Iessi
Center for Gender-Specific Medicine, Oncology Unit, Istituto Superiore di Sanità, Rome, Italy

Albert Jeltsch
Department of Biochemistry, University of Stuttgart, Institute of Biochemistry and Technical Biochemistry, Stuttgart, Germany

Gelina S. Kopeina
Faculty of Medicine, MV Lomonosov Moscow State University, Moscow, Russia

Walter Malorni
Center for Gender-Specific Medicine, Oncology Unit, Istituto Superiore di Sanità; Department of Biology, University of Rome Tor Vergata, Rome, Italy

Valeria Manganelli
Department of Experimental Medicine, Sapienza University, Rome, Italy

Matteo Marconi
Center for Gender-Specific Medicine, Oncology Unit, Istituto Superiore di Sanità, Rome, Italy

Paola Matarrese
Center for Gender-Specific Medicine, Oncology Unit, Istituto Superiore di Sanità, Rome, Italy

Jan B. Parys
KU Leuven, Laboratory of Molecular and Cellular Signaling, Department of Cellular and Molecular Medicine and Leuven Kanker Instituut (LKI), Leuven, Belgium

Nikolay V. Pervushin
Faculty of Medicine, MV Lomonosov Moscow State University, Moscow, Russia

Hamsa Puthalakath
Department of Biochemistry and Genetics, La Trobe University, Bundoora, VIC, Australia

Markus Rehm
University of Stuttgart, Institute of Cell Biology and Immunology; University of Stuttgart, Stuttgart Research Center Systems Biology; University of Stuttgart, Stuttgart Centre for Simulation Science, Stuttgart, Germany

Uris Ros
Interfaculty Institute of Biochemistry, Eberhard-Karls-Universität Tübingen, Tübingen, Germany

Nicolas Rosa
KU Leuven, Laboratory of Molecular and Cellular Signaling, Department of Cellular and Molecular Medicine and Leuven Kanker Instituut (LKI), Leuven, Belgium

Viacheslav V. Senichkin
Faculty of Medicine, MV Lomonosov Moscow State University, Moscow, Russia

Flore Sneyers
KU Leuven, Laboratory of Molecular and Cellular Signaling, Department of Cellular and Molecular Medicine and Leuven Kanker Instituut (LKI), Leuven, Belgium

Maurizio Sorice
Department of Experimental Medicine, Sapienza University, Rome, Italy

Daniela Stöhr
University of Stuttgart, Institute of Cell Biology and Immunology; University of Stuttgart, Stuttgart Research Center Systems Biology, Stuttgart, Germany

Boris Zhivotovsky
Faculty of Medicine, MV Lomonosov Moscow State University, Moscow, Russia; Institute of Environmental Medicine, Karolinska Institutet, Stockholm, Sweden

Preface: Life through death—Key role of cellular suicide for colonial and organismal homeostasis

Johan Spetz[a,b,*], Lorenzo Galluzzi[c,d,e,f,g,*]
[a]John B. Little Center for Radiation Sciences, Harvard School of Public Health, Boston, MA, United States
[b]Laboratory of Systems Pharmacology, Harvard Program in Therapeutic Science, Department of Systems Biology, Harvard Medical School, Boston, MA, United States
[c]Department of Radiation Oncology, Weill Cornell Medical College, New York, NY, United States
[d]Sandra and Edward Meyer Cancer Center, New York, NY, United States
[e]Caryl and Israel Englander Institute for Precision Medicine, New York, NY, United States
[f]Department of Dermatology, Yale School of Medicine, New Haven, CT, United States
[g]Université de Paris, Paris, France
[*]Corresponding authors: e-mail address: spetz@hsph.harvard.edu; deadoc80@gmail.com

The term "regulated cell death" (RCD) is commonly employed to indicate a form of cell death that is controlled by a dedicated molecular machinery, and hence can be modulated (at least to some degree) with specific pharmacological or genetic interventions (Galluzzi et al., 2015, 2018a). RCD can occur in the context of two diametrically opposed scenarios. On the one hand, RCD can represent the ultimate response of cells to microenvironmental perturbations that are excessively intense or prolonged for cells to adapt and recover homeostasis (Spetz et al., 2019). On the other hand, RCD can contribute to purely physiological programs of development and adult tissue homeostasis (Bock and Tait, 2020). The expression "programmed cell death" (PCD) is often used to refer to these latter variants of RCD that occur at a predetermined point in time, irrespective of potentially cytotoxic alterations of the intracellular or extracellular microenvironment (Fuchs and Steller, 2011, 2015).

For multicellular organisms, the advantages provided by genetically encoded programs for cell death are obvious. On the one hand, PCD can be harnessed as a morphogenetic mechanism that shapes first (and maintains then) tissue and organ architecture (Tuzlak et al., 2016). On the other hand, stress-driven RCD constitutes a robust mechanism for the elimination first (and replacement then) of postmitotic cells that are damaged beyond recovery, and hence functionless and/or prone to constitute a threat for the entire organism (e.g., prone to malignant transformation) (Cao and Tait, 2018). Less obvious are the advantages provided by RCD to unicellular organisms, such as bacteria and yeast. However, both prokaryotes and unicellular eukaryotes have rudimentary RCD machineries that allow individual cells to suicide (Carmona-Gutierrez et al., 2018). This happens, as an example, when nutrients for the entire colony are limiting, and the demise of old cells increases the chances of younger, fitter cells to survive dwindling nutrient levels and ultimately reconstitute the colony (Fabrizio et al., 2004; Herker et al., 2004).

In higher eukaryotes, the molecular machinery for RCD has considerably diversified (Galluzzi et al., 2017a; Stockwell et al., 2017; Vanden Berghe et al., 2014; Weinlich et al., 2017) to enable variants of RCD that (1) occur according to diverse kinetics (Buque et al., 2019; Grootjans et al., 2016) and (2) are accompanied by the emission of specific signals that engage or suppress immune reactions (Galluzzi et al., 2017b; Rodriguez-Ruiz et al., 2019; Vanpouille-Box et al., 2018). Thus, dying mammalian cells can deliver signals in the form of damage-associated molecular patterns (DAMPs) or cytokines that are responsible for immune modulation at sites of active RCD (Galluzzi et al., 2018b; Green et al., 2009; Krysko et al., 2012). Many of the enzymes that have long been considered as pure RCD regulators, such as apoptotic caspases, have now been shown to mediate multipronged immunomodulatory functions (Galluzzi et al., 2016; Rodriguez-Ruiz et al., 2019; Rongvaux et al., 2014; White et al., 2014), and to support tissue regeneration, at least in some settings (Cheng et al., 2019; Huang et al., 2011; Ichim and Tait, 2016).

Given the key role of RCD in the maintenance of organismal homeostasis and the control of immune responses against dying cells, it is not surprising that deregulated RCD contributes to the pathology of a variety of human disorders. On the one hand, uncontrolled RCD associated with inflammatory reactions is involved in the pathogenesis of ischemic, cardiac, renal and neurodegenerative conditions, among others (Del Re et al., 2019; Fang et al., 2019; Hetz and Saxena, 2017; Kers et al., 2016; Quinlan et al., 2017). On the other hand, insufficient RCD (as such or in the context of immunosuppressive signaling) is etiologically linked to autoimmunity, proliferative syndromes, and cancer (Cao and Tait, 2018; Czabotar et al., 2014; Khan and Ghazanfar, 2018; Nagata and Tanaka, 2017; Singh et al., 2019; Spetz et al., 2019). In summary, RCD in its various forms stands out as a key mechanism for the preservation of colonial and organismal homeostasis, thus constituting an attractive target for the development of pharmacological interventions against a plethora of human disorders for more than 2 decades (Degterev et al., 2019; Huang and Ichikawa, 2008; Naderer and Fulcher, 2018; Valentin et al., 2018). As of today, however, very few drugs that target the molecular machinery for RCD have successfully entered clinical development, including the BCL2 inhibitor venetoclax (currently approved for the therapy of some hematological tumors) (Jain et al., 2019), and the pan-caspase inhibitor emricasan (which was recently granted fast track status by the US Food and Drug Administration for the therapy of cirrhosis) (Mehta et al., 2018; Shiffman et al., 2019). It is tempting to speculate, yet remains to be formally demonstrated, that translating RCD modulation from the bench to the bedside has been difficult because RCD has been often investigated (especially when the molecular machinery for RCD has begun to emerge) irrespective of its immunological context.

In *Cell Death Regulation in Health and Disease* (a special issue of the *International Review of Cell and Molecular Biology* series) a group of world-renowned experts in the field gathered to discuss mechanistic features and physiopathological implications of RCD in a variety of model organisms. Each of the chapters of *Cell Death Regulation in Health and Disease* addresses a specific aspect of cell death biology, ranging from RCD control in prokaryotes and lower eukaryotes to the contribution of cell death deregulation to various human disorders. The book (which comes in three separate volumes, i.e., parts A, B, and C) is equally addressed to experienced investigators who may want to acquire in-depth insights into a specific aspect of cell death biology, as well as to students or neophytes who are moving their first steps in this broad field of research.

Acknowledgments

L.G. is supported by a Breakthrough Level 2 grant from the US Department of Defense (DoD), Breast Cancer Research Program (BRCP) (#BC180476P1), by the 2019 Laura Ziskin Prize in Translational Research (#ZP-6177, PI: Formenti) from the Stand up to Cancer (SU2C), by a Mantle Cell Lymphoma Research Initiative (MCL-RI, PI: Chen-Kiang) grant from the Leukemia and Lymphoma Society (LLS), by a startup grant from the Department of Radiation Oncology at Weill Cornell Medicine (New York, USA), by a Rapid Response Grant from the Functional Genomics Initiative (New York, USA), by industrial collaborations with Lytix (Oslo, Norway) and Phosplatin (New York, USA), and by donations from Phosplatin (New York, USA), the Luke Heller TECPR2 Foundation (Boston, USA) and Sotio a.s. (Prague, Czech Republic).

Conflict of interest statement

LG received consulting fees from OmniSEQ, Astra Zeneca, Inzen and the Luke Heller TECPR2 Foundation, and he is member of the Scientific Advisory Committee of Boehringer Ingelheim, The Longevity Labs and OmniSEQ.

References

Bock, F.J., Tait, S.W.G., 2020. Mitochondria as multifaceted regulators of cell death. Nat. Rev. Mol. Cell Biol. 21, 85–100.

Buque, A., et al., 2019. Apoptotic caspases cut down the immunogenicity of radiation. Onco. Targets. Ther. 8 (11), e1655364.

Cao, K., Tait, S.W.G., 2018. Apoptosis and cancer: force awakens, phantom menace, or both? Int. Rev. Cell Mol. Biol. 337, 135–152.

Carmona-Gutierrez, D., et al., 2018. Guidelines and recommendations on yeast cell death nomenclature. Microb. Cell 5 (1), 4–31.

Cheng, J., et al., 2019. The caspase-3/PKCdelta/Akt/VEGF-A signaling pathway mediates tumor repopulation during radiotherapy. Clin. Cancer Res. 25 (12), 3732–3743.

Czabotar, P.E., et al., 2014. Control of apoptosis by the BCL-2 protein family: implications for physiology and therapy. Nat. Rev. Mol. Cell Biol. 15 (1), 49–63.

Degterev, A., et al., 2019. Targeting RIPK1 for the treatment of human diseases. Proc. Natl. Acad. Sci. U. S. A. 116 (20), 9714–9722.

Del Re, D.P., et al., 2019. Fundamental mechanisms of regulated cell death and implications for heart disease. Physiol. Rev. 99 (4), 1765–1817.

Fabrizio, P., et al., 2004. Superoxide is a mediator of an altruistic aging program in Saccharomyces cerevisiae. J. Cell Biol. 166 (7), 1055–1067.

Fang, X., et al., 2019. Ferroptosis as a target for protection against cardiomyopathy. Proc. Natl. Acad. Sci. U. S. A. 116 (7), 2672–2680.

Fuchs, Y., Steller, H., 2011. Programmed cell death in animal development and disease. Cell 147 (4), 742–758.

Fuchs, Y., Steller, H., 2015. Live to die another way: modes of programmed cell death and the signals emanating from dying cells. Nat. Rev. Mol. Cell Biol. 16 (6), 329–344.

Galluzzi, L., et al., 2015. Essential versus accessory aspects of cell death: recommendations of the NCCD 2015. Cell Death Differ. 22 (1), 58–73.

Galluzzi, L., et al., 2016. Caspases connect cell-death signaling to organismal homeostasis. Immunity 44 (2), 221–231.

Galluzzi, L., et al., 2017a. Necroptosis: mechanisms and relevance to disease. Annu. Rev. Pathol. 12, 103–130.

Galluzzi, L., et al., 2017b. Immunogenic cell death in cancer and infectious disease. Nat. Rev. Immunol. 17 (2), 97–111.

Galluzzi, L., et al., 2018a. Molecular mechanisms of cell death: recommendations of the nomenclature committee on cell death 2018. Cell Death Differ. 25 (3), 486–541.

Galluzzi, L., et al., 2018b. Linking cellular stress responses to systemic homeostasis. Nat. Rev. Mol. Cell Biol. 19 (11), 731–745.

Green, D.R., et al., 2009. Immunogenic and tolerogenic cell death. Nat. Rev. Immunol. 9 (5), 353–363.

Grootjans, S., et al., 2016. A real-time fluorometric method for the simultaneous detection of cell death type and rate. Nat. Protoc. 11 (8), 1444–1454.

Herker, E., et al., 2004. Chronological aging leads to apoptosis in yeast. J. Cell Biol. 164 (4), 501–507.

Hetz, C., Saxena, S., 2017. ER stress and the unfolded protein response in neurodegeneration. Nat. Rev. Neurol. 13 (8), 477–491.

Huang, D., Ichikawa, K., 2008. Drug discovery targeting apoptosis. Nat. Rev. Drug Discov. 7 (12), 1041.

Huang, Q., et al., 2011. Caspase 3-mediated stimulation of tumor cell repopulation during cancer radiotherapy. Nat. Med. 17 (7), 860–866.

Ichim, G., Tait, S.W., 2016. A fate worse than death: apoptosis as an oncogenic process. Nat. Rev. Cancer 16 (8), 539–548.

Jain, N., et al., 2019. Ibrutinib and venetoclax for first-line treatment of CLL. N. Engl. J. Med. 380 (22), 2095–2103.
Kers, J., et al., 2016. An overview of pathways of regulated necrosis in acute kidney injury. Semin. Nephrol. 36 (3), 139–152.
Khan, U., Ghazanfar, H., 2018. T lymphocytes and autoimmunity. Int. Rev. Cell Mol. Biol. 341, 125–168.
Krysko, D.V., et al., 2012. Immunogenic cell death and DAMPs in cancer therapy. Nat. Rev. Cancer 12 (12), 860–875.
Mehta, G., et al., 2018. A placebo-controlled, multicenter, double-blind, phase 2 randomized trial of the pan-caspase inhibitor emricasan in patients with acutely decompensated cirrhosis. J. Clin. Exp. Hepatol. 8 (3), 224–234.
Naderer, T., Fulcher, M.C., 2018. Targeting apoptosis pathways in infections. J. Leukoc. Biol. 103 (2), 275–285.
Nagata, S., Tanaka, M., 2017. Programmed cell death and the immune system. Nat. Rev. Immunol. 17 (5), 333–340.
Quinlan, S., et al., 2017. MicroRNAs in neurodegenerative diseases. Int. Rev. Cell Mol. Biol. 334, 309–343.
Rodriguez-Ruiz, M.E., et al., 2019. Apoptotic caspases inhibit abscopal responses to radiation and identify a new prognostic biomarker for breast cancer patients. Onco. Targets. Ther. 8 (11), e1655964.
Rongvaux, A., et al., 2014. Apoptotic caspases prevent the induction of type I interferons by mitochondrial DNA. Cell 159 (7), 1563–1577.
Shiffman, M., et al., 2019. Randomised clinical trial: emricasan versus placebo significantly decreases ALT and caspase 3/7 activation in subjects with non-alcoholic fatty liver disease. Aliment. Pharmacol. Ther. 49 (1), 64–73.
Singh, R., et al., 2019. Regulation of apoptosis in health and disease: the balancing act of BCL-2 family proteins. Nat. Rev. Mol. Cell Biol. 20 (3), 175–193.
Spetz, J., et al., 2019. T cells and regulated cell death: kill or be killed. Int. Rev. Cell Mol. Biol. 342, 27–71.
Stockwell, B.R., et al., 2017. Ferroptosis: a regulated cell death nexus linking metabolism, redox biology, and disease. Cell 171 (2), 273–285.
Tuzlak, S., et al., 2016. Interrogating the relevance of mitochondrial apoptosis for vertebrate development and postnatal tissue homeostasis. Genes Dev. 30 (19), 2133–2151.
Valentin, R., et al., 2018. The rise of apoptosis: targeting apoptosis in hematologic malignancies. Blood 132 (12), 1248–1264.
Vanden Berghe, T., et al., 2014. Regulated necrosis: the expanding network of non-apoptotic cell death pathways. Nat. Rev. Mol. Cell Biol. 15 (2), 135–147.
Vanpouille-Box, C., et al., 2018. Cytosolic DNA sensing in organismal tumor control. Cancer Cell 34 (3), 361–378.
Weinlich, R., et al., 2017. Necroptosis in development, inflammation and disease. Nat. Rev. Mol. Cell Biol. 18 (2), 127–136.
White, M.J., et al., 2014. Apoptotic caspases suppress mtDNA-induced STING-mediated type I IFN production. Cell 159 (7), 1549–1562.

CHAPTER ONE

Bcl-2 family proteins, beyond the veil

Jason Andrew Glab, Zhipeng Cao, Hamsa Puthalakath*
Department of Biochemistry and Genetics, La Trobe University, Bundoora, VIC, Australia
*Corresponding author: e-mail address: h.puthalakath@latrobe.edu.au

Contents

Abstract

Apoptosis is an important part of both health and disease and is often regulated by the BCL-2 family of proteins. These proteins are either pro- or anti-apoptotic, existing in a delicate balance during homeostasis. They are best known for their role in regulating the activation of caspases and the execution of a cell in response to a variety of stimuli. However, it is often forgotten that these BCL-2 family proteins also have important roles to play in cell maintenance that are not associated with apoptosis. These include roles in regulating processes such as cell cycle progression, mitochondrial function, autophagy, intracellular calcium concentration, glucose and lipid metabolism, and the unfolded protein response. In addition to these established alternate functions, further discoveries are being made that have potential therapeutic benefits in diseases such as cancer. BOK, a BCL-2 family protein thought comparable to multidomain pro-apoptotic proteins BAX and BAK, has recently been identified as a key player in metabolism of and resistance to the commonly used chemotherapeutic 5-FU. As a result of such findings, which could see the potential use of BOK as a biomarker for 5-FU sensitivity or mimetic molecules as a resensitization strategy, new targets and mechanisms of pathology may arise from further investigation into the realm of alternate functions of BCL-2 family proteins.

International Review of Cell and Molecular Biology, Volume 351
ISSN 1937-6448
https://doi.org/10.1016/bs.ircmb.2019.12.001

1. Introduction

Apoptosis, a form of programmed cell death, is vital for the development and maintenance of healthy tissue in the body, as well as a key factor in many diseases. It is regulated by two main pathways; the receptor pathway (via extrinsic regulation) and the mitochondrial pathway (via intrinsic regulation). The mitochondrial apoptotic pathway is controlled by proteins of the B cell lymphoma 2 (BCL-2) family, which either instigate or prevent the activation of deadly caspases, depending on whether they belong to pro- or anti-apoptotic subgroups. The presence of proteins regulating programmed cell death was realized in *C. elegans*, where depletion of ced-3 and ced-4 resulted in the survival of cells that would otherwise die during development (Ellis and Horvitz, 1986). The importance such proteins in mammalian biology was realized in the late 1980s and early 1990s with the discovery of heightened expression of the *Bcl-2* gene in chronic lymphocytic leukemia variants and homology with the Epstein-Barr virus protein BHRF, which was later discovered to be a functional homolog of BCL-2 (Cleary et al., 1986; Hanada et al., 1993; Henderson et al., 1996). The role of BCL-2 in apoptosis was not identified until 1988, when it was demonstrated that increased expression of BCL-2 both in vitro and in vivo promoted growth and survival of cells (McDonnell et al., 1989; Vaux et al., 1988). The theory was further advanced, demonstrating its evolutionarily conserved status for regulating cell death through the expression of human BCL-2 in mutant *C. elegans* (Vaux et al., 1992). Further research delved deeper into its conservation, revealing the same role in *S. cerevisiae* (Kane et al., 1993). Following this initial understanding of BCL-2's role in regulating apoptosis, additional members of the family were identified, and their roles elucidated. This included additional anti-apoptotic members, such as BCL-xL and MCL-1, as well as pro-apoptotic members BIM, PUMA, BAX and BAK (Boise et al., 1993; Kozopas et al., 1993; Lindsten et al., 2000; O'Connor et al., 1998; Yu et al., 2001).

However, in addition to their roles in regulating apoptosis, the members of the BCL-2 family have also been found to be key players in entirely different cellular processes. Such varied alternate roles include regulation of mitochondrial processes, calcium homeostasis at the ER, metabolism, autophagy and the DNA damage response (summarized in Table 1). In the following sections of this review, we will be discussing in further detail these additional roles performed by the BCL-2 family proteins.

Table 1 Bcl-2 family members—divided into antiapoptotic, pro-apoptotic and multidomain pro-apoptotic—and their alternate nonapoptotic roles.

	Cell cycle regulation	Mitochondrial function	Autophagy	Calcium regulation	Glucose and lipid metabolism	Unfolded protein response	Nucleic acid synthesis
A1			★				
Bcl-2	★	★		★		★	
Bcl-B			★				
Bcl-W	★	★					
Bcl-xL	★		★	★		★	
Mcl-1	★	★	★			★	
Bad	★				★	★	
Bcl-G							
Bcl-Rambo		★	★				
Bid					★		
Bik							
Bim			★		★	★	
Bmf					★		
Bnip3					★		
Noxa			★		★		
Puma						★	
Bak				★		★	
Bax			★	★		★	
Bok	★					★	★

2. Cell cycle regulation

The cell cycle is required for the accurate replication of a cell's genetic material in preparation for its division (Alberts et al., 2002). This process is controlled by cyclin-dependent kinases (CDKs), which remain inactive until bound to by cyclin subunits (Barnum and O'Connell, 2014). Hence, regulation of cell cycle progression is also mediated by synthesis of cyclin and its degradation by ubiquitin-dependent proteolysis. Due to their role in removing aberrant or damaged cells, proteins regulating apoptosis have previously been linked with managing cell cycle progression (Field et al., 1996; Li et al., 1998). One such example is the tumor suppressor protein, p53, which is responsible for the transcriptional activation of both pro-apoptotic proteins (e.g., Puma, Noxa) and the cell cycle inhibitor p21 (el-Deiry et al., 1993; Nakano and Vousden, 2001; Oda et al., 2000). In addition to these apoptotic regulatory proteins, a number of BCL-2 family proteins have been found to play important roles in regulating cell cycle progression.

Myeloid Cell Leukemia 1 (MCL-1), discovered in the ML-1 cell line, is commonly known for its anti-apoptotic activity (Kozopas et al., 1993). However, it has also been shown to play a role in cell cycle regulation. MCL-1 was found to interact with proliferating cell nuclear antigen (PCNA), a protein that forms a homotrimer around the DNA to enable initiation of replication by DNA polymerase δ (Fujise et al., 2000; Jiang et al., 1995). This interaction was found to block DNA synthesis by preventing the binding of replication factor C (RF-C) to PCNA, which is required for its trimerization. MCL-1 has also been found to induce terminal mitosis and differentiation of neural precursor cells (NPCs), a process vital to brain development (Hasan et al., 2013). The cell cycle regulator $p27^{Kip1}$, which is responsible for the timing of this key mechanism in brain layer formation and size determination, was shown to potentially be stabilized by MCL-1 to enable its phosphorylation by Cdk5 and subsequent entry into terminal mitosis by the containing neuron. Other BCL-2 family members, namely BCL-2 and BCL-xL, have also been linked with cell cycle regulation through interaction with $p27^{Kip1}$ (Greider et al., 2002; Vairo et al., 2000). Both were found to prevent entry into the cell cycle in vitro, although the physiological relevance of this finding is yet to be determined. Further links between MCL-1 and cell cycle mechanisms have been made following mitotic arrest, which interlink with its anti-apoptotic activity (Millman and Pagano, 2011; Zhou et al., 2013).

As mentioned earlier, BCL-2 has also been identified as a regulator of cell cycle progression. This was first posed in 1996, when it was shown that BCL-2 expression could increase the length of the G1 phase by 30–60% (Mazel et al., 1996). As a result, cells would remain at the G1 to S phase transition control point and avoid initiating apoptosis. This was shown to be the result of BCL-2 activity in response to reduced intracellular ROS levels (Deng et al., 2003). Also found to regulate cell cycle entry are BAD and BCL-xL, with their heterodimerization resulting in bypass of the G0/G1 checkpoint (Chattopadhyay et al., 2001). This is in contrast to the roles of the pro-apoptotic proteins BAX and NOXA, both of which have displayed acceleratory effects on proliferation despite their more commonly known pro-apoptotic function (Brady et al., 1996; Knudson et al., 2001; Zhong et al., 2014). As such, the potential roles for BCL-2 family proteins in cell cycle control are likely yet to be fully understood, as indicated by conflicting roles suggested for BCL-W in spermatogenesis (Print et al., 1998; Yan et al., 2003). Pro-apoptotic family member BID has also been shown to regulate cell cycle entry and arrest in response to DNA damage in hematopoietic stem cells (HSCs), interacting with its receptor MTCH2 at the mitochondria to regulate ROS levels and cell quiescence (Kamer et al., 2005; Maryanovich et al., 2015; Zinkel et al., 2005). From this, we can conclude that the BCL-2 family proteins have a varied repertoire of roles to play in cell cycle regulation, of which our understanding is continuing to grow.

3. Mitochondrial function

Mitochondria are commonly known by the apt but clichéd label "powerhouse of the cell." However, in the context of the BCL-2 family, mitochondria are more famously known for their central role in the induction of apoptosis. As part of the intrinsic apoptotic pathway, initiated in response to various cellular stress stimuli, mitochondria play an essential role in the execution of a cell (Wang and Youle, 2009). Though the usual involvement of the BCL-2 family proteins in the function of the mitochondria is mediating the formation of pores in the outer membrane to initiate the activation of downstream caspase enzymes, they have been found to perform a number of other roles in the regulation of mitochondrial function.

BCL-2 family proteins have been found to play roles in both normal structural and functional features of mitochondria. Shown to be important to the structural integrity was the anti-apoptotic protein BCL-W, which was found to play a key role in the morphology of mitochondria in the dendrites of Purkinje cells (Liu and Shio, 2008). Knockout of *Bcl-w* from these cells resulted in elongated mitochondria, in turn resulting in dysfunctional mitochondrial fission. Another anti-apoptotic protein, MCL-1, was also found to influence mitochondrial morphology (Wang et al., 2013). The absence of *Mcl-1* induced morphological abnormalities and reduced the respiration capacity of mitochondria in cardiomyocytes, causing their death, independent of BAX and BAK. A truncated isoform of MCL-1 imported into the mitochondrial matrix, known as MCL-1^{Matrix}, has also been shown to have a vital role in the maintenance of mitochondrial membrane structure and ensuring efficient mitochondrial respiration (Perciavalle et al., 2012). While not key to mitochondrial morphology in cardiomyocytes, BAX and BAK were found to influence mitochondrial fusion through regulation of the GTPase Mitofusin-2 in other cell types (Karbowski et al., 2006). BOK has also been identified as a key player in the maintenance of mitochondrial morphology and function in vitro, with its depletion from cells significantly reducing the rate of mitochondrial fusion (Schulman et al., 2019). Further roles of BCL-2 family proteins in a purely mitochondrial function mediatory capacity have been found for BCL-xL and BCL-2. The former has been shown to be important in lowering membrane potential and preventing mitochondrial swelling in response to cellular stress, promoting cell survival in a role more indirect than its usual anti-apoptotic activity (Vander Heiden et al., 1997). BCL-xL has also been reported to regulate metabolic rate by interacting with mitochondrial F_1F_0 ATP synthetase in neuronal cells (Alavian et al., 2011). BCL-2, on the other hand, has been found to function as a pro-oxidant and positive mediator of mitochondrial respiration, despite the potential for increasing DNA damage due to the presence of excess ROS (Chen and Pervaiz, 2007; Steinman, 1995). In addition to structural and functional regulation of mitochondria, BCL-2 family proteins have also been demonstrated to have essential roles in their quality control by mediating mitophagy. This has been found in the cases of BCL-RAMBO, BCL-xL and MCL-1 (Hollville et al., 2014; Murakawa et al., 2015). In summary, BCL-2 family proteins can be said to have major roles to play in the structure, function and turnover of mitochondria, in addition to their well-known activities in regulating the mitochondrial pathway of apoptosis.

4. Autophagy

Autophagy is important for the removal of damaged organelles and aberrant proteins, among other things, from the intracellular space

(Glick et al., 2010). Hence, it can be considered a survival mechanism for the cell, as its dysregulation can result in cell death and lead to disease (e.g., Parkinson's disease and Crohn's disease.). Regulating the process are key protein machinery, including Beclin-1 and members of the Atg family (Glick et al., 2010). Also, among these important proteins are various members of the BCL-2 family.

Autophagy can be induced in response to cellular stresses, such as ATP unavailability and hypoxia, as a result of mTOR inhibition (Glick et al., 2010). In addition to regulating the formation of the autophagosome, mTOR is also a translational regulator of MCL-1 (Jung et al., 2010; Mills et al., 2008). As a result of mTOR inhibition during nutrient starvation, resulting in lowered ATP availability, MCL-1 is degraded and autophagy is promoted in differentiated cells (i.e., mature neurons) (Germain et al., 2011). However, the balance between autophagy and apoptosis in the absence of MCL-1 is reliant on the relative levels of Beclin-1 and BAX. This effect may be cell type-specific though, as loss of *Mcl-1* in cardiac cells causes autophagy impairment and culminates in heart failure (Thomas et al., 2013).

Although MCL-1 has displayed some pro-autophagy activity, it and most other BCL-2 family proteins have also been found to have the opposite effect. Autophagy-suppressive effects of MCL-1 and BCL-xL, were observed in Human Ovarian Surface Epithelial (HOSE) cells expressing H-Ras^{V12} (Elgendy et al., 2011). This was overcome through its displacement from Beclin-1 by fellow BCL-2 family protein NOXA. Other BCL-2 proteins have also been found to interact with Beclin-1, resulting in the inhibition of autophagy. Overexpression of BCL-B was found to inhibit Beclin-1 activity by binding to its BH3 domain, desensitizing cells to a variety of autophagic stimuli (Robert et al., 2012). Conversely, upon knockdown of *Bcl-B*, cells were sensitized to nutrient starvation and autophagic cell death. Similar results were also seen from A1, which impacted the ability of in vitro infected macrophages to effect bacterial growth through autophagy (Kathania et al., 2011). BCL-2, through mutation of key phosphorylation sites required for its interaction with Beclin-1, was shown to have an important role in the regulation of exercise-induced autophagy (He et al., 2012). Disruption of BCL-2-mediated inhibition of Beclin-1 in mice was also found to decrease age-related pathologies and improve longevity through increased autophagy (Fernandez et al., 2018). An insight into the potential mechanism of regulating Beclin-1 by BCL-2 family proteins was revealed through BIM, whose expression levels were also inversely correlated with Beclin-1 activity (Luo et al., 2012). BIM was found to facilitate the binding of Beclin-1 to the dynein motor complex, mislocalizing it within the cell. These reports reveal the important role BCL-2 family proteins play in regulating and initiating autophagy.

5. Calcium regulation

Calcium is an important regulatory and signaling molecule in a variety of cellular processes. Its activity within the cell is dependent on its concentration and localization, as its presence is required for life but in excess can prove lethal (Clapham, 2007). Levels of calcium are regulated by ion pumps and organelles such as the endoplasmic reticulum (ER), which contains proteins that sequester free calcium ions (Carafoli, 1988). The cellular functions that it regulates include cell growth, differentiation, protein folding, metabolism, contractility and neuronal excitability. As a result it is also linked with a number of diseases, both physical (e.g., Alzheimer's, Parkinson's and heart disease) and psychological (e.g., schizophrenia and bipolar disorder) in nature (Berridge, 2017). Therefore, any role to play for BCL-2 proteins could be vital to further understand basic physiology and disease states.

Potential involvement of BCL-2 family proteins in calcium regulation, in a nonapoptotic manner, came about with the discovery of their ability to localize at the ER (Krajewski et al., 1993). The first BCL-2 family protein linked with ER calcium regulation was BCL-2, which was initially found to prevent its release following thapsigargin and hydrogen peroxide treatment (Distelhorst et al., 1996; Lam et al., 1994). However, more recently it was shown that BCL-2 had entirely the opposite role, inducing calcium release from the ER through interactions with the inositol triphosphate receptor (IP3R) upon its overexpression (Chen et al., 2004; Pinton et al., 2000). This process is negatively regulated by BAX and BAK, the double knockout of which was shown to increase calcium leakage via the IP3R (Oakes et al., 2005). BCL-xL has also similar shown similar interactions to BCL-2 with IP3R, resulting in increased calcium release (White et al., 2005). In addition to interacting with IP3R, overexpression of BCL-xL has also been shown to downregulate the former through decreased induction of its transcription factor NFATc2 (Li et al., 2002). As a result, T cell receptor-activated calcium flux was found to have lower calcium release capacity.

In addition to interactions with the IP3R, BCL-2 family members have also been found to target ryanodine receptors (RyR). BCL-2 was found to interact with RyR through binding of the BH4 domain, leading to a lessened release of calcium upon its exogenous expression or electroporative insertion of the BH4 domain in HEK293 cell models (Vervliet et al., 2014). Introduction of a BH4 domain peptide into hippocampal neurons further supported this, decreasing RyR-mediated calcium release. A similar result was also observed from BCL-xL, binding to RyR3 through not only its BH4 domain but also its BH3 domain (Vervliet et al., 2015). This was observed in both HEK293 cells and rat hippocampal neurons. From these reports, it is clear that proteins of the BCL-2 family are involved in calcium regulation within the cell in a non-apoptotic setting.

6. Glucose and lipid metabolism

Glucose and lipid metabolism are associated in many ways, particularly in the mitochondria. It is reasonable to speculate that BCL-2 family proteins may be involved in the regulation of glucose and lipid metabolism, as most are located at the mitochondria. BAD was first shown to participate in glycolysis through the observation of a holoenzyme complex that includes itself, protein kinase K and protein phosphatase 1 catalytic units, Wiskott-Aldrich family member WAVE-1 and glucokinase (Danial et al., 2003). *Bad*$^{-/-}$ cells lack this complex, leading to reduced glucokinase activity and glucose intolerance (Danial et al., 2003). Furthermore, activation of glucokinase was shown to be dependent on the phosphorylation of a serine residue in the BAD BH3 domain (Ser155 in mBAD; Ser118 in hBAD) (Polzien et al., 2009). A cell-permeable BAD BH3 phospho-mimic was capable of restoring glucose-driven mitochondrial respiration in *Bad*$^{-/-}$ cells and increase functional β-cell mass in diabetes (Danial et al., 2008; Ljubicic et al., 2015). More recently, SAD-A, a serine/threonine protein kinase of the AMPK subfamily, was found to promote the phosphorylation of BAD S155 to regulate β-cell function (Wang et al., 2019). Thus, it is clear that BAD is involved in regulating glucose metabolism. Other BCL-2 proteins are also involved in regulating glucose metabolism. *Bmf* deficiency was found to exacerbate hyperglycemia in mice and result in a diminished glucose-stimulated insulin response in vivo (Pfeiffer et al., 2015). Loss of *Bim* was shown to increase oxygen consumption, improve insulin sensitivity and decrease fasting blood glucose (Wali et al., 2017). Phosphorylated NOXA (Ser13) was found to stimulate glucose consumption and could enhance glucose turnover via the pentose phosphate pathway, while its dephosphorylation increased cell sensitivity to glucose starvation (Lowman et al., 2010).

Beyond regulating the metabolism of glucose, the BCL-2 family has also been implicated in regulating that of lipids. *Bnip3*$^{-/-}$ mice displayed increased hepatocellular respiration in the presence of glucose and increased expression of lipogenic enzymes, resulting in increased lipid synthesis in the liver (Glick et al., 2012). Conversely, loss of *Bnip3* caused reduced fatty acid oxidation and defective glucose output under fasting conditions (Glick et al., 2012). Truncated BID (tBID) displayed lipid transfer activity and inhibited fatty acid oxidation via inhibition of carnitine palmitoyltransferase-1 (CPT-1) (Giordano et al., 2005). Interestingly, loss of mitochondrial carrier homolog 2 (MTCH2), a tBID-interacting protein, resulted in increased energy utilization, mitochondrial oxidative phosphorylation and protection from diet-induced obesity (Buzaglo-Azriel et al., 2016). Recently, BIM was also shown to be involved in lipid metabolism. $Bim^{-/-}$ cells presented with a higher mitochondrial oxygen consumption rate, while $Bim^{-/-}$ mice displayed increased lipid oxidation, reduced adiposity and lower hepatic lipid content (Wali et al., 2017). The detailed mechanism of how BIM regulates these processes is still unclear. MCL-1—more specifically MCL-1^{Matrix}—has also recently been implicated in lipid metabolism through its interaction with the mitochondrial fatty acid pathway regulator very long-chain acyl-CoA dehydrogenase (VLCAD) via its BH3 domain (Escudero et al., 2018). Through its depletion, the role of MCL-1^{Matrix} was confirmed by the accumulation of VLCAD substrates. Based on such observations, BCL-2 family proteins can be intrinsically linked with the metabolism of lipids and glucose within a cell.

7. Regulation of unfolded protein response

In addition to their calcium regulatory role in the ER, the BCL-2 family also mediate protein quality and quantity in the ER. Most proteins located in the ER, Golgi apparatus, lysosomes and plasma membrane are translated by ribosomes on rough ER and folded with protein-folding machines in the ER (Sevier and Kaiser, 2002). However, many misfolded and incompletely folded proteins arise, despite the contribution of these protein-folding machines. Accumulation of misfolded proteins eventually leads to the activation of a signal transduction pathway known as the unfolded protein response (UPR) (Schröder and Kaufman, 2005). The UPR is initiated by the stimulation of three ER stress sensors located on ER membrane; protein kinase RNA-like ER kinase (PERK), Inositol-requiring enzyme 1 (IRE1α) and activating transcription factor 6 (ATF6) (Wang and Kaufman, 2016). Once activated, the UPR induces an adaptive response to restore protein-folding homeostasis by reducing protein synthesis and enhancing the degradation of misfolded proteins. If this process fails it will eventually initiate apoptosis (Shore et al., 2011).

PERK activation leads to the phosphorylation of its downstream target—elongation initiation factor 2α (eIF2α), resulting in the subsequent translational upregulation of activating transcription factor 4 (ATF4) (Harding et al., 2000). Increased expression of ATF4 induces the expression of downstream genes, such as the transcription factor C/EBP homologous protein (CHOP) (Harding et al., 2000). Two different groups have shown the involvement of BOK in the PERK pathway (Carpio et al., 2015; Llambi et al., 2016). *Bok*$^{-/-}$ cells were shown to have a defective response to ER stress stimuli (Carpio et al., 2015). Further analysis found that activation of ATF4 and CHOP were diminished in *Bok*$^{-/-}$ cells and mice (Carpio et al., 2015). Furthermore, inhibition of PERK with a specific inhibitor resulted in BOK-induced cell death when stimulated with the ER stress agent tunicamycin (Llambi et al., 2016). These results suggest that BOK is involved in PERK activation. However, many subsequent studies have disputed the role of Bok in UPR response (Fernandez-Marrero et al., 2016; Schulman et al., 2019).

IRE1α contains a kinase domain and an endoribonuclease domain, requiring the former activity in order to activate the latter (Tirasophon et al., 2000). The kinase domain can also activate c-Jun amino-terminal kinases (JNK), which, in turn, can phosphorylate BAD, BCL-2 and BCL-xL to suppress apoptosis (Urano et al., 2000; Yang and Yao, 2015; Yu et al., 2004). BAX and BAK were shown to directly interact with the cytosolic domain of IRE1α, which is essential for its activation (Hetz et al., 2006). As a result, *Bax*$^{-/-}$/*Bak*$^{-/-}$ cells have decreased endoribonuclease activity under ER stress (Hetz et al., 2006). BCL-2 was also shown to regulate IRE1α activity through an unknown mechanism (Chonghaile et al., 2015). Other BCL-2 family members (i.e., BIM and PUMA) were originally reported to be involved in upstream regulation of the UPR pathway through IRE1α, however, another group demonstrated that they function downstream of the pathway (Herold et al., 2014; Rodriguez et al., 2012).

ATF6 has been reported to be upstream of several BCL-2 family proteins, such as MCL-1 and PUMA (Galehdar et al., 2010; Morishima et al., 2011). However, none of these are identified as upstream regulators of ATF6. Therefore, the ATF6 arm of the UPR may not be regulated by BCL-2 family proteins, while the PERK- and IRE1α-regulated UPR are.

8. Recent findings

The main functions of most BCL-2 family proteins have been well documented, while some of them have recently been found to have additional novel functions. BCL-RAMBO was shown to induce apoptosis via interaction with adenine nucleotide translocator (Kim et al., 2012). Later studies reported that murine BCL-RAMBO was a mammalian Atg32 homolog, which interacts with LC3, mediating mitophagy and mitochondrial fragmentation (Murakawa et al., 2015; Nakazawa et al., 2016). Recently, voltage-dependent anion channels (VDAC)—also known as mitochondrial porin—were identified as another BCL-RAMBO-interacting protein by a comprehensive screen using a *Drosophila* model, in which human BCL-RAMBO was ectopically expressed in eye imaginal discs (Matsubara et al., 2019). Human BCL-RAMBO and VDAC were shown to cooperatively promote caspase activation in 293T cells (Matsubara et al., 2019). Moreover, this group also found that BCL-RAMBO induces the perinuclear accumulation of fragmented mitochondria in A549 cells (Matsubara et al., 2019).

For a long time, BOK has been considered as a BAK/BAK-like protein with a redundant role in regulating mitochondrial permeability transition during apoptosis. This was based on ectopic over-expression (Hsu et al., 1997), on the finding that genomic deletion of *Bok* had minimal or no phenotype—despite its broad expression pattern (Ke et al., 2012)—and on the developmental abnormalities observed in a compound $Bok^{-/-}$/$Bax^{-/-}$/$Bak^{-/-}$ background (Ke et al., 2018). Recently, our group demonstrated that BOK is a regulator of uridine metabolism and chemosensitivity through its interaction with uridine monophosphate synthetase (UMPS) (Srivastava et al., 2019). This is distinct from previous links of BCL-2 family proteins to UMP synthesis, which lie in the metabolism of glucose and lipids using its downstream derivatives. Apart from its role as a nucleotide in RNA, uridine is a versatile molecule that exerts multitargeted effects because it can be used to produce other biologically active molecules. For example, uridine diphosphate glucose (UDPG), is a basic building block for glycogen biosynthesis (Roach et al., 2012); UDP-GlcNAc, is a donor substrate for protein glycosylation (Hawkins et al., 1997); uridine catabolism produces acetyl-CoA, which is an important molecule in energy metabolism and in the biosynthesis of the neurotransmitter acetyl choline (Connolly and Duley, 1999). Acetyl-CoA is also a donor substrate for protein lysine acetylation (de Cabo and Navas, 2009). For these reasons, uridine has

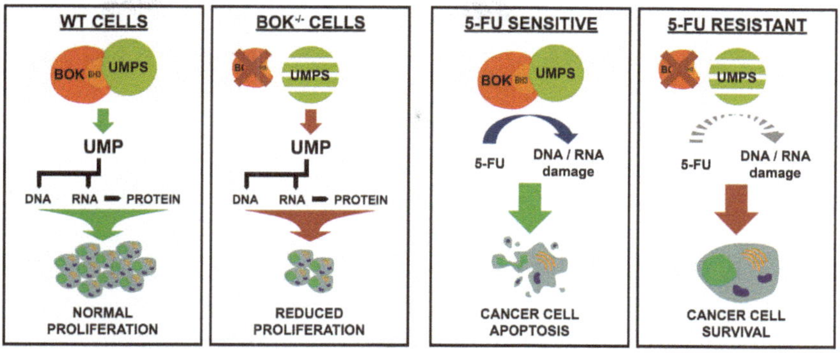

Fig. 1 Bok regulates UMPS activity and chemosensitivity. Bok enhances UMPS activity and cellular proliferation and is involved in 5-FU conversion. Reduced expression of Bok in cancer cells leads to reduced proliferation and chemo-resistance.

the ability to affect a wide range of biological processes (Fig. 1). In fact, uridine administration in C57B/6 mice is associated with altered liver protein glycosylation profile, reduced phosphorylation level of insulin signaling proteins, reduced liver hemin levels and reduced ability for insulin-stimulated blood glucose removal during an insulin tolerance test (Urasaki et al., 2014). Furthermore, there is positive correlation between plasma uridine concentration and insulin resistance in humans (Dudzinska et al., 2013; Yamamoto et al., 2010). If this interaction is crucial for UMPS activity, why wouldn't this manifest in an observable phenotype in *Bok*$^{-/-}$ mice? The answer is two-fold i.e. in addition to de novo synthesis, the salvage pathway, converts uridine into uracil via uridine phosphorylase (UPase) (Ahmed, 1984) and more importantly via food intake (Holguin et al., 2008). Therefore, mice grown under ideal conditions (no limit on food intake) is unlikely to exhibit any phenotype. Consistent with this notion, we have seen a distinctive phenotype in CCl_4-induced liver fibrosis model in *Bok*$^{-/-}$ mice, liver being an organ with high demand for uridine biosynthesis (Srivastava et al., 2019). Our findings on the role of Bok in uridine metabolism and cellular proliferation is in agreement with finding that a significant proliferation defect was observed in diethylnitrosamine (DEN)-induced hepatocarcinogenesis in *Bok*$^{-/-}$ mice (Rabachini et al., 2018); 5-FU sensitivity of pancreatic cancer (Schober et al., 2015) and the role of BOK in erythropoiesis (Kang et al., 2019).

9. Conclusion and future perspectives

Since the discovery of the prototypic member BCL-2's role in mitochondrial apoptotic process (Vaux et al., 1988), this family of proteins has been one of the most well characterized, both in terms of their structure and their physiological function. Our understanding of their relative abundance, affinity in various tissues and types of cancer has led to the development of the new age cancer therapeutic "BH3 mimetics" (Merino et al., 2018). Our understanding of the nonapoptotic functions of this family has also been increasing steadily over the last two decades. These nonapoptotic functions are widely varied, with some of them possessing therapeutic implications of similar significance to BH3 mimetics in cancer. For example, the role of BAD in in regulating glucokinase activity, respiration, ATP production and as a sentinel to sense glucose levels in pancreatic islets (Danial et al., 2003, 2008). Understanding the structural basis of this interaction led to the development of "BAD mimetics," which have potential as a new generation of glucokinase activators for treating type 2 diabetes (Szlyk et al., 2014). Our own recent finding on the role of BOK in 5-FU metabolism and chemoresistance (Srivastava et al., 2019) could lead to the development of BOK as a biomarker for stratifying cancer patients for 5-FU sensitivity. Furthermore, understanding the structural basis of the interaction between BOK and UMPS could lead to the development of "BOK mimetics" to resensitize recalcitrant cancers to 5-FU, one of the most commonly used chemotherapeutic drugs in the world.

Acknowledgments

We wish to acknowledge the financial help from RFA (Understanding Disease), La Trobe University. J.G. and Z.C. are supported by La Trobe University's postgraduate scholarships.

References

Ahmed, N.K., 1984. Enzymes of the de novo and salvage pathways for pyrimidine biosynthesis in normal colon, colon carcinoma, and xenografts. Cancer 54, 1370–1373.

Alavian, K.N., Li, H., Collis, L., Bonanni, L., Zeng, L., Sacchetti, S., Lazrove, E., Nabili, P., Flaherty, B., Graham, M., Chen, Y., Messerli, S.M., Mariggio, M.A., Rahner, C., McNay, E., Shore, G.C., Smith, P.J., Hardwick, J.M., Jonas, E.A., 2011. Bcl-xL regulates metabolic efficiency of neurons through interaction with the mitochondrial F1FO ATP synthase. Nat. Cell Biol. 13, 1224–1233.

Alberts, B., Johnson, A., Lewis, J., Raff, M., Roberts, K., Walter, P. (Eds.), 2002. An overview of the cell cycle. In: Molecular Biology of the Cell. Garland Science, New York.

Barnum, K.J., O'Connell, M.J., 2014. Cell cycle regulation by checkpoints. Methods Mol. Biol. 1170, 29–40.

Berridge, M.J., 2017. Calcium signalling in health and disease. Biochem. Biophys. Res. Commun. 485, 5.

Boise, L.H., Gonzalez-Garcia, M., Postema, C.E., Ding, L., Lindsten, T., Turka, L.A., Mao, X., Nunez, G., Thompson, C.B., 1993. bcl-x, a bcl-2-related gene that functions as a dominant regulator of apoptotic cell death. Cell 74, 597–608.

Brady, H.J., Gil-Gomez, G., Kirberg, J., Berns, A.J., 1996. Bax alpha perturbs T cell development and affects cell cycle entry of T cells. EMBO J. 15, 6991–7001.
Buzaglo-Azriel, L., Kuperman, Y., Tsoory, M., Zaltsman, Y., Shachnai, L., Zaidman, S.L., Bassat, E., Michailovici, I., Sarver, A., Tzahor, E., Haran, M., Vernochet, C., Gross, A., 2016. Loss of muscle MTCH2 increases whole-body energy utilization and protects from diet-induced obesity. Cell Rep. 14, 1602–1610.
Carafoli, E., 1988. The role of calcium in the control of cell function. In: Lemasters, J.J., Hackenbrock, C.R., Thurman, R.G., Westerhoff, H.V. (Eds.), Integration of Mitochondrial Function. Springer, Boston, MA, pp. 475–485.
Carpio, M.A., Michaud, M., Zhou, W., Fisher, J.K., Walensky, L.D., Katz, S.G., 2015. BCL-2 family member BOK promotes apoptosis in response to endoplasmic reticulum stress. Proc. Natl. Acad. Sci. U. S. A. 112, 7201–7206.
Chattopadhyay, A., Chiang, C.W., Yang, E., 2001. BAD/BCL-[X(L)] heterodimerization leads to bypass of G0/G1 arrest. Oncogene 20, 4507–4518.
Chen, Z.X., Pervaiz, S., 2007. Bcl-2 induces pro-oxidant state by engaging mitochondrial respiration in tumor cells. Cell Death Differ. 14, 1617.
Chen, R., Valencia, I., Zhong, F., McColl, K.S., Roderick, H.L., Bootman, M.D., Berridge, M.J., Conway, S.J., Holmes, A.B., Mignery, G.A., Velez, P., Distelhorst, C.W., 2004. Bcl-2 functionally interacts with inositol 1,4,5-trisphosphate receptors to regulate calcium release from the ER in response to inositol 1,4,5-trisphosphate. J. Cell Biol. 166, 193–203.
Chonghaile, T.N., Gupta, S., John, M., Szegezdi, E., Logue, S.E., Samali, A., 2015. BCL-2 modulates the unfolded protein response by enhancing splicing of X-box binding protein-1. Biochem. Biophys. Res. Commun. 466, 40–45.
Clapham, D.E., 2007. Calcium signaling. Cell 131, 1047–1058.
Cleary, M.L., Smith, S.D., Sklar, J., 1986. Cloning and structural analysis of cDNAs for bcl-2 and a hybrid bcl-2/immunoglobulin transcript resulting from the t(14;18) translocation. Cell 47, 19–28.
Connolly, G.P., Duley, J.A., 1999. Uridine and its nucleotides: biological actions, therapeutic potentials. Trends Pharmacol. Sci. 20, 218–225.
Danial, N.N., Gramm, C.F., Scorrano, L., Zhang, C.Y., Krauss, S., Ranger, A.M., Datta, S.R., Greenberg, M.E., Licklider, L.J., Lowell, B.B., Gygi, S.P., Korsmeyer, S.J., 2003. BAD and glucokinase reside in a mitochondrial complex that integrates glycolysis and apoptosis. Nature 424, 952–956.
Danial, N.N., Walensky, L.D., Zhang, C.-Y., Choi, C.S., Fisher, J.K., Molina, A.J.A., Datta, S.R., Pitter, K.L., Bird, G.H., Wikstrom, J.D., Deeney, J.T., Robertson, K., Morash, J., Kulkarni, A., Neschen, S., Kim, S., Greenberg, M.E., Corkey, B.E., Shirihai, O.S., Shulman, G.I., Lowell, B.B., Korsmeyer, S.J., 2008. Dual role of proapoptotic BAD in insulin secretion and beta cell survival. Nat. Med. 14, 144–153.
de Cabo, R., Navas, P., 2009. "AcCoA" lade for energy and life span. Cell Metab. 9, 305–306.
Deng, X., Gao, F., May, W.S., 2003. Bcl2 retards G1/S cell cycle transition by regulating intracellular ROS. Blood 102, 3179–3185.
Distelhorst, C.W., Lam, M., McCormick, T.S., 1996. Bcl-2 inhibits hydrogen peroxide-induced ER Ca2+ pool depletion. Oncogene 12, 2051–2055.
Dudzinska, W., Lubkowska, A., Jakubowska, K., Suska, M., Skotnicka, E., 2013. Insulin resistance induced by maximal exercise correlates with a post-exercise increase in uridine concentration in the blood of healthy young men. Physiol. Res. 62, 163–170.
el-Deiry, W.S., Tokino, T., Velculescu, V.E., Levy, D.B., Parsons, R., Trent, J.M., Lin, D., Mercer, W.E., Kinzler, K.W., Vogelstein, B., 1993. WAF1, a potential mediator of p53 tumor suppression. Cell 75, 817–825.

Elgendy, M., Sheridan, C., Brumatti, G., Martin, S.J., 2011. Oncogenic Ras-induced expression of Noxa and Beclin-1 promotes autophagic cell death and limits clonogenic survival. Mol. Cell 42, 23–35.

Ellis, H.M., Horvitz, H.R., 1986. Genetic control of programmed cell death in the nematode C. elegans. Cell 44, 817–829.

Escudero, S., Zaganjor, E., Lee, S., Mill, C.P., Morgan, A.M., Crawford, E.B., Chen, J., Wales, T.E., Mourtada, R., Luccarelli, J., Bird, G.H., Steidl, U., Engen, J.R., Haigis, M.C., Opferman, J.T., Walensky, L.D., 2018. Dynamic regulation of long-chain fatty acid oxidation by a noncanonical interaction between the MCL-1 BH3 helix and VLCAD. Mol. Cell 69. 729–743.e727.

Fernandez, A.F., Sebti, S., Wei, Y., Zou, Z., Shi, M., McMillan, K.L., He, C., Ting, T., Liu, Y., Chiang, W.C., Marciano, D.K., Schiattarella, G.G., Bhagat, G., Moe, O.W., Hu, M.C., Levine, B., 2018. Disruption of the beclin 1-BCL2 autophagy regulatory complex promotes longevity in mice. Nature 558, 136–140.

Fernandez-Marrero, Y., Ke, F., Echeverry, N., Bouillet, P., Bachmann, D., Strasser, A., Kaufmann, T., 2016. Is BOK required for apoptosis induced by endoplasmic reticulum stress? Proc. Natl. Acad. Sci. U. S. A. 113, E492–E493.

Field, S.J., Tsai, F.-Y., Kuo, F., Zubiaga, A.M., Kaelin, W.G., Livingston, D.M., Orkin, S.H., Greenberg, M.E., 1996. E2F-1 functions in mice to promote apoptosis and suppress proliferation. Cell 85, 549–561.

Fujise, K., Zhang, D., Liu, J., Yeh, E.T., 2000. Regulation of apoptosis and cell cycle progression by MCL1. Differential role of proliferating cell nuclear antigen. J. Biol. Chem. 275, 39458–39465.

Galehdar, Z., Swan, P., Fuerth, B., Callaghan, S.M., Park, D.S., Cregan, S.P., 2010. Neuronal apoptosis induced by endoplasmic reticulum stress is regulated by ATF4–CHOP-mediated induction of the Bcl-2 homology 3-only member PUMA. J. Neurosci. 30, 16938–16948.

Germain, M., Nguyen, A.P., Le Grand, J.N., Arbour, N., Vanderluit, J.L., Park, D.S., Opferman, J.T., Slack, R.S., 2011. MCL-1 is a stress sensor that regulates autophagy in a developmentally regulated manner. EMBO J. 30, 395–407.

Giordano, A., Calvani, M., Petillo, O., Grippo, P., Tuccillo, F., Melone, M.A., Bonelli, P., Calarco, A., Peluso, G., 2005. tBid induces alterations of mitochondrial fatty acid oxidation flux by malonyl-CoA-independent inhibition of carnitine palmitoyltransferase-1. Cell Death Differ. 12, 603–613.

Glick, D., Barth, S., Macleod, K.F., 2010. Autophagy: cellular and molecular mechanisms. J. Pathol. 221, 3–12.

Glick, D., Zhang, W., Beaton, M., Marsboom, G., Gruber, M., Simon, M.C., Hart, J., Dorn, G.W., Brady, M.J., Macleod, K.F., 2012. BNip3 regulates mitochondrial function and lipid metabolism in the liver. Mol. Cell. Biol. 32, 2570–2584.

Greider, C., Chattopadhyay, A., Parkhurst, C., Yang, E., 2002. BCL-x(L) and BCL2 delay Myc-induced cell cycle entry through elevation of p27 and inhibition of G1 cyclin-dependent kinases. Oncogene 21, 7765–7775.

Hanada, M., Delia, D., Aiello, A., Stadtmauer, E., Reed, J.C., 1993. bcl-2 gene hypomethylation and high-level expression in B-cell chronic lymphocytic leukemia. Blood 82, 1820–1828.

Harding, H.P., Novoa, I., Zhang, Y., Zeng, H., Wek, R., Schapira, M., Ron, D., 2000. Regulated translation initiation controls stress-induced gene expression in mammalian cells. Mol. Cell 6, 1099–1108.

Hasan, S.M.M., Sheen, A.D., Power, A.M., Langevin, L.M., Xiong, J., Furlong, M., Day, K., Schuurmans, C., Opferman, J.T., Vanderluit, J.L., 2013. Mcl1 regulates the terminal mitosis of neural precursor cells in the mammalian brain through $p27^{Kip1}$. Development 140, 3118–3127.

Hawkins, M., Angelov, I., Liu, R., Barzilai, N., Rossetti, L., 1997. The tissue concentration of UDP-N-acetylglucosamine modulates the stimulatory effect of insulin on skeletal muscle glucose uptake. J. Biol. Chem. 272, 4889–4895.

He, C., Bassik, M.C., Moresi, V., Sun, K., Wei, Y., Zou, Z., An, Z., Loh, J., Fisher, J., Sun, Q., Korsmeyer, S., Packer, M., May, H.I., Hill, J.A., Virgin, H.W., Gilpin, C., Xiao, G., Bassel-Duby, R., Scherer, P.E., Levine, B., 2012. Exercise-induced BCL2-regulated autophagy is required for muscle glucose homeostasis. Nature 481, 511–515.

Henderson, G.S., Brown, K.A., Perkins, S.L., Abbott, T.M., Clayton, F., 1996. bcl-2 is down-regulated in atypical endometrial hyperplasia and adenocarcinoma. Mod. Pathol. 9, 430–438.

Herold, M.J., O'Reilly, L.A., Lin, A., Srivastava, R., Doerflinger, M., Bouillet, P., Strasser, A., Puthalakath, H., 2014. Evidence against upstream regulation of the unfolded protein response (UPR) by pro-apoptotic BIM and PUMA. Cell Death Dis. 5, e1354.

Hetz, C., Bernasconi, P., Fisher, J., Lee, A.H., Bassik, M.C., Antonsson, B., Brandt, G.S., Iwakoshi, N.N., Schinzel, A., Glimcher, L.H., Korsmeyer, S.J., 2006. Proapoptotic BAX and BAK modulate the unfolded protein response by a direct interaction with IRE1alpha. Science (New York, N.Y.) 312, 572–576.

Holguin, S., Martinez, J., Chow, C., Wurtman, R., 2008. Dietary uridine enhances the improvement in learning and memory produced by administering DHA to gerbils. FASEB J. 22, 3938–3946.

Hollville, E., Carroll, R.G., Cullen, S.P., Martin, S.J., 2014. Bcl-2 family proteins participate in mitochondrial quality control by regulating Parkin/PINK1-dependent mitophagy. Mol. Cell 55, 451–466.

Hsu, S.Y., Kaipia, A., McGee, E., Lomeli, M., Hsueh, A.J., 1997. Bok is a pro-apoptotic Bcl-2 protein with restricted expression in reproductive tissues and heterodimerizes with selective anti-apoptotic Bcl-2 family members. Proc. Natl. Acad. Sci. U. S. A. 94, 12401–12406.

Jiang, Y., Zhang, S.-J., Wu, S.-M., Lee, M.Y.W.T., 1995. Immunoaffinity purification of DNA polymerase δ1. Arch. Biochem. Biophys. 320, 297–304.

Jung, C.H., Ro, S.H., Cao, J., Otto, N.M., Kim, D.H., 2010. mTOR regulation of autophagy. FEBS Lett. 584, 1287–1295.

Kamer, I., Sarig, R., Zaltsman, Y., Niv, H., Oberkovitz, G., Regev, L., Haimovich, G., Lerenthal, Y., Marcellus, R.C., Gross, A., 2005. Proapoptotic BID is an ATM effector in the DNA-damage response. Cell 122, 593–603.

Kane, D.J., Sarafian, T.A., Anton, R., Hahn, H., Gralla, E.B., Valentine, J.S., Ord, T., Bredesen, D.E., 1993. Bcl-2 inhibition of neural death: decreased generation of reactive oxygen species. Science (New York, N.Y.) 262, 1274–1277.

Kang, S.H., Perales, O., Michaud, M., Katz, S.G., 2019. BOK promotes erythropoiesis in a mouse model of myelodysplastic syndrome. Ann. Hematol. 98, 2089–2096.

Karbowski, M., Norris, K.L., Cleland, M.M., Jeong, S.Y., Youle, R.J., 2006. Role of Bax and Bak in mitochondrial morphogenesis. Nature 443, 658–662.

Kathania, M., Raje, C.I., Raje, M., Dutta, R.K., Majumdar, S., 2011. Bfl-1/A1 acts as a negative regulator of autophagy in mycobacteria infected macrophages. Int. J. Biochem. Cell Biol. 43, 573–585.

Ke, F., Voss, A., Kerr, J.B., O'Reilly, L.A., Tai, L., Echeverry, N., Bouillet, P., Strasser, A., Kaufmann, T., 2012. BCL-2 family member BOK is widely expressed but its loss has only minimal impact in mice. Cell Death Differ. 19, 915–925.

Ke, F.F.S., Vanyai, H.K., Cowan, A.D., Delbridge, A.R.D., Whitehead, L., Grabow, S., Czabotar, P.E., Voss, A.K., Strasser, A., 2018. Embryogenesis and adult life in the absence of intrinsic apoptosis effectors BAX, BAK, and BOK. Cell 173. 1217–1230 e1217.

Kim, J.-Y., So, K.-J., Lee, S., Park, J.-H., 2012. Bcl-rambo induces apoptosis via interaction with the adenine nucleotide translocator. FEBS Lett. 586, 3142–3149.

Knudson, C.M., Johnson, G.M., Lin, Y., Korsmeyer, S.J., 2001. Bax accelerates tumorigenesis in p53-deficient mice. Cancer Res. 61, 659–665.

Kozopas, K.M., Yang, T., Buchan, H.L., Zhou, P., Craig, R.W., 1993. MCL1, a gene expressed in programmed myeloid cell differentiation, has sequence similarity to BCL2. Proc. Natl. Acad. Sci. U. S. A. 90, 3516–3520.

Krajewski, S., Tanaka, S., Takayama, S., Schibler, M.J., Fenton, W., Reed, J.C., 1993. Investigation of the subcellular distribution of the bcl-2 oncoprotein: residence in the nuclear envelope, endoplasmic reticulum, and outer mitochondrial membranes. Cancer Res. 53, 4701–4714.

Lam, M., Dubyak, G., Chen, L., Nuñez, G., Miesfeld, R.L., Distelhorst, C.W., 1994. Evidence that BCL-2 represses apoptosis by regulating endoplasmic reticulum-associated Ca2+ fluxes. Proc. Natl. Acad. Sci. U. S. A. 91, 6569–6573.

Li, F., Ambrosini, G., Chu, E.Y., Plescia, J., Tognin, S., Marchisio, P.C., Altieri, D.C., 1998. Control of apoptosis and mitotic spindle checkpoint by survivin. Nature 396, 580–584.

Li, C., Fox, C.J., Master, S.R., Bindokas, V.P., Chodosh, L.A., Thompson, C.B., 2002. Bcl-X(L) affects Ca(2+) homeostasis by altering expression of inositol 1,4,5-trisphosphate receptors. Proc. Natl. Acad. Sci. U. S. A. 99, 9830–9835.

Lindsten, T., Ross, A.J., King, A., Zong, W.X., Rathmell, J.C., Shiels, H.A., Ulrich, E., Waymire, K.G., Mahar, P., Frauwirth, K., Chen, Y., Wei, M., Eng, V.M., Adelman, D.M., Simon, M.C., Ma, A., Golden, J.A., Evan, G., Korsmeyer, S.J., MacGregor, G.R., Thompson, C.B., 2000. The combined functions of proapoptotic Bcl-2 family members bak and bax are essential for normal development of multiple tissues. Mol. Cell 6, 1389–1399.

Liu, Q.A., Shio, H., 2008. Mitochondrial morphogenesis, dendrite development, and synapse formation in cerebellum require both Bcl-w and the glutamate receptor delta2. PLoS Genet. 4, e1000097.

Ljubicic, S., Polak, K., Fu, A., Wiwczar, J., Szlyk, B., Chang, Y., Alvarez-Perez, J.C., Bird, G.H., Walensky, L.D., Garcia-Ocaña, A., Danial, N.N., 2015. Phospho-BAD BH3 mimicry protects β cells and restores functional β cell mass in diabetes. Cell Rep. 10, 497–504.

Llambi, F., Wang, Y.M., Victor, B., Yang, M., Schneider, D.M., Gingras, S., Parsons, M.J., Zheng, J.H., Brown, S.A., Pelletier, S., Moldoveanu, T., Chen, T., Green, D.R., 2016. BOK is a non-canonical BCL-2 family effector of apoptosis regulated by ER-associated degradation. Cell 165, 421–433.

Lowman, X.H., McDonnell, M.A., Kosloske, A., Odumade, O.A., Jenness, C., Karim, C.B., Jemmerson, R., Kelekar, A., 2010. The proapoptotic function of Noxa in human leukemia cells is regulated by the kinase Cdk5 and by glucose. Mol. Cell 40, 823–833.

Luo, S., Garcia-Arencibia, M., Zhao, R., Puri, C., Toh, P.P., Sadiq, O., Rubinsztein, D.C., 2012. Bim inhibits autophagy by recruiting Beclin 1 to microtubules. Mol. Cell 47, 359–370.

Maryanovich, M., Zaltsman, Y., Ruggiero, A., Goldman, A., Shachnai, L., Zaidman, S.L., Porat, Z., Golan, K., Lapidot, T., Gross, A., 2015. An MTCH2 pathway repressing mitochondria metabolism regulates haematopoietic stem cell fate. Nat. Commun. 6, 7901.

Matsubara, H., Tanaka, R., Tateishi, T., Yoshida, H., Yamaguchi, M., Kataoka, T., 2019. The human Bcl-2 family member Bcl-rambo and voltage-dependent anion channels manifest a genetic interaction in Drosophila and cooperatively promote the activation of effector caspases in human cultured cells. Exp. Cell Res. 381, 223–234.

Mazel, S., Burtrum, D., Petrie, H.T., 1996. Regulation of cell division cycle progression by bcl-2 expression: a potential mechanism for inhibition of programmed cell death. J. Exp. Med. 183, 2219–2226.

McDonnell, T.J., Deane, N., Platt, F.M., Nunez, G., Jaeger, U., McKearn, J.P., Korsmeyer, S.J., 1989. bcl-2-immunoglobulin transgenic mice demonstrate extended B cell survival and follicular lymphoproliferation. Cell 57, 79–88.

Merino, D., Kelly, G.L., Lessene, G., Wei, A.H., Roberts, A.W., Strasser, A., 2018. BH3-mimetic drugs: blazing the trail for new cancer medicines. Cancer Cell 34, 879–891.

Millman, S.E., Pagano, M., 2011. MCL1 meets its end during mitotic arrest. EMBO Rep. 12, 384–385.

Mills, J.R., Hippo, Y., Robert, F., Chen, S.M., Malina, A., Lin, C.J., Trojahn, U., Wendel, H.G., Charest, A., Bronson, R.T., Kogan, S.C., Nadon, R., Housman, D.E., Lowe, S.W., Pelletier, J., 2008. mTORC1 promotes survival through translational control of Mcl-1. Proc. Natl. Acad. Sci. U. S. A. 105, 10853–10858.

Morishima, N., Nakanishi, K., Nakano, A., 2011. Activating transcription factor-6 (ATF6) mediates apoptosis with reduction of myeloid cell leukemia sequence 1 (Mcl-1) protein via induction of WW domain binding protein 1. J. Biol. Chem. 286, 35227–35235.

Murakawa, T., Yamaguchi, O., Hashimoto, A., Hikoso, S., Takeda, T., Oka, T., Yasui, H., Ueda, H., Akazawa, Y., Nakayama, H.J.N.C., 2015. Bcl-2-like protein 13 is a mammalian Atg32 homologue that mediates mitophagy and mitochondrial fragmentation. 6, 7527.

Nakano, K., Vousden, K.H., 2001. PUMA, a novel proapoptotic gene, is induced by p53. Mol. Cell 7, 683–694.

Nakazawa, M., Matsubara, H., Matsushita, Y., Watanabe, M., Vo, N., Yoshida, H., Yamaguchi, M., Kataoka, T.J.P.O., 2016. The human Bcl-2 family member Bcl-rambo localizes to mitochondria and induces apoptosis and morphological aberrations in Drosophila. PLoS One 11, e0157823.

Oakes, S.A., Scorrano, L., Opferman, J.T., Bassik, M.C., Nishino, M., Pozzan, T., Korsmeyer, S.J., 2005. Proapoptotic BAX and BAK regulate the type 1 inositol trisphosphate receptor and calcium leak from the endoplasmic reticulum. Proc. Natl. Acad. Sci. U. S. A. 102, 105–110.

O'Connor, L., Strasser, A., O'Reilly, L.A., Hausmann, G., Adams, J.M., Cory, S., Huang, D.C., 1998. Bim: a novel member of the Bcl-2 family that promotes apoptosis. EMBO J. 17, 384–395.

Oda, E., Ohki, R., Murasawa, H., Nemoto, J., Shibue, T., Yamashita, T., Tokino, T., Taniguchi, T., Tanaka, N., 2000. Noxa, a BH3-only member of the Bcl-2 family and candidate mediator of p53-induced apoptosis. Science (New York, N.Y.) 288, 1053–1058.

Perciavalle, R.M., Stewart, D.P., Koss, B., Lynch, J., Milasta, S., Bathina, M., Temirov, J., Cleland, M.M., Pelletier, S., Schuetz, J.D., Youle, R.J., Green, D.R., Opferman, J.T., 2012. Anti-apoptotic MCL-1 localizes to the mitochondrial matrix and couples mitochondrial fusion to respiration. Nat. Cell Biol. 14, 575–583.

Pfeiffer, S., Halang, L., Düssmann, H., Byrne, M.M., Prehn, J.H.M., 2015. BH3-only protein bmf is required for the maintenance of glucose homeostasis in an in vivo model of HNF1α-MODY diabetes. Cell Death Dis. 1, 15041.

Pinton, P., Ferrari, D., Magalhães, P., Schulze-Osthoff, K., Di Virgilio, F., Pozzan, T., Rizzuto, R., 2000. Reduced loading of intracellular Ca^{2+} stores and downregulation of capacitative Ca^{2+} Influx in Bcl-2–overexpressing cells. J. Cell Biol. 148, 857–862.

Polzien, L., Baljuls, A., Rennefahrt, U.E.E., Fischer, A., Schmitz, W., Zahedi, R.P., Sickmann, A., Metz, R., Albert, S., Benz, R., Hekman, M., Rapp, U.R., 2009. Identification of novel in vivo phosphorylation sites of the human proapoptotic protein BAD: pore-forming activity of BAD is regulated by phosphorylation. J. Biol. Chem. 284, 28004–28020.

Print, C.G., Loveland, K.L., Gibson, L., Meehan, T., Stylianou, A., Wreford, N., de Kretser, D., Metcalf, D., Köntgen, F., Adams, J.M., Cory, S., 1998. Apoptosis regulator Bcl-w is essential for spermatogenesis but appears otherwise redundant. Proc. Natl. Acad. Sci. U. S. A. 95, 12424–12431.

Rabachini, T., Fernandez-Marrero, Y., Montani, M., Loforese, G., Sladky, V., He, Z., Bachmann, D., Wicki, S., Villunger, A., Stroka, D., Kaufmann, T., 2018. BOK promotes chemical-induced hepatocarcinogenesis in mice. Cell Death Differ. 25, 706–718.

Roach, P.J., Depaoli-Roach, A.A., Hurley, T.D., Tagliabracci, V.S., 2012. Glycogen and its metabolism: some new developments and old themes. Biochem. J. 441, 763–787.

Robert, G., Gastaldi, C., Puissant, A., Hamouda, A., Jacquel, A., Dufies, M., Belhacene, N., Colosetti, P., Reed, J.C., Auberger, P., Luciano, F., 2012. The anti-apoptotic Bcl-B protein inhibits BECN1-dependent autophagic cell death. Autophagy 8, 637–649.

Rodriguez, D.A., Zamorano, S., Lisbona, F., Rojas-Rivera, D., Urra, H., Cubillos-Ruiz, J.R., Armisen, R., Henriquez, D.R., Cheng, E.H., Letek, M., Vaisar, T., Irrazabal, T., Gonzalez-Billault, C., Letai, A., Pimentel-Muinos, F.X., Kroemer, G., Hetz, C., 2012. BH3-only proteins are part of a regulatory network that control the sustained signalling of the unfolded protein response sensor IRE1alpha. EMBO J. 31, 2322–2335.

Schober, M., Esenofsky, R., Faissner, R., Krug, S., Weidenauer, C., Hagmann, W., Haas, S., Heuchel, R., Löhr, M., 2015. Acquired 5-fluorouracil resistance in human pancreatic carcinoma cells—a paradigm for chemoresistance mechanisms in pancreatic cancer. JOP 16, 256–265.

Schröder, M., Kaufman, R.J., 2005. The mammalian unfolded protein response. Annu. Rev. Biochem. 74, 739–789.

Schulman, J.J., Szczesniak, L.M., Bunker, E.N., Nelson, H.A., Roe, M.W., Wagner 2nd, L.E., Yule, D.I., Wojcikiewicz, R.J.H., 2019. Bok regulates mitochondrial fusion and morphology. Cell Death Differ. 26, 2682–2694.

Sevier, C.S., Kaiser, C.A., 2002. Formation and transfer of disulphide bonds in living cells. Nat. Rev. Mol. Cell Biol. 3, 836.

Shore, G.C., Papa, F.R., Oakes, S.A., 2011. Signaling cell death from the endoplasmic reticulum stress response. Curr. Opin. Cell Biol. 23, 143–149.

Srivastava, R., Cao, Z., Nedeva, C., Naim, S., Bachmann, D., Rabachini, T., Gangoda, L., Shahi, S., Glab, J., Menassa, J., Osellame, L., Nelson, T., Fernandez-Marrero, Y., Brown, F., Wei, A., Ke, F., O'Reilly, L., Doerflinger, M., Allison, C., Kueh, A., Ramsay, R., Smith, B.J., Mathivanan, S., Kaufmann, T., Puthalakath, H., 2019. BCL-2 family protein BOK is a positive regulator of uridine metabolism in mammals. Proc. Natl. Acad. Sci. U. S. A. 116, 15469–15474.

Steinman, H.M., 1995. The Bcl-2 oncoprotein functions as a pro-oxidant. J. Biol. Chem. 270, 3487–3490.

Szlyk, B., Braun, C.R., Ljubicic, S., Patton, E., Bird, G.H., Osundiji, M.A., Matschinsky, F.M., Walensky, L.D., Danial, N.N., 2014. A phospho-BAD BH3 helix activates glucokinase by a mechanism distinct from that of allosteric activators. Nat. Struct. Mol. Biol. 21, 36–42.

Thomas, R.L., Roberts, D.J., Kubli, D.A., Lee, Y., Quinsay, M.N., Owens, J.B., Fischer, K.M., Sussman, M.A., Miyamoto, S., Gustafsson, A.B., 2013. Loss of MCL-1 leads to impaired autophagy and rapid development of heart failure. Genes Dev. 27, 1365–1377.

Tirasophon, W., Lee, K., Callaghan, B., Welihinda, A., Kaufman, R.J., 2000. The endoribonuclease activity of mammalian IRE1 autoregulates its mRNA and is required for the unfolded protein response. Genes Dev. 14, 2725–2736.
Urano, F., Wang, X., Bertolotti, A., Zhang, Y., Chung, P., Harding, H.P., Ron, D., 2000. Coupling of stress in the ER to activation of JNK protein kinases by transmembrane protein kinase IRE1. Science (New York, N.Y.) 287, 664–666.
Urasaki, Y., Pizzorno, G., Le, T.T., 2014. Uridine affects liver protein glycosylation, insulin signaling, and heme biosynthesis. PLoS One 9 e99728.
Vairo, G., Soos, T.J., Upton, T.M., Zalvide, J., DeCaprio, J.A., Ewen, M.E., Koff, A., Adams, J.M., 2000. Bcl-2 retards cell cycle entry through $p27^{Kip1}$, pRB Relative p130, and altered E2F regulation. Mol. Cell. Biol. 20, 4745–4753.
Vander Heiden, M.G., Chandel, N.S., Williamson, E.K., Schumacker, P.T., Thompson, C.B., 1997. Bcl-xL regulates the membrane potential and volume homeostasis of mitochondria. Cell 91, 627–637.
Vaux, D.L., Cory, S., Adams, J.M., 1988. Bcl-2 gene promotes haemopoietic cell survival and cooperates with c-myc to immortalize pre-B cells. Nature 335, 440–442.
Vaux, D.L., Weissman, I.L., Kim, S.K., 1992. Prevention of programmed cell death in Caenorhabditis elegans by human bcl-2. Science (New York, N.Y.) 258, 1955–1957.
Vervliet, T., Decrock, E., Molgo, J., Sorrentino, V., Missiaen, L., Leybaert, L., De Smedt, H., Kasri, N.N., Parys, J.B., Bultynck, G., 2014. Bcl-2 binds to and inhibits ryanodine receptors. J. Cell Sci. 127, 2782–2792.
Vervliet, T., Lemmens, I., Vandermarliere, E., Decrock, E., Ivanova, H., Monaco, G., Sorrentino, V., Nadif Kasri, N., Missiaen, L., Martens, L., De Smedt, H., Leybaert, L., Parys, J.B., Tavernier, J., Bultynck, G., 2015. Ryanodine receptors are targeted by anti-apoptotic Bcl-XL involving its BH4 domain and Lys87 from its BH3 domain. Sci. Rep. 5, 9641.
Wali, J.A., Galic, S., Tan, C.Y.R., Gurzov, E.N., Frazier, A.E., Connor, T., Ge, J., Pappas, E.G., Stroud, D., Varanasi, L.C., Selck, C., Ryan, M.T., Thorburn, D.R., Kemp, B.E., Krishnamurthy, B., Kay, T.W.H., McGee, S.L., Thomas, H.E., 2017. Loss of BIM increases mitochondrial oxygen consumption and lipid oxidation, reduces adiposity and improves insulin sensitivity in mice. Cell Death Differ. 25, 217.
Wang, M., Kaufman, R.J., 2016. Protein misfolding in the endoplasmic reticulum as a conduit to human disease. Nature 529, 326.
Wang, C., Youle, R.J., 2009. The role of mitochondria in apoptosis. Annu. Rev. Genet. 43, 95–118.
Wang, X., Bathina, M., Lynch, J., Koss, B., Calabrese, C., Frase, S., Schuetz, J.D., Rehg, J.E., Opferman, J.T., 2013. Deletion of MCL-1 causes lethal cardiac failure and mitochondrial dysfunction. Genes Dev. 27, 1351–1364.
Wang, K., Liu, D., Zhang, Y., Chang, X., Xu, R., Pang, J., Li, K., Sun, P., Zhu, Y., Han, X., 2019. SAD-A, a downstream mediator of GLP-1 signaling, promotes the phosphorylation of Bad S155 to regulate in vitro β-cell functions. Biochem. Biophys. Res. Commun. 509, 76–81.
White, C., Li, C., Yang, J., Petrenko, N.B., Madesh, M., Thompson, C.B., Foskett, J.K., 2005. The endoplasmic reticulum gateway to apoptosis by Bcl-XL modulation of the InsP3R. Nat. Cell Biol. 7, 1021–1028.
Yamamoto, T., Inokuchi, T., Ka, T., Yamamoto, A., Takahashi, S., Tsutsumi, Z., Tamada, D., Okuda, C., Moriwaki, Y., 2010. Relationship between plasma uridine and insulin resistance in patients with non-insulin-dependent diabetes mellitus. Nucleosides Nucleotides Nucleic Acids 29, 504–508.
Yan, W., Huang, J.X., Lax, A.S., Pelliniemi, L., Salminen, E., Poutanen, M., Toppari, J., 2003. Overexpression of Bcl-W in the testis disrupts spermatogenesis: revelation of a role of BCL-W in male germ cell cycle control. Mol. Endocrinol. 17, 1868–1879.

Yang, J., Yao, S., 2015. JNK-Bcl-2/Bcl-xL-Bax/Bak pathway mediates the crosstalk between matrine-induced autophagy and apoptosis via interplay with Beclin 1. Int. J. Mol. Sci. 16, 25744–25758.

Yu, J., Zhang, L., Hwang, P.M., Kinzler, K.W., Vogelstein, B., 2001. PUMA induces the rapid apoptosis of colorectal cancer cells. Mol. Cell 7, 673–682.

Yu, C., Minemoto, Y., Zhang, J., Liu, J., Tang, F., Bui, T.N., Xiang, J., Lin, A., 2004. JNK suppresses apoptosis via phosphorylation of the proapoptotic Bcl-2 family protein BAD. Mol. Cell 13, 329–340.

Zhong, J.X., Zhou, L., Li, Z., Wang, Y., Gui, J.F., 2014. Zebrafish Noxa promotes mitosis in early embryonic development and regulates apoptosis in subsequent embryogenesis. Cell Death Differ. 21, 1013.

Zhou, T., Li, G., Cao, B., Liu, L., Cheng, Q., Kong, H., Shan, C., Huang, X., Chen, J., Gao, N., 2013. Downregulation of Mcl-1 through inhibition of translation contributes to benzyl isothiocyanate-induced cell cycle arrest and apoptosis in human leukemia cells. Cell Death Dis. 4, e515.

Zinkel, S.S., Hurov, K.E., Ong, C., Abtahi, F.M., Gross, A., Korsmeyer, S.J., 2005. A role for proapoptotic BID in the DNA-damage response. Cell 122, 579–591.

CHAPTER TWO

Mcl-1 as a "barrier" in cancer treatment: Can we target it now?

Nikolay V. Pervushin[a], Viacheslav V. Senichkin[a], Boris Zhivotovsky[a,b,*], Gelina S. Kopeina[a,*]
[a]Faculty of Medicine, MV Lomonosov Moscow State University, Moscow, Russia
[b]Institute of Environmental Medicine, Karolinska Institutet, Stockholm, Sweden
*Corresponding authors: e-mail address: boris.zhivotovsky@ki.se; lirroster@gmail.com

Contents

Abstract

During the last two decades, the study of Mcl-1, an anti-apoptotic member of the Bcl-2 family, attracted researchers due to its important role in cancer cell survival and tumor development. The significance of Mcl-1 protein in resistance to chemotherapeutics

International Review of Cell and Molecular Biology, Volume 351
ISSN 1937-6448
https://doi.org/10.1016/bs.ircmb.2020.01.002

makes it an attractive target in cancer therapy. Here, we discuss the diverse possibilities for indirect Mcl-1 inhibition through its downregulation, for example, via targeting for proteasomal degradation or blockage of translation and transcription. We also provide an overview of the direct blocking of protein-protein interactions with pro-apoptotic Bcl-2 family proteins, including examples of the most promising regulators of Mcl-1 and selective BH3-mimetics, which at present are under clinical evaluation. Moreover, several approaches for the co-targeting of Mcl-1 and other proteins (e.g., CDKs) are also presented. In addition, we highlight the broad spectrum of problems that accompanied the discovery and development of effective Mcl-1 inhibitors.

1. Introduction

Programmed cell death is a natural barrier that prevents cancer development. Apoptosis, the most described type of programmed cell death, is one of the key processes for eliminating damaged or unfavorable cells, which could be threatening to an organism's health. Dysregulation or evasion of this physiological process is one of the important hallmarks of cancer (Hanahan and Weinberg, 2011).

Proteins of the Bcl-2 (B-cell lymphoma 2) family have a great significance in the regulation of the intrinsic pathway of apoptosis. This family of proteins is divided into three subsets according to their functions in the apoptotic process and the number of Bcl-2 homology (BH) domains (Czabotar et al., 2007). The anti-apoptotic group contains proteins (Bcl-2, Bcl-xL, Mcl-1, Bfl-1, etc.) that have four BH domains and prevent apoptosis through binding and neutralizing both effector proteins and regulatory pro-apoptotic proteins (Czabotar et al., 2014). The subset of pro-apoptotic effectors consists of Bak, Bax, and Bok, which also contain four BH domains (BH1-BH4), and these proteins are responsible for mitochondrial outer membrane permeabilization (MOMP). The subset of regulatory pro-apoptotic members also includes proteins (Bim, Bad, Puma, etc.) that contain only one BH3 domain and thus are called "BH3-only proteins." According to the mechanisms of action, this subset can be subdivided into two groups: activators and sensitizers. The latter (e.g., Bad) can only displace Bax/Bak from complexes with anti-apoptotic proteins after binding to them, while the former (e.g., Bim) can directly bind Bax/Bak in order to facilitate their oligomerization (Czabotar et al., 2014; Elkholi et al., 2011; Kodama et al., 2011).

The intrinsic apoptotic pathway is activated in response to different stimuli (for example, genotoxic stress, viral infections, etc.) and leads to Bax/Bak oligomerization. This event causes pore formation in the outer mitochondrial membrane (OMM) and the subsequent release of cytochrome *c* and other apoptosis-related proteins from the intermembrane space of mitochondria to the cytosol (Elmore, 2007). Being in cytosol, cytochrome *c* binds to cytosolic proteins Apaf-1 and pro-caspase-9 and in the presence of dATP forms the apoptosome complex leading to the caspase-9 activation. Finally, the active form of this protease promotes activation of downstream executioner caspases, which eventually leads to cell death (Letai, 2011) (Fig. 1).

The interplay between BH3-only proteins and pro- or anti-apoptotic members of the Bcl-2 family occurs through electrostatic and hydrophobic interactions between the BH3 domain and the special BH3 binding site. This site represents the hydrophobic groove formed by BH1-3 domains, which is responsible for selective interactions between Bcl-2 family proteins and their partners (Czabotar et al., 2014). Different cancer types are often characterized by an increased expression of anti-apoptotic proteins that leads to evasion of apoptosis and facilitates malignant transformation (Chonghaile et al., 2011; Glaser et al., 2012).

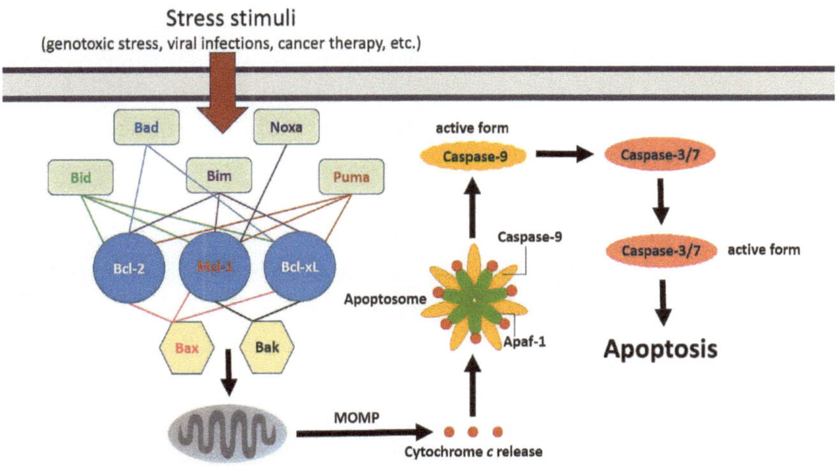

Fig. 1 Bcl-2 family members act as gatekeepers of the intrinsic pathway of apoptosis. Lines indicate protein-protein interactions (PPIs), and different colors of lines demonstrate various sets of inhibitory PPIs between anti-apoptotic proteins (Bcl-2, Mcl-1, and Bcl-xL) and pro-apoptotic (Bax—pink, Bak—black) or BH3-only (Bid—green, Bad—blue, Bim—violet, Noxa—dark blue, and Puma—brown) proteins of the Bcl-2 family. MOMP—mitochondrial outer membrane permeabilization.

Myeloid cell leukemia-1 (Mcl-1) has some remarkable characteristics (Senichkin et al., 2019). Thus, the depletion of this protein causes embryonic lethality in the peri-implantation period (Rinkenberger et al., 2000). Moreover, Mcl-1 plays a key role in the development and regulation of the homeostasis of hematopoietic stem cells, lymphocytes, neurons, and other cell types (Arbour et al., 2008; Dzhagalov et al., 2014; Opferman et al., 2005). Three variants of alternative splicing (Mcl-1L, Mcl-1S and Mcl-1ES) have been described. The most abundant isoform of the protein is Mcl-1L (long), which is usually named Mcl-1 (hereafter Mcl-1L is referred to as Mcl-1). This protein is located in the OMM and exhibits an anti-apoptotic activity. Two other isoforms of the protein, Mcl-1S (short) and Mcl-1ES (extra-short), exert pro-apoptotic functions but they are poorly studied (De Blasio et al., 2018) and beyond the scope of this review.

Mcl-1 with a molecular weight 37.2kDa is larger than other anti-apoptotic proteins (Bcl-2 and Bcl-xL) due to the presence of a N-terminal domain that contains two proline/glutamic acid/serine/threonine rich (PEST) regions (Germain and Duronio, 2007). PEST is a common motif for proteins with short half-life; therefore, this protein is stable during 1–4h depending on cellular context (Thomas et al., 2010). Mcl-1 is highly regulated at transcriptional (via signal transducer and activator of transcription 3 (STAT3), p38 mitogen-activated protein kinase (MAPK), phosphoinositide-3-kinase/RAC-alpha serine/threonine-protein kinase (PI3K/AKT) signaling pathways), translational (more than 10 miRNAs have been found to influence Mcl-1 translation), and post-translational (via deubiquitinases and E3 ubiquitin ligases (Akgul, 2009; Becker et al., 2014; Cui and Placzek, 2018; Gomez-Bougie et al., 2011; Mojsa et al., 2014)) levels.

The presence of *MCL1* amplification has been proven in different types of human cancer (Wei et al., 2012). Moreover, high level of Mcl-1 promotes cell survival and leads to malignant transformation through inhibition of apoptosis in several hematological (e.g., acute myeloid leukemia (AML) and multiple myeloma (MM)) and solid (e.g., lung, breast, ovarian, and cervical) types of cancer (Brotin et al., 2010; Ding et al., 2007; Song et al., 2005). Drug resistance of cancer cells is frequently associated with overexpression of Mcl-1 (Li et al., 2015; Shi et al., 2017; Wertz et al., 2011). Consistently, siRNA-mediated knockdown of Mcl-1 leads to tumor growth inhibition and the death of cells overexpressing Mcl-1 (Akgul, 2009; Chen et al., 2009; Konopleva et al., 2006; Lucas et al., 2012). Taken together, one can conclude that Mcl-1 is a promising target for cancer treatment, but its great physiological significance in the maintenance of cell survival complicates the development of "ideal" inhibitors. The scope of this review is to summarize the most promising approaches and achievements in the inhibition of the anti-apoptotic protein Mcl-1 with natural or chemical compounds that have been developed during the last several years.

2. Agents that influence the regulation of Mcl-1 expression

Mcl-1 is an attractive target for indirect inhibition due to its multistep regulation and short half-life (Senichkin et al., 2020). Many natural compounds have been studied to find molecules with potent biological activity, which could be used in cancer therapy. Among them there are natural compounds that might effectively suppress Mcl-1 expression via different mechanisms. Therefore, here we focused on several approaches of Mcl-1 indirect inhibition, which may be promising for further study (Fig. 2).

2.1 Agents that regulate Mcl-1 transcription

Among various chemical compounds, the most well-known regulators of Mcl-1 transcription are flavonoids and their hemi-synthetic or synthetic derivatives. Emodin is the first example of them, acting through Interleukin-6 (IL-6) signaling cascade in human MM. IL-6 is involved in

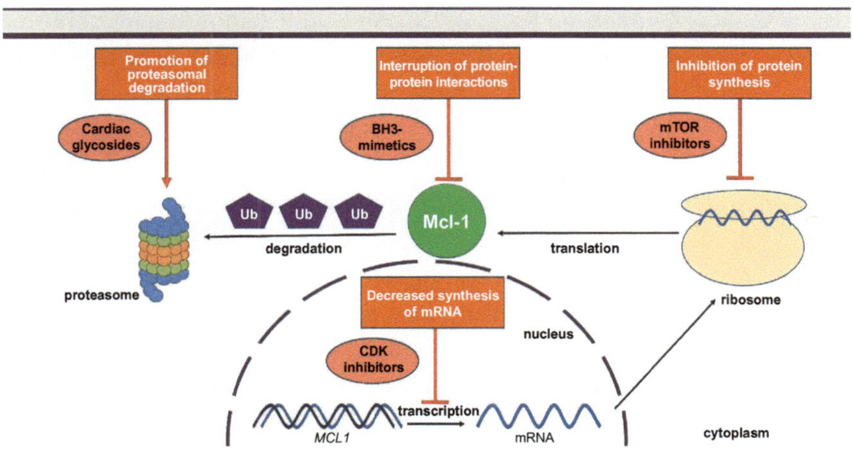

Fig. 2 How can we target Mcl-1 now? There are two principal approaches in the targeting of Mcl-1: direct inhibition through interruption of protein-protein interactions via small molecule inhibitors (BH3-mimetics) and indirect inhibition through down-regulation of Mcl-1 via targeting of proteasomal degradation (e.g., cardiac glycosides) or blockage of translation (e.g., mTOR inhibitors) and transcription (e.g., CDK inhibitors). Ub—ubiquitin.

different signaling cascades (MAPK pathway, PI3K/AKT pathway, and STAT3 pathway) and plays a key role in the regulation of inflammation and proliferation. IL-6-mediated activation of STAT3 causes upregulation of c-Myc, Bcl-xL, and Mcl-1. Emodin was found to block the IL-6 pathway in MM cells that results in a decrease in the Mcl-1 level without significantly reducing the levels of other anti-apoptotic proteins (Muto et al., 2007).

Wogonin and other natural flavonoids with similar structures (e.g., apigenin, chrysin, and luteolin) are inhibitors of cyclin-dependent kinase 9 (CDK9). CDK9 inhibition leads to the decreased synthesis of mRNA, the subsequent downregulation of the short-lived proteins, including Mcl-1, and induction of apoptosis in cancer cells. Furthermore, these anti-cancer agents enhance apoptosis induced by ABT-263 (selective inhibitor of Bcl-2 and Bcl-xL) via decreased Mcl-1 level in different types of tumors. Hence, these flavonoids allow the use of ABT-263 at lower doses, decreasing the risk of its side effects, such as dose-dependent development of thrombocytopenia due to Bcl-xL inhibition. Finally, the combination of ABT-263 with wogonin demonstrated tumor regression *in vivo* that was shown in xenografted human leukemic cells (Polier et al., 2015).

Alvocidib (flavopiridol) is a hemi-synthetic flavonoid derivative and the first CDK inhibitor which has been studied in clinical trials (Kadia et al., 2019). The mechanisms of its anti-cancer activity include suppression of several transcriptional targets, such as *c-MYC* and *MCL1*, via inhibition of various CDKs (Zeidner and Karp, 2015). Alvocidib demonstrated promising results in pre-clinical studies and has been tested both alone and in combination with chemotherapy in clinical trials during the last two decades (more than 60 trials were carried out) (Kadia et al., 2019; Zeidner and Karp, 2015). However, this drug has not yet been approved for cancer treatment due to limited therapeutic efficacy in patients. Nevertheless, alvocidib is still being studied in current clinical trials. For instance, the preliminary efficacy of alvocidib in combination with venetoclax (also known as ABT-199) is being evaluated in phase I trials in patients with relapsed/refractory (R/R) AML (NCT03441555). Additionally, the clinical response to alvocidib combined with cytarabine + mitoxantrone (C + M) in comparison with C + M alone is under phase II clinical trials in R/R AML patients with registered Mcl-1 dependence (NCT02520011). Hopefully, biomarker-driven studies of alvocidib will be more successful than the earlier ones.

Voruciclib, a compound with a structure similar to alvocidib, inhibits CDK9 and has less off-target effects in comparison with alvocidib.

Voruciclib successfully repressed *MCL1* expression in both *in vitro* and *in vivo* models of diffuse large B-cell lymphoma (DLBCL). Furthermore, this compound leads to enhanced inhibition of tumor growth in combination with venetoclax in a xenograft mouse model. Importantly, this combination has good oral bioavailability in contrast with the combination of venetoclax with either alvocidib or dinaciclib (other CDK inhibitors). Hence, voruciclib, which is being evaluated in phase I clinical trials, might become a new drug for cancer treatment (Dey et al., 2017).

AZD4573, a novel selective CDK9 inhibitor, demonstrates anti-cancer activity in various hematological cancer models *in vitro* and leads to tumor regression in different xenograft models. The main mechanism of anti-tumor activity of this inhibitor is associated with Mcl-1 suppression in a time- and dose-dependent way. Moreover, venetoclax enhances the cytotoxicity of AZD4573 in both *in vitro* and *in vivo* models (Cidado et al., 2019). Notably, AZD4573 is being tested in a phase I clinical trial in patients with several hematological malignancies (NCT03263637).

SY-1365, a selective CDK7 inhibitor, leads to the downregulation of Mcl-1 in different cancer cell lines, and the response to this compound is associated with Mcl-1 dependence. This inhibitor leads to the suppression of cell growth in a large number of cancer types *in vitro* and possesses anti-tumor activity in several hematologic (AML) and solid (ovarian cancer) xenograft models. Additionally, SY-1365 synergizes with venetoclax both *in vitro* and *in vivo* (Hu et al., 2019). Based on positive pre-clinical results SY-1365 has become the first selective CDK7 inhibitor progressed into clinical trials (NCT03134638).

Taken together, there is a lot of evidence that short-lived Mcl-1 represents an attractive target for CDK inhibitors (Fig. 2). These compounds affecting multiple cellular pathways could be more effective compared to the selective inhibition of Mcl-1. Moreover, downregulation of Mcl-1 via CDK inhibition in combination with venetoclax could boost the sensitivity of cancer cells to Bcl-2 inhibition, possibly leading to increased therapeutic efficacy in patients with various malignancies, such as AML. Thus, Mcl-1 dependence serves as a predictive biomarker for treatment by CDK inhibitors, which have therapeutic potential in cancer therapy.

2.2 Agents that act via regulation of Mcl-1 translation

As mentioned above, Mcl-1 is tightly regulated at translational level by various miRNAs or the modulation of translational activity of ribosomes.

Consequently, several miRNAs may represent potential biomarkers, indicating increased levels of Mcl-1. Additionally, drugs targeting mTOR (mammalian target of rapamycin) or other modulators of protein synthesis may be used as a possible approach to selectively downregulate short-lived Mcl-1 (Fig. 2). Thus, sorafenib, a broad spectrum tyrosine kinase inhibitor also known as Nexavar, is approved for the therapy of different types of cancer. This drug was reported to inhibit Mcl-1 translation in leukemia cells (Huber et al., 2011; Rahmani et al., 2005). In addition, sorafenib treatment led to improved efficacy of other anti-cancer drugs through downregulation of Mcl-1. Sorafenib augmented apoptosis *in vitro* and prolonged the survival of mice in models of AML in combination with all-trans retinoic acid (ATRA). ATRA in combination with chemotherapeutics is used as the first-line therapy in acute promyelocytic leukemia (APL) patients and leads to diverse effects on the levels of anti-apoptotic proteins of the Bcl-2 family. While causing transcriptional downregulation of Bcl-2, ATRA results in a time-dependent upregulation of Mcl-1 due to activation of the translation and stabilization of Mcl-1 through ribosomal S6 kinase (p90RSK) activation and glycogen synthase kinase 3 beta (GSK3β) inactivation, respectively (Wang et al., 2016). Thus, the concomitant suppression of both Bcl-2 and Mcl-1 demonstrates the benefits of co-administration of ATRA and sorafenib, which prevented ATRA-induced upregulation of Mcl-1.

Meanwhile, recent reports demonstrate the contribution of Mcl-1 to sorafenib resistance *in vitro*, whereas Ras-association domain family member 6 (RASSF6) has been found to activate c-Jun N-terminal kinase (JNK) that led to overcoming the resistance and proteasomal degradation of Mcl-1 (Liang et al., 2018; Tutusaus et al., 2018). Further studies are required to assess how Mcl-1 can regulate the sensitivity of tumor cells to sorafenib. Since sorafenib resistance is commonly observed in patients and limits drug efficacy, this issue is of great significance.

The use of mTOR inhibitors (such as BEZ235 or AZD8055), which effectively downregulate Mcl-1 in triple negative breast cancer (TNBC) cells, is another example of Mcl-1 regulation at the translational level. The suppression of Mcl-1 synthesis by BEZ235 or AZD8055 through blocking translation enhanced ABT-263-induced apoptosis. Moreover, the combination of BEZ235 and ABT263 demonstrated high efficacy in a xenograft mouse model of TNBC (Li et al., 2018). Taken together, these data show that the combination of mTOR inhibitors and BH3 mimetics might hopefully be a good therapeutic strategy in the treatment of TNBC and other malignancies.

Norcantharidin (NCTD) has shown anti-cancer activity in several cancer cell types. Recently, NCTD was found to upregulate miR-320d, a negative regulator of Mcl-1, and Mcl-1 suppression resulted in mitochondrial dysfunction and activation of apoptosis in prostate cancer cells. Furthermore, NCTD strongly reduced tumor growth of prostate cancer PC3 xenografts (Lin et al., 2017a). Thus, NCTD treatment could be a novel option for prostate cancer treatment via translational regulation of Mcl-1. These data highlight therapeutic opportunities in the therapy of Mcl-1-dependent tumors via the inhibition of translation.

2.3 Agents that act through the regulation of Mcl-1 proteasomal degradation

The ubiquitin-proteasome system plays an important role in the rapid degradation of Mcl-1. The rate of ubiquitination is determined by the activity of E3 ubiquitin ligases and deubiquitinases. Several E3 ubiquitin ligases (e.g., Mule, Parkin, and Trim17) and deubiquitinases (e.g., USP9X and Ku70) were reported to regulate Mcl-1 level. For instance, Mule was found to be involved in Noxa-mediated Mcl-1 downregulation (Mojsa et al., 2014). Furthermore, GSK-3β-mediated phosphorylation of Mcl-1 promotes its ubiquitination and consequent proteasomal degradation (Mojsa et al., 2014). At the same time, DNA repair protein Ku70 and several other proteins (such as DUB3 and USP9X (Wu et al., 2019)) act as deubiquitinases, stabilizing Mcl-1 (Wang et al., 2014). Thus, drugs targeting Mcl-1 via inhibition of its deubiquitinases and/or activation of proteasomal degradation of Mcl-1 serve as an attractive tool for the inhibition of Mcl-1 (Fig. 2).

Several cardiac glycosides, which are often used in treatment of cardiovascular diseases, possess potent anti-cancer activities. For example, UNBS1450 effectively decreased Mcl-1 levels, but not Bcl-2 or Bcl-xL, with subsequent induction of apoptosis in different adherent and non-adherent cancer cell lines. Overexpression of a mutated form of Mcl-1, which cannot be ubiquitinated, or the use of the proteasome inhibitor MG132 prevented UNBS1450-induced apoptosis (Cerella et al., 2015). Other glycosides with different chemical structures (cardenolides, such as ouabain, digitoxin, and digoxin, or bufadienolides, such as bufalin, cinobufagin, and proscillaridin A) can also downregulate Mcl-1 (Chanvorachote and Pongrakhananon, 2012; Pongrakhananon et al., 2014; Waiwut et al., 2012). For example, digitoxin and its synthetic derivative D6-MA leads to a decrease in Mcl-1 level through GSK-3β-dependent proteasomal degradation in non-small-cell lung carcinoma (NSCLC) (Pongrakhananon et al., 2014). Similar results were demonstrated for bufalin in NSCLC. Both the proteasome inhibitor MG132 and GSK-3β inhibition significantly prevented the bufalin-induced downregulation of Mcl-1, confirming a post-translational mechanism of Mcl-1 regulation by cardiac glycosides (Kang et al., 2017). In addition, bufalin was found to overcome acquired resistance to osimertinib (a third-generation epidermal growth factor receptor tyrosine kinase inhibitor approved by the FDA) by inhibiting Ku70-mediated Mcl-1 upregulation in NSCLC cells (Cao et al., 2019).

Hence, the indirect inhibition of Mcl-1 via blocking of the specific deubiquitinases (e.g., Ku70) might have significant therapeutic potential in cancer therapy. In comparison with small molecule inhibitors of Mcl-1, it could be more selective for cancer cells characterized by dysregulation of deubiquitinases and could protect healthy cells from excessive toxicity mediated by the direct inhibition of Mcl-1.

PaTrin-2, O^6-methylguanine-DNA methyltransferase (MGMT) inhibitor, could be another example of indirect inhibition of Mcl-1. MGMT is a transcriptional activator of deubiquitinating enzyme 3 (DUB3), which prevents the proteasomal degradation of Mcl-1. PaTrin-2 downregulates Mcl-1 in ovarian cancer via blocking MGMT. Moreover, histone deacetylase inhibitors (HDACis) could significantly enhance PaTrin-2-induced Mcl-1 suppression. Thus, co-administration of PaTrin-2 and HDACis could be a novel and selective approach for Mcl-1 inhibition in chemoresistant cancer cells, which are characterized by dysregulation of MGMT-DUB3-Mcl-1 axis (Wu et al., 2019).

Mcl-1 could be a drug target for diseases other than cancer. For instance, the high level of Mcl-1 has been found in rheumatoid arthritis synovial fibroblasts (RASFs), which are responsible for the pathogenesis of rheumatoid arthritis (RA). It was shown that ursolic acid (UA) selectively induces Noxa and leads to RASFs apoptosis caused by the downregulation of Mcl-1 via proteasomal degradation (Kim et al., 2018). These data illustrate the possibility of targeting Mcl-1 for the treatment of non-cancer diseases. Taken together, targeting of Mcl-1 through the ubiquitin-proteasome system possesses the therapeutic potential both in cancer and non-cancer diseases. Taking into consideration that cancer cells could be more sensitive to indirect inhibition of Mcl-1 than normal cells, the stimulation of its proteasomal degradation can not only improve the outcome of anti-cancer therapy but also avoid the side effects of the treatment.

It should be noted that the whole indirect inhibitors of Mcl-1 mentioned above possess multi-target mechanisms of action, and these drugs exert their anti-cancer activity via different ways, which not necessarily involve "pure" Mcl-1 targeting. Accordingly, the potential risks of undesirable toxicity due to non-selectivity of these indirect inhibitors of Mcl-1 should be considered. For instance, the limited clinical utility of flavopiridol is associated with gastrointestinal side effects, which are probably linked to higher affinity of flavopiridol to several non-CDK targets, in particular intestinal cell kinase (ICK) and male germ cell-associated kinase (MAK) (Dey et al., 2017; Lanasa et al., 2015).

3. Mcl-1 inhibitors: Problems for clinical usage

Despite the discovery of various regulators of Mcl-1 transcription, translation, and degradation, which lead to the downregulation of this protein, small molecule inhibitors represent the most attractive option for targeting Mcl-1. Fortunately, the significant progress has been achieved over the last several years toward the discovery of selective small molecule inhibitors of Mcl-1, which are being evaluated in current clinical trials (Fig. 3). However, it has been a very long road since the first steps of developing of Mcl-1 inhibitors.

3.1 Difficulties in molecular targeting of Mcl-1

As mentioned above, Mcl-1 protein has several characteristics. For example, it is essential for embryogenesis (Rinkenberger et al., 2000), it maintains the survival of different types of cells (Arbour et al., 2008; Dzhagalov et al., 2014; Opferman et al., 2005), and it possesses essential non-apoptotic functions (Fujise et al., 2000; Jamil et al., 2008; Wang et al., 2014). Its characteristics may lead to undesirable toxicity due to Mcl-1 inhibition. Moreover, Mcl-1 has different structure of the BH3-binding groove in comparison with other anti-apoptotic proteins of the Bcl-2 family.

These structural characteristics of Mcl-1 complicate the selection of ligands as potent Mcl-1 inhibitors. Large lipophilic molecules with an acidic group to mimic the salt-bridge interaction with Arg-263 were the only available compounds, which invaded the large hydrophobic groove of Mcl-1. Unfortunately, the large size and high lipophilicity of the potential

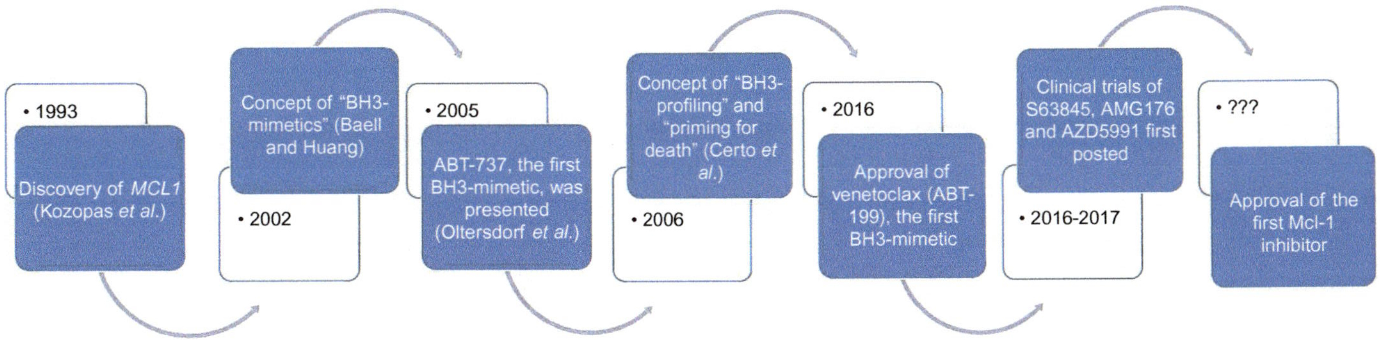

Fig. 3 When will the first Mcl-1 inhibitor be approved? Timeline of the main events in the development of BH3-mimetics and small molecules targeting Mcl-1.

Mcl-1 inhibitors have resulted in unsatisfying drug-like properties and created new challenges for scientists. The use of advanced and innovative methods (e.g., NMR-based molecular design) has allowed the development of more successful small molecule inhibitors, which are discussed below.

3.2 Stability of molecules and poor cell uptake

Generally, two main types of approved drugs exist: macromolecular biologics and small molecules. The former contain a greater surface area for contact with targets, but they are not able to pass through cell membranes that limits the possibility of their use. An alternative approach is the use of peptides instead of whole proteins, but they have low efficacy *in vivo* due to their susceptibility to proteolysis and inadequate cellular uptake. These disadvantages have been overcome by using synthetic modifications, such as peptide stapling, which resulted in the enhanced stability of these drug candidates. The class of stapled peptides contains a great number of compounds, which disrupt different PPIs, including those of Bcl-2 family members.

The first mention of "stapled peptides" was in 2000 (Schafmeister et al., 2000). Bid was the first protein that was modified by "stapling," and a Bid "stabilized α-helix of Bcl-2 domains" ($SAHB_A$) successfully induced apoptosis through the binding and activation of Bax and inhibited tumor growth in a xenograft mouse model of leukemia (Walensky et al., 2004, 2006). Later, in 2010, the first stapled peptide which was able to inhibit Mcl-1 was developed. Mcl-1 $SAHB_D$ treatment of cells led to the activation of the intrinsic pathway of apoptosis in a Bak-dependent manner (Stewart et al., 2010). In 2012, a set of stapled Noxa BH3-peptides with ubiquitin-conjugation was released. The conjugated peptides showed higher activity and increased stability in mouse serum compared with non-conjugated peptides (Muppidi et al., 2012, 2014). Two years later, three Mcl-1-selective peptides (MS), possessing higher affinity in comparison with Noxa-derived peptides, were engineered. Further optimization of MS-1 promoted the development of peptides, which had improved drug-like properties (Foight et al., 2014; Rezaei Araghi et al., 2016). In 2018, a new study of MS-1 was presented. After screening a set of constructions, SAH-MS1-14 and SAH-MS1-18 were selected for further consideration. These peptides demonstrated improved cellular uptake and selectively induced apoptosis via the disruption of Mcl-1/Bak complexes in Mcl-1-dependent cell lines (Rezaei Araghi et al., 2018). Stapled peptides targeting Mcl-1 have not yet been tested in clinical trials. Nevertheless, these investigations have revealed that the isolated BH3-peptide of Mcl-1 represents a unique selective inhibitor of Mcl-1 protein. It should be noted that small molecule inhibitor MIM1 has been selected for further study after a stapled peptide-based screen (Cohen et al., 2012). Taken together, these data highlight the significance of stapled peptides as an experimental model for the development of small molecules.

3.3 Low affinity and non-selectivity

During the evolution of Mcl-1 inhibitors, the values of their affinities to Mcl-1 have changed from micromolar to subnanomolar. The first compounds reported to inhibit Mcl-1 possessed low affinity to this protein (for instance, IC50 of YC137 was 2.47 μM (Zhai et al., 2006)) and usually demonstrated poor selectivity. Further development of Mcl-1 inhibitors led to the emergence of a new class of drugs: pan-Bcl-2 inhibitors (e.g., Obatoclax and AT-101), which possessed increased affinity to Mcl-1, but still showed low selectivity. Promiscuous binding of pan-Bcl-2 inhibitors to different Bcl-2 family members complicated the accurate assessment of direct Mcl-1 inhibition for anti-cancer activity of these drugs. Moreover, pan-Bcl-2 inhibitors had various off-target effects mentioned below.

3.4 Off-target effects

Careful examination of putative Mcl-1 inhibitors, which demonstrated high efficacy in pre-clinical studies, revealed that many of these drugs had various off-target effects not associated with the inhibition of Mcl-1. These off-target effects limited the clinical utility of numerous potential Mcl-1 inhibitors and hampered their translation to clinical trials.

Gossypol and its derivatives were regarded as promising compounds for the inhibition of Mcl-1. Indeed, (−)-gossypol (AT-101), TW-37, and apogossypolone (ApoG2) demonstrated strong anti-tumor activity in preclinical assays. For example, TW-37, first reported in 2006, was presented as a new pan-Bcl-2 inhibitor (K_i=260, 290 and 1100 nM for Mcl-1, Bcl-2 and Bcl-xL, respectively). This compound has been shown to inhibit cell growth and effectively induced apoptosis in prostate cancer cells PC3 and lymphoma cells WSU-DLCL2 through the disruption of complexes of anti-apoptotic proteins with Bax or Bak. Moreover, treatment with this agent led to the activation of apoptosis in a Bak-dependent manner in Mcl-1-dependent cells. In addition, TW-37 demonstrated higher efficacy in combination with cyclophosphamide-doxorubicin-vincristine-prednisone (CHOP) application than CHOP alone, in *in vitro* and *in vivo* models of lymphoma (Brumatti and Ekert, 2013; Mohammad et al., 2007; Wang et al., 2006). Recently, it was discovered that TW-37 increased the chemosensitivity of nasopharyngeal carcinoma *in vivo* and significantly inhibited colorectal cancer and neuroblastoma cell growth as a single agent, both *in vitro* and *in vivo* (Klenke et al., 2019; Lei et al., 2017; Lu et al., 2017). Notably, TW-37-induced apoptosis was partially dependent on Noxa (Varadarajan et al., 2013).

Nevertheless, TW-37 is not the only Mcl-1 inhibitor with off-target effects. It was reported that gossypol could block IL-6 pro-survival signaling cascade in multiple myeloma cells, resulting in the downregulation of Mcl-1 levels (Sadahira et al., 2014). Moreover, gossypol was able to activate Phospholipase A2 (PLA2)-mediated and Ras homolog gene family, member A (RhoA)-mediated signaling cascades, resulting in the activation of pro-apoptotic proteins, such as Bim and Bid (Li et al., 2014; Soderquist et al., 2014b; Spijkers-Hagelstein et al., 2014). Apogossypolone inhibits c-Myc signaling (Hu et al., 2009), while both gossypol and ApoG2 activate autophagy or autophagic cell death through Autophagy related 5 (Atg5) and Beclin-1-dependent mechanisms (Cheng et al., 2013; He et al., 2014; Voss et al., 2010).

Pan-Bcl-2 inhibitor, obatoclax, exhibited anti-tumor activity as a single agent in different types of hematological cancers *in vitro* and suppressed tumor growth in different xenograft models (Champa et al., 2014; Li et al., 2008). Moreover, obatoclax enhanced cisplatin-induced cell death in both cisplatin-resistant and cisplatin-sensitive cell lines (Crawford et al., 2011; Konopleva et al., 2008; Nguyen et al., 2007; Palve et al., 2014; Timme et al., 2013). However, several studies have demonstrated diverse off-target effects of obatoclax, including cell cycle aberrations (S/G2 and G1/G0 cell cycle arrest, p38/p21 (waf1/Cip1) pathway-mediated cell cycle block, and downregulation of cyclin D) (Broecker-Preuss et al., 2015; Koehler et al., 2014; Konopleva et al., 2008; Zhong et al., 2014); unfolded protein response (UPR) stress (Wroblewski et al., 2013); inhibition of autophagy and impaired lysosomal function (Basit et al., 2013; Liang et al., 2015; Stamelos et al., 2016; Sulkshane and Teni, 2016) and induction of necroptotic cell death (Basit et al., 2013; Bonapace et al., 2010; Sulkshane and Teni, 2016). Thus, a broad spectrum of off-target effects of obatoclax and low therapeutic efficacy demonstrated in clinical trials (see below) do not allow the consideration of this drug as a promising Mcl-1 inhibitor.

S1 inhibitor demonstrates anti-tumor activity and activates Bak/Bax-dependent apoptosis in a time- and dose-dependent manner, via the disruption of Mcl-1 and Bcl-2 complexes with pro-apoptotic proteins.

However, further investigations revealed that S1 might be a Noxa inducer. Moreover, this compound causes accumulation of reactive oxygen species (ROS) and, consequently, endoplasmic reticulum (ER) stress. Finally, induction of autophagy in response to ER stress and disruption of Bcl-2/Beclin-1 complexes may also contribute to S1-induced cell death (Albershardt et al., 2011; Soderquist et al., 2014a; Zhang et al., 2007, 2011).

Another inhibitor, A-1210477, exhibits partial disruption of Bcl-2/Bim complexes in a concentration of 10 μM (Leverson et al., 2015). Furthermore, there is new evidence of its off-target effects: the presence of DRP1-dependent mitochondrial fragmentation and partially DRP1-dependent apoptosis (Milani et al., 2017). A-1210477 did not induce Noxa expression in contrast with MIM1 and UMI-77; however, it caused Bak/Bax-independent apoptosis at high concentrations (>10 μM) (Mallick et al., 2019).

MIM1 was found to induce apoptosis in leukemia cells in a Bax/Bak-dependent manner (Cohen et al., 2012). However, this molecule indirectly downregulated Mcl-1 via the induction of Noxa expression following an unfolded protein response (UPR) pathway that was found in NB4, Jurkat, and U937 cell lines. Furthermore, MIM1 sensitized NB4 cells to venetoclax in a Noxa-dependent manner (Mallick et al., 2019). Likewise, another inhibitor of Mcl-1, UMI-77, induced Noxa expression (Lin et al., 2017b). Nevertheless, the increase of Noxa levels, which is demonstrated for various putative Mcl-1 inhibitors, might be considered as a partial "on-target" effect, because this protein selectively inhibits Mcl-1 and Bfl-1, but not other members of the Bcl-2 family. Noxa is regulated by multiple stimuli, and the upregulation of this protein could be exploited to overcome Mcl-1-dependent resistance to apoptosis in cancer therapy (Ding et al., 2019; Guikema et al., 2017).

3.5 Bad bioavailability

A-1210477 was developed by Abbvie's research team and demonstrated high selectivity for Mcl-1 in comparison with other Bcl-2 proteins (for Mcl-1, $K_i = 0.45$ nM, over 100-fold higher affinity than for Bcl-2 and Bcl-xL). This inhibitor disrupted Mcl-1/Bim complexes in H929 cells and effectively killed Mcl-1-dependent cell lines (as shown for several NSCLC cell lines) (Leverson et al., 2015). Moreover, A-1210477 inhibited *in vivo* tumor growth of esophageal squamous cell carcinoma induced by 4-nitroquinoline oxide (Lin et al., 2017b). Nevertheless, A-1210477 has poor pharmacological properties (exhibited significant serum protein binding) that limited further study and application (Leverson et al., 2015).

3.6 The "controversial" drugs

Marinopyrrole A (maritoclax), a natural compound with antibiotic activity from Streptomyces (Haste et al., 2011), was presented as a putative Mcl-1 inhibitor in 2012. This agent led to a decreased Mcl-1 rate through proteasomal degradation and induced apoptosis in Mcl-1-dependent leukemia and melanoma cell lines (Doi et al., 2012; Pandey et al., 2013). However, further investigations revealed that marinopyrrole A was equally effective in several Bcl-2- and Mcl-1-dependent cells. Moreover, this compound did not affect Mcl-1 expression levels in some types of cell lines (Eichhorn et al., 2013). In addition, this agent caused less cell death in wild type mouse embryonic fibroblasts (MEFs) in comparison with Mcl-1-deficient MEFs (Varadarajan et al., 2015). It should be noted that the mechanism of action of marinopyrrole A is probably different from that of other Mcl-1 inhibitors. This drug is considered to facilitate proteasomal degradation of Mcl-1, presumably acting on this protein in a Noxa-like manner (Song et al., 2016). Indeed, if marinopyrrole A directly promotes proteasomal decay of Mcl-1, this compound might represent a prototype of drugs which target Mcl-1 for degradation.

3.7 Failure of clinical studies

Several compounds that are able to inhibit Mcl-1 were investigated in earlier clinical trials. First, gossypol (AT-101) was evaluated both as a single agent and in combination with other anti-cancer drugs. Notably, more than 20 clinical trials were carried out with gossypol. Unfortunately, most of the results were disappointing (Ready et al., 2011; Schelman et al., 2014; Sonpavde et al., 2012; Stein et al., 2019; Xie et al., 2019). Similarly, clinical trials with obatoclax either alone or in combination with other drugs have shown low therapeutic efficacy (Arellano et al., 2014; Brown et al., 2015; Chiappori et al., 2012, 2014; Goy et al., 2014; Langer et al., 2014; Paik et al., 2010, 2011). Perhaps, concomitant suppression of various pro-survival proteins by pan-Bcl-2 inhibitors limits the possibility of using doses which efficiently neutralize individual anti-apoptotic members of the Bcl-2 family. In this context, the combination of selective BH3-mimetics represents a more favorable approach for cancer therapy in comparison with the use of pan-Bcl-2 inhibitors, since different doses of various compounds could be applied to achieve efficacy while staying within the therapeutic window.

It is worth noting that several combinations of selective BH3-mimetics seem to be tolerable, while others demonstrate high toxicity. Indeed, the combined use of BH3-mimetics to Bcl-xL and Mcl-1 leads to acute liver toxicity in mice (Weeden et al., 2018). Moreover, the inhibition of Mcl-1 could exacerbate thrombocytopenia caused by Bcl-xL inhibition (Tse et al., 2008) as Mcl-1 also plays a critical role for homeostasis of megakaryocytes and platelet production (Debrincat et al., 2012; Opferman et al., 2005). At the same time, concomitant inhibition of Mcl-1 and Bcl-2 does not lead to excessive toxicity and is being evaluated in clinical trials.

3.8 Humanized Mcl-1 mice for testing on-target toxicity

As mentioned above, Mcl-1 plays an important role in the development and regulation of the homeostasis of various types of cells (Arbour et al., 2008; Dzhagalov et al., 2014; Opferman et al., 2005). Hence, testing Mcl-1 inhibitors *in vivo* was required to address their potential on-target toxicity. However, there are some differences in the amino acid sequences between mouse and human Mcl-1. As a result, several Mcl-1 inhibitors have a lower affinity for mouse Mcl-1 compared with its human counterpart. For example, highly potent and selective inhibitors of Mcl-1, S63845 and AMG-176, have approximately 6- and 1000-fold lower affinities, respectively, for mouse versus human Mcl-1. These differences in the affinity for Mcl-1 might lead to an underestimation of the possible toxicity of Mcl-1 inhibitors in human trials. To solve this problem, humanized Mcl-1 mice (huMcl-1) were generated for testing Mcl-1 inhibitors. This approach has enabled careful pre-clinical evaluation of S63845, which demonstrated a reduction of the maximum tolerated dose from 40 mg/kg in wild type mice to 12.5 mg/kg in huMcl-1 mice (Brennan et al., 2018). AMG-176 has been tested in huMcl-1 mice due to failure of this compound to inhibit mouse Mcl-1 (Caenepeel et al., 2018). The treatment of huMcl-1 mice with S63845 or AMG-176 have demonstrated that a favorable safety window could be achieved for these compounds.

4. The latest achievements in the development of small molecule inhibitors of Mcl-1

Mcl-1 inhibitors belong to the class of BH3-mimetics. This class of therapeutics includes drugs that mimic the activity of BH3-only proteins, inhibiting the function of anti-apoptotic members of the Bcl-2 family. Cancer cells are often resistant to traditional chemotherapies like DNA-damaging agents due to the increased expression of anti-apoptotic Bcl-2 family proteins. Meanwhile, various stresses force cancer cells to increase expression of pro-apoptotic factors, such as BH3-only proteins. As a result, cancer cells are "primed" for apoptosis, i.e., the inhibition of anti-apoptotic Bcl-2 family proteins would result in the efficient derepression of their pro-apoptotic counterparts, leading to apoptosis induction. Hence, cancer cells are more prone to apoptosis induced by BH3-mimetics than various normal unprimed or apoptosis-refractory cells, which do not express high levels of pro-apoptotic proteins (Montero and Letai, 2018). At the same time, various types of normal cells, such as hematopoietic ones, are also highly primed for apoptosis. Importantly, BH3-mimetics demonstrate higher efficacy in those tumor cells, which depend predominantly on the targeted protein for their survival. Thus, these agents demonstrate a significant potential in hematological cancers but are not sufficiently effective in different types of solid tumors, which often survive due to high levels of various anti-apoptotic proteins. It is necessary to mention that selection of the right strategy for cancer therapy could be provided by a BH3 profiling assay based on the evaluation of cell sensitivity to different drugs *ex vivo* (Montero and Letai, 2016; Villalobos-Ortiz et al., 2019). BH3 profiling is a functional assay of the efficacy of synthetic BH3 peptides by measuring their ability to cause MOMP. This method could help to predict the clinical response to therapy, including to BH3-mimetics. Below we summarize the latest data concerning new Mcl-1 inhibitors, which may have clinical significance in the future.

4.1 S63845

The research teams at Servier and Vernalis have discovered the first potent and selective inhibitor of Mcl-1 ($K_i = 1.2\,nM$) possessing *in vivo* activity, S63845. This compound activates Bak/Bax-dependent apoptosis and promotes the death of different hematological cell lines *in vitro* and *in vivo* (Kotschy et al., 2016). Moreover, this inhibitor enhanced apoptosis in combination with venetoclax in *in vitro* and *in vivo* AML and T-cell acute lymphoblastic leukemia (T-ALL) models (Algarín et al., 2019; Li et al., 2019; Moujalled et al., 2018) and in combination with docetaxel in TNBC and trastuzumab or lapatinib in human epidermal growth factor receptor 2 (HER2)-amplified breast cancer models (Merino et al., 2017). Nowadays, a modified derivative of S63845—S64315/MIK665 (administered by infusion)—is undergoing phase I clinical trials in patients with relapsed and/or refractory lymphoma, MM, AML, or myelodysplastic syndrome (NCT02979366 and NCT02992483), and as a combined therapy with venetoclax in patients with AML (NCT03672695).

4.2 AMG-176

A new Mcl-1 inhibitor AMG-176 was recently introduced by Amgen. This molecule has subnanomolar affinity ($K_i < 1$ nM) to Mcl-1 and effectively disrupted complexes between Mcl-1 and its pro-apoptotic binding partners of the Bcl-2 family, leading to Bak/Bax-dependent cell death. Various hematological cell lines were found to be highly sensitive to AMG-176. This inhibitor demonstrated anti-tumor activity in *in vivo* models of MM and AML. Moreover, it enhanced anti-cancer activity of the proteasome inhibitor, carfilzomib, in an orthotopic model of MM and synergized with venetoclax in *in vitro* and *in vivo* AML models. Furthermore, trametinib increased the efficacy of AMG-176 in KRAS-mutant NSCLC models both *in vitro* and *in vivo* via induction of Bim accumulation by MEK inhibition. AMG-176 is being evaluated in clinical studies as a single agent in patients with relapsed or refractory MM and AML (NCT02675452) and as a combined therapy with venetoclax in patients with AML and NHL (Non-Hodgkin lymphoma) (NCT03797261). Importantly, its intravenous administration has been used in clinical investigations in contrast to oral administration in pre-clinical studies. Meanwhile, AMG-397 (other Mcl-1 inhibitor) is also being studied in patients with MM, AML, or NHL (NCT03465540), and it is the first Mcl-1 inhibitor given *per os* (Caenepeel et al., 2018; Nangia et al., 2018). Despite the advantages of infusion therapy (quick onset of drug action; precise control of extent and duration of drug exposure; reduced interpatient variability), the oral route is more convenient, painless, and safe for patients. Hence, AMG-397 is compared favorably with other Mcl-1 inhibitors due to the oral administration.

4.3 AZD5991

A new Mcl-1 inhibitor, which possesses a subnanomolar affinity ($K_i = 0.2$ nM) to Mcl-1, was developed by Astra Zeneca. Similar to the inhibitors described previously, AZD5991 stimulates the intrinsic pathway of apoptosis in a Bak/Bax-dependent manner. AZD5991 activity was shown *in vitro* in different hematological cell lines, some types of NSCLC and breast cancer. There is also some evidence of *in vivo* activity of AZD5991: the inhibition of tumor growth in several MM and AML (including disseminated forms) xenografted models. Venetoclax enhanced sensitivity to AZD5991 in the AML models (including those resistant to either therapy alone). Likewise, the proteasome inhibitor bortezomib synergized with AZD5991 in MM models, both *in vitro* and *in vivo*.

Furthermore, AZD5991 demonstrated efficacy as a single agent *in vivo* in PDX models of T-cell lymphomas (TCL). Nowadays, AZD5991 is being evaluated in phase I clinical trials in patients with relapsed or refractory MM, NHL, TCL, chronic lymphocytic leukemia (CLL), small lymphocytic leukemia, or Richter's syndrome (NCT03218683) (Koch et al., 2019; Tron et al., 2018).

4.4 VU661013

Recently, Vanderbilt University and Boehringer Ingelheim reported a new Mcl-1 inhibitor—VU661013. This molecule possesses subnanomolar affinity ($K_i = 0.097 \pm 0.030$ nM) to Mcl-1 and demonstrates potent cytotoxic activity against several AML cell lines. Venetoclax enhanced the efficiency of VU661013 in AML cell lines. VU661013 is also active *in vivo* in AML xenografts, both as a single agent and in combination with venetoclax. Although VU661013 is not yet in clinical testing, Vanderbilt University and Boehringer Ingelheim have recently announced its introduction into clinical trials (Ramsey et al., 2018).

It is notable that the search for new Mcl-1 inhibitors continues, and several new compounds have recently been presented (Lee et al., 2019; Lu et al., 2019; Szlávik et al., 2019).

4.5 BH3-mimetics to Mcl-1: Summary

The small molecule inhibitors of Mcl-1 have several common features. As BH3-mimetics, these compounds cause a disruption of the complexes between Mcl-1 and its pro-apoptotic binding partners. Moreover, treatment with these drugs leads to the increased level of Mcl-1 in a dose-dependent manner. The accumulation of Mcl-1 is not associated with increased Mcl-1 mRNA levels and probably occurs due to prevention of Mcl-1 proteasomal degradation. At the same time, decreased levels of Mcl-1 protein were observed in sensitive cell lines, as evidenced for AZD5991, due to the cleavage of Mcl-1 by active caspases. Furthermore, it was shown that the efficacy of inhibitors does not correlate with *MCL1* mRNA expression but there is an inverse correlation between the efficacy of BH3-mimetics to Mcl-1, such as AMG-176 and S63845, and the level of *BCL2L1* mRNA (encoding Bcl-xL) (Caenepeel et al., 2018; Kotschy et al., 2016; Tron et al., 2018). Therefore, high levels of the Bcl-xL protein might limit the sensitivity of cancer cells to Mcl-1 inhibition and serve as mechanism of drug resistance. In addition, genomic analysis also revealed the significance of high levels of Bak as a predictor of activity of Mcl-1 inhibitors, as evidenced for AMG-176 (Caenepeel et al., 2018).

Finally, it should be noted that combinations of Mcl-1 inhibitors with other BH3-mimetics (in particular, with venetoclax) or chemotherapies (e.g., with docetaxel) have been found in several *in vivo* studies to increase efficacy or overcome resistance in comparison with monotherapy. It could be explained by the redundancy of anti-apoptotic Bcl-2 family proteins in different types of tumor cells. In addition, the simultaneous attacking of various targets in cancer cells along with the inhibition of Mcl-1 could improve the efficacy of the current therapeutic strategies. The combinations of BH3-mimetics selective to Mcl-1 and venetoclax are under clinical investigation and are expected to provide new options for the treatment of patients with various hematological malignancies.

5. Conclusions

Mcl-1 plays a major role in drug resistance in cancer therapy. There are two principal approaches in the targeting of Mcl-1: direct inhibition (BH3-mimetics) and indirect inhibition (e.g., CDK inhibitors and mTOR inhibitors). The direct targeting via BH3-mimetics leads to the potent and selective inhibition of Mcl-1, and now several small molecule inhibitors are under clinical evaluation. However, these drugs could have a lack of therapeutic efficacy as monotherapy and cause expected on-target toxicities. At the same time, targeting Mcl-1 indirectly might be more effective in some cases due to the influence on multiple cell signaling pathways (e.g., CDK inhibitors) and less toxic due to more selective killing of cancer cells (e.g., PaTrin-2). In any case, it is essential to seek novel drug combinations with both direct and indirect Mcl-1 inhibitors in order to improve therapeutic efficacy.

We are convinced that the significance of studying Mcl-1 will not decrease in the future and, hopefully, the compounds targeting Mcl-1 will demonstrate high efficacy in clinical trials. One of the most important challenges will be the suitable optimization of dosing regimens (parenteral versus oral route) for the safe and efficient treatment of patients. Undoubtedly, clinical approval of the first Mcl-1 inhibitor will be a great scientific breakthrough, providing the chance of a new life for many patients all over the world.

Acknowledgments

This work was supported by the Grant from the Russian Science Foundation (19-15-00125). The work in the authors' laboratories is also supported by the Grants from the Russian Foundation for Basic Research (18-29-09005, 20-015-00500A), the Swedish (190345) and Stockholm (181301) Cancer Societies. We apologize to those authors whose primary works could not be cited owing to space limitations.

References

Akgul, C., 2009. Mcl-1 is a potential therapeutic target in multiple types of cancer. Cell. Mol. Life Sci. 66, 1326–1336.

Albershardt, T.C., Salerni, B.L., Soderquist, R.S., Bates, D.J.P., Pletnev, A.A., Kisselev, A.F., Eastman, A., 2011. Multiple BH3 mimetics antagonize antiapoptotic MCL1 protein by inducing the endoplasmic reticulum stress response and up-regulating BH3-only protein NOXA. J. Biol. Chem. 286, 24882–24895.

Algarín, E.M., Díaz-Tejedor, A., Mogollón, P., Hernández-García, S., Corchete, L.A., San-Segundo, L., Martín-Sánchez, M., González-Méndez, L., Schoumacher, M., Banquet, S., Kraus-Berthier, L., Kloos, I., Derreal, A., Halilovic, E., Maacke, H., Gutiérrez, N.C., Mateos, M.-V., Paíno, T., Garayoa, M., Ocio, E.M., 2019. Preclinical evaluation of the simultaneous inhibition of MCL-1 and BCL-2 with the combination of S63845 and venetoclax in multiple myeloma. Haematologica. https://doi.org/10.3324/haematol.2018.212308. [Epub ahead of print].

Arbour, N., Vanderluit, J.L., Le Grand, J.N., Jahani-Asl, A., Ruzhynsky, V.A., Cheung, E.C.C., Kelly, M.A., MacKenzie, A.E., Park, D.S., Opferman, J.T., Slack, R.S., 2008. Mcl-1 is a key regulator of apoptosis during CNS development and after DNA damage. J. Neurosci. 28, 6068–6078.

Arellano, M.L., Borthakur, G., Berger, M., Luer, J., Raza, A., 2014. A phase II, multicenter, open-label study of obatoclax mesylate in patients with previously untreated myelodysplastic syndromes with anemia or thrombocytopenia. Clin. Lymphoma Myeloma Leuk. 14, 534–539.

Basit, F., Cristofanon, S., Fulda, S., 2013. Obatoclax (GX15-070) triggers necroptosis by promoting the assembly of the necrosome on autophagosomal membranes. Cell Death Differ. 20, 1161–1173.

Becker, T.M., Boyd, S.C., Mijatov, B., Gowrishankar, K., Snoyman, S., Pupo, G.M., Scolyer, R.A., Mann, G.J., Kefford, R.F., Zhang, X.D., Rizos, H., 2014. Mutant B-RAF-Mcl-1 survival signaling depends on the STAT3 transcription factor. Oncogene 33, 1158–1166.

Bonapace, L., Bornhauser, B.C., Schmitz, M., Cario, G., Ziegler, U., Niggli, F.K., Schäfer, B.W., Schrappe, M., Stanulla, M., Bourquin, J.P., 2010. Induction of autophagy-dependent necroptosis is required for childhood acute lymphoblastic leukemia cells to overcome glucocorticoid resistance. J. Clin. Invest. 120, 1310–1323.

Brennan, M.S., Chang, C., Tai, L., Lessene, G., Strasser, A., Dewson, G., Kelly, G.L., Herold, M.J., 2018. Humanized mcl-1 mice enable accurate preclinical evaluation of MCL-1 inhibitors destined for clinical use. Blood 132, 1573–1583.

Broecker-Preuss, M., Viehof, J., Jastrow, H., Becher-Boveleth, N., Fuhrer, D., Mann, K., 2015. Cell death induction by the BH3 mimetic GX15-070 in thyroid carcinoma cells. J. Exp. Clin. Cancer Res. 34.

Brotin, E., Meryet-Figuière, M., Simonin, K., Duval, R.E., Villedieu, M., Leroy-Dudal, J., Saison-Behmoaras, E., Gauduchon, P., Denoyelle, C., Poulain, L., 2010. Bcl-xL and MCL-1 constitute pertinent targets in ovarian carcinoma and their concomitant inhibition is sufficient to induce apoptosis. Int. J. Cancer 126, 885–895.

Brown, J.R., Tesar, B., Yu, L., Werner, L., Takebe, N., Mikler, E., Reynolds, H.M., Thompson, C., Fisher, D.C., Neuberg, D., Freedman, A.S., 2015. Obatoclax in combination with fludarabine and rituximab is well-tolerated and shows promising clinical activity in relapsed chronic lymphocytic leukemia. Leuk. Lymphoma 56, 3336–3342.

Brumatti, G., Ekert, P.G., 2013. Seeking a MCL-1 inhibitor. Cell Death Differ. 435, 677–681.

Caenepeel, S., Brown, S.P., Belmontes, B., Moody, G., Keegan, K.S., Chui, D., Whittington, D.A., Huang, X., Poppe, L., Cheng, A.C., Cardozo, M., Houze, J., Li, Y., Lucas, B., Paras, N.A., Wang, X., Taygerly, J.P., Vimolratana, M., Zancanella, M., Zhu, L., Cajulis, E., Osgood, T., Sun, J., Damon, L., Egan, R.K.,

Greninger, P., McClanaghan, J.D., Gong, J., Moujalled, D., Pomilio, G., Beltran, P., Benes, C.H., Roberts, A.W., Huang, D.C., Wei, A., Canon, J., Coxon, A., Hughes, P.E., 2018. AMG 176, a selective MCL1 inhibitor, is effective in hematologic cancer models alone and in combination with established therapies. Cancer Discov. 8, 1582–1597.

Cao, F., Gong, Y.B., Kang, X.H., Lu, Z.H., Wang, Y., Zhao, K.L., Miao, Z.H., Liao, M.J., Xu, Z.Y., 2019. Degradation of MCL-1 by bufalin reverses acquired resistance to osimertinib in EGFR-mutant lung cancer. Toxicol. Appl. Pharmacol. 379.

Cerella, C., Muller, F., Gaigneaux, A., Radogna, F., Viry, E., Chateauvieux, S., Dicato, M., Diederich, M., 2015. Early downregulation of Mcl-1 regulates apoptosis triggered by cardiac glycoside UNBS1450. Cell Death Dis. 6, e1782.

Champa, D., Russo, M.A., Liao, X.H., Refetoff, S., Ghossein, R.A., Di Cristofano, A., 2014. Obatoclax overcomes resistance to cell death in aggressive thyroid carcinomas by countering Bcl2a1 and Mcl1 overexpression. Endocr. Relat. Cancer 21, 755–767.

Chanvorachote, P., Pongrakhananon, V., 2012. Ouabain downregulates Mcl-1 and sensitizes lung cancer cells to TRAIL-induced apoptosis. Am. J. Physiol-Cell Physiol. 304, 263–272.

Chen, W., Bai, L., Wang, X., Xu, S., Belinsky, S.A., Lin, Y., 2009. Acquired activation of the Akt/cyclooxygenase-2/Mcl-1 pathway renders lung cancer cells resistant to apoptosis. Mol. Pharmacol. 77, 416–423.

Cheng, P., Ni, Z., Dai, X., Wang, B., Ding, W., Rae Smith, A., Xu, L., Wu, D., He, F., Lian, J., 2013. The novel BH-3 mimetic apogossypolone induces Beclin-1- and ROS-mediated autophagy in human hepatocellular carcinoma cells. Cell Death Dis. 4, e489.

Chiappori, A.A., Schreeder, M.T., Moezi, M.M., Stephenson, J.J., Blakely, J., Salgia, R., Chu, Q.S., Ross, H.J., Subramaniam, D.S., Schnyder, J., Berger, M.S., 2012. A phase I trial of pan-Bcl-2 antagonist obatoclax administered as a 3-h or a 24-h infusion in combination with carboplatin and etoposide in patients with extensive-stage small cell lung cancer. Br. J. Cancer 106, 839–845.

Chiappori, A., Williams, C., Northfelt, D.W., Adams, J.W., Malik, S., Edelman, M.J., Rosen, P., Van Echo, D.A., Berger, M.S., Haura, E.B., 2014. Obatoclax mesylate, a pan-bcl-2 inhibitor, in combination with docetaxel in a phase 1/2 trial in relapsed non-small-cell lung cancer. J. Thorac. Oncol. 9, 121–125.

Chonghaile, T.N., Sarosiek, K.A., Vo, T.-T., Ryan, J.A., Tammareddi, A., Moore, V.D.G., Deng, J., Anderson, K.C., Richardson, P., Tai, Y.-T., Mitsiades, C.S., Matulonis, U.A., Drapkin, R., Stone, R., DeAngelo, D.J., McConkey, D.J., Sallan, S.E., Silverman, L., Hirsch, M.S., Carrasco, D.R., Letai, A., 2011. Pretreatment mitochondrial priming correlates with clinical response to cytotoxic chemotherapy. Science 334, 1129–1133.

Cidado, J., Boiko, S., Proia, T., Ferguson, D., Criscione, S.W., San Martin, M., Pop-Damkov, P., Su, N., Roamio Franklin, V.N., Sekhar Reddy Chilamakuri, C., D'Santos, C.S., Shao, W., Saeh, J.C., Koch, R., Weinstock, D.M., Zinda, M., Fawell, S.E., Drew, L., 2019. AZD4573 is a highly selective CDK9 inhibitor that suppresses Mcl-1 and induces apoptosis in hematological cancer cells. Clin. Cancer Res. https://doi.org/10.1158/1078-0432.CCR-19-1853. [Epub ahead of print].

Cohen, N.A., Stewart, M.L., Gavathiotis, E., Tepper, J.L., Bruekner, S.R., Koss, B., Opferman, J.T., Walensky, L.D., 2012. A competitive stapled peptide screen identifies a selective small molecule that overcomes MCL-1-dependent leukemia cell survival. Chem. Biol. 19, 1175–1186.

Crawford, N., Chacko, A.D., Savage, K.I., McCoy, F., Redmond, K., Longley, D.B., Fennell, D.A., 2011. Platinum resistant cancer cells conserve sensitivity to BH3 domains and obatoclax induced mitochondrial apoptosis. Apoptosis 16, 311–320.

Cui, J., Placzek, W.J., 2018. Post-transcriptional regulation of anti-apoptotic BCL2 family members. Int. J. Mol. Sci. 19, 308.

Czabotar, P.E., Lee, E.F., van Delft, M.F., Day, C.L., Smith, B.J., Huang, D.C.S., Fairlie, W.D., Hinds, M.G., Colman, P.M., 2007. Structural insights into the degradation of Mcl-1 induced by BH3 domains. Proc. Natl. Acad. Sci. U. S. A. 104, 6217–6222.

Czabotar, P.E., Lessene, G., Strasser, A., Adams, J.M., 2014. Control of apoptosis by the BCL-2 protein family: implications for physiology and therapy. Nat. Rev. Mol. Cell Biol. 15, 49–63.

De Blasio, A., Vento, R., Di Fiore, R., 2018. Mcl-1 targeting could be an intriguing perspective to cure cancer. J. Cell. Physiol. 233, 8482–8498.

Debrincat, M.A., Josefsson, E.C., James, C., Henley, K.J., Ellis, S., Lebois, M., Betterman, K.L., Lane, R.M., Rogers, K.L., White, M.J., Roberts, A.W., Harvey, N.L., Metcalf, D., Kile, B.T., 2012. Mcl-1 and Bcl-x L coordinately regulate megakaryocyte survival. Blood 119, 5850–5858.

Dey, J., Deckwerth, T.L., Kerwin, W.S., Casalini, J.R., Merrell, A.J., Grenley, M.O., Burns, C., Ditzler, S.H., Dixon, C.P., Beirne, E., Gillespie, K.C., Kleinman, E.F., Klinghoffer, R.A., 2017. Voruciclib, a clinical stage oral CDK9 inhibitor, represses MCL-1 and sensitizes high-risk diffuse large B-cell lymphoma to BCL2 inhibition. Sci. Rep. 7, 18007.

Ding, Q., He, X., Xia, W., Hsu, J.M., Chen, C.T., Li, L.Y., Lee, D.F., Yang, J.Y., Xie, X., Liu, J.C., Hung, M.C., 2007. Myeloid cell leukemia-1 inversely correlates with glycogen synthase kinase-3β activity and associates with poor prognosis in human breast cancer. Cancer Res. 67, 4564–4571.

Ding, X., Zhang, Y., Huang, T., Xu, G., Peng, C., Chen, G., Kong, B., Friess, H., Shen, S., Lv, Y., Roberts, L.R., Wang, L., Zou, X., 2019. Targeting sphingosine kinase 2 suppresses cell growth and synergizes with BCL2/BCL-XL inhibitors through NOXA-mediated MCL1 degradation in cholangiocarcinoma. Am. J. Cancer Res. 9, 546.

Doi, K., Li, R., Sung, S.S., Wu, H., Liu, Y., Manieri, W., Krishnegowda, G., Awwad, A., Dewey, A., Liu, X., Amin, S., Cheng, C., Qin, Y., Schonbrunn, E., Daughdrill, G., Loughran, T.P., Sebti, S., Wang, H.G., 2012. Discovery of marinopyrrole A (maritoclax) as a selective Mcl-1 antagonist that overcomes ABT-737 resistance by binding to and targeting Mcl-1 for proteasomal degradation. J. Biol. Chem. 287, 10224–10235.

Dzhagalov, I., Dunkle, A., He, Y.-W., 2014. The anti-apoptotic Bcl-2 family member Mcl-1 promotes T lymphocyte survival at multiple stages. J. Immunol. 181, 521–528.

Eichhorn, J.M., Alford, S.E., Hughes, C.C., Fenical, W., Chambers, T.C., 2013. Purported Mcl-1 inhibitor marinopyrrole a fails to show selective cytotoxicity for Mcl-1-dependent cell lines. Cell Death Dis. 4, e880.

Elkholi, R., Floros, K.V., Chipuk, J.E., 2011. The role of BH3-only proteins in tumor cell development, signaling, and treatment. Genes and Cancer 2, 523–537.

Elmore, S., 2007. Apoptosis: a review of programmed cell death. Toxicol. Pathol. 35, 495–516.

Foight, G.W., Ryan, J.A., Gullá, S.V., Letai, A., Keating, A.E., 2014. Designed BH3 peptides with high affinity and specificity for targeting Mcl-1 in cells. ACS Chem. Biol. 9, 1962–1968.

Fujise, K., Zhang, D., Liu, J., Yeh, E.T.H., 2000. Regulation of apoptosis and cell cycle progression by MCL1. J. Biol. Chem. 275, 39458–39465.

Germain, M., Duronio, V., 2007. The N terminus of the anti-apoptotic BCL-2 homologue MCL-1 regulates its localization and function. J. Biol. Chem. 282, 32233–32242.

Glaser, S.P., Lee, E.F., Trounson, E., Bouillet, P., Wei, A., Fairlie, W.D., Izon, D.J., Zuber, J., Rappaport, A.R., Herold, M.J., Alexander, W.S., Lowe, S.W., Robb, L., Strasser, A., 2012. Anti-apoptotic Mcl-1 is essential for the development and sustained growth of acute myeloid leukemia. Genes Dev. 26, 120–125.

Gomez-Bougie, P., Ménoret, E., Juin, P., Dousset, C., Pellat-Deceunynck, C., Amiot, M., 2011. Noxa controls mule-dependent Mcl-1 ubiquitination through the regulation of the Mcl-1/USP9X interaction. Biochem. Biophys. Res. Commun. 413, 460–464.

Goy, A., Hernandez-Ilzaliturri, F.J., Kahl, B., Ford, P., Protomastro, E., Berger, M., 2014. A phase I/II study of the pan Bcl-2 inhibitor obatoclax mesylate plus bortezomib for relapsed or refractory mantle cell lymphoma. Leuk. Lymphoma 55, 2761–2768.

Guikema, J.E., Amiot, M., Eldering, E., 2017. Exploiting the pro-apoptotic function of NOXA as a therapeutic modality in cancer. Expert Opin. Ther. Targets 8, 767–779.

Hanahan, D., Weinberg, R.A., 2011. Hallmarks of cancer: the next generation. Cell 144, 646–674.

Haste, N.M., Hughes, C.C., Tran, D.N., Fenical, W., Jensen, P.R., Nizet, V., Hensler, M.E., 2011. Pharmacological properties of the marine natural product marinopyrrole A against methicillin-resistant Staphylococcus aureus. Antimicrob. Agents Chemother. 55, 3305–3312.

He, J.H., Liao, X.L.I., Wang, W., Li, D.D., Chen, W.D., Deng, R., Yang, D., Han, Z.E.P., Jiang, J.W., Zhu, X.F., 2014. Apogossypolone, a small-molecule inhibitor of Bcl-2, induces radiosensitization of nasopharyngeal carcinoma cells by stimulating autophagy. Int. J. Oncol. 45, 1099–1108.

Hu, Z.Y., Sun, J., Zhu, X.F., Yang, D., Zeng, Y.X., 2009. ApoG2 induces cell cycle arrest of nasopharyngeal carcinoma cells by suppressing the c-Myc signaling pathway. J. Transl. Med. 7, 74.

Hu, S., Marineau, J.J., Rajagopal, N., Hamman, K.B., Choi, Y.J., Schmidt, D.R., Ke, N., Johannessen, L., Bradley, M.J., Orlando, D.A., Alnemy, S.R., Ren, Y., Ciblat, S., Winter, D.K., Kabro, A., Sprott, K.T., Hodgson, J.G., Fritz, C.C., Carulli, J.P., Di Tomaso, E., Olson, E.R., 2019. Discovery and characterization of SY-1365, a selective, covalent inhibitor of CDK7. Cancer Res. 79, 3479–3491.

Huber, S., Oelsner, M., Decker, T., Zum Büschenfelde, C.M., Wagner, M., Lutzny, G., Kuhnt, T., Schmidt, B., Oostendorp, R.A.J., Peschel, C., Ringshausen, I., 2011. Sorafenib induces cell death in chronic lymphocytic leukemia by translational downregulation of Mcl-1. Leukemia 25, 838–847.

Jamil, S., Mojtabavi, S., Hojabrpour, P., Cheah, S., Duronio, V., 2008. An essential role for MCL-1 in ATR-mediated CHK1 phosphorylation. Mol. Biol. Cell 19, 3212–3220.

Kadia, T.M., Kantarjian, H.M., Konopleva, M., 2019. Myeloid cell leukemia-1 dependence in acute myeloid leukemia: a novel approach to patient therapy. Oncotarget 10, 1250–1265.

Kang, X.H., Zhang, J.H., Zhang, Q.Q., Cui, Y.H., Wang, Y., Kou, W.Z., Miao, Z.H., Lu, P., Wang, L.F., Xu, Z.Y., Cao, F., 2017. Degradation of Mcl-1 through GSK-3β activation regulates apoptosis induced by bufalin in non-small cell lung cancer H1975 cells. Cell. Physiol. Biochem. 41, 2067–2076.

Kim, E.Y., Sudini, K., Singh, A.K., Haque, M., Leaman, D., Khuder, S., Ahmed, S., 2018. Ursolic acid facilitates apoptosis in rheumatoid arthritis synovial fibroblasts by inducing SP1-mediated Noxa expression and proteasomal degradation of Mcl-1. FASEB J. 32, 6174–6185.

Klenke, S., Akdeli, N., Stelmach, P., Heukamp, L., Schulte, J.H., Bachmann, H.S., 2019. The small molecule Bcl-2/Mcl-1 inhibitor TW-37 shows single-agent cytotoxicity in neuroblastoma cell lines. BMC Cancer 19, 243.

Koch, R., Christie, A.L., Crombie, J.L., Palmer, A.C., Plana, D., Shigemori, K., Morrow, S.N., Van Scoyk, A., Wu, W., Brem, E.A., Paul Secrist, J., Drew, L., Schuller, A.G., Cidado, J., Letai, A., Weinstock, D.M., 2019. Biomarker-driven strategy for MCL1 inhibition in T-cell lymphomas. Blood 133, 566–575.

Kodama, T., Takehara, T., Hikita, H., Shimizu, S., Shigekawa, M., Li, W., Miyagi, T., Hosui, A., Tatsumi, T., Ishida, H., Kanto, T., Hiramatsu, N., Yin, X.M.,

Hayashi, N., 2011. BH3-only activator proteins Bid and Bim are dispensable for Bak/Bax-dependent thrombocyte apoptosis induced by Bcl-xL deficiency molecular requisites for the mitochondrial pathway to apoptosis in platelets. J. Biol. Chem. 286, 13905–13913.

Koehler, B.C., Scherr, A.L., Lorenz, S., Elssner, C., Kautz, N., Welte, S., Jaeger, D., Urbanik, T., Schulze-Bergkamen, H., 2014. Pan-Bcl-2 inhibitor obatoclax delays cell cycle progression and blocks migration of colorectal cancer cells. PLoS One 9, e106571.

Konopleva, M., Contractor, R., Tsao, T., Samudio, I., Ruvolo, P.P., Kitada, S., Deng, X., Zhai, D., Shi, Y.X., Sneed, T., Verhaegen, M., Soengas, M., Ruvolo, V.R., McQueen, T., Schober, W.D., Watt, J.C., Jiffar, T., Ling, X., Marini, F.C., Harris, D., Dietrich, M., Estrov, Z., McCubrey, J., May, W.S., Reed, J.C., Andreeff, M., 2006. Mechanisms of apoptosis sensitivity and resistance to the BH3 mimetic ABT-737 in acute myeloid leukemia. Cancer Cell 10, 375–388.

Konopleva, M., Watt, J., Contractor, R., Tsao, T., Harris, D., Estrov, Z., Bornmann, W., Kantarjian, H., Viallet, J., Samudio, I., Andreeff, M., 2008. Mechanisms of antileukemic activity of the novel Bcl-2 homology domain-3 mimetic GX15-070 (Obatoclax). Cancer Res. 68, 3413–3420.

Kotschy, A., Szlavik, Z., Murray, J., Davidson, J., Maragno, A.L., Le Toumelin-Braizat, G., Chanrion, M., Kelly, G.L., Gong, J.N., Moujalled, D.M., Bruno, A., Csekei, M., Paczal, A., Szabo, Z.B., Sipos, S., Radics, G., Proszenyak, A., Balint, B., Ondi, L., Blasko, G., Robertson, A., Surgenor, A., Dokurno, P., Chen, I., Matassova, N., Smith, J., Pedder, C., Graham, C., Studeny, A., Lysiak-Auvity, G., Girard, A.M., Gravé, F., Segal, D., Riffkin, C.D., Pomilio, G., Galbraith, L.C.A., Aubrey, B.J., Brennan, M.S., Herold, M.J., Chang, C., Guasconi, G., Cauquil, N., Melchiore, F., Guigal-Stephan, N., Lockhart, B., Colland, F., Hickman, J.A., Roberts, A.W., Huang, D.C.S., Wei, A.H., Strasser, A., Lessene, G., Geneste, O., 2016. The MCL1 inhibitor S63845 is tolerable and effective in diverse cancer models. Nature 538, 477–482.

Lanasa, M.C., Andritsos, L., Brown, J.R., Gabrilove, J., Caligaris-Cappio, F., Ghia, P., Larson, R.A., Kipps, T.J., Leblond, V., Milligan, D.W., Janssens, A., Johnson, A.J., Heerema, N.A., Bühler, A., Stilgenbauer, S., Devin, J., Hallek, M., Byrd, J.C., Grever, M.R., 2015. Final results of EFC6663: a multicenter, international, phase 2 study of alvocidib for patients with fludarabine-refractory chronic lymphocytic leukemia. Leuk. Res. 39, 495–500.

Langer, C.J., Albert, I., Ross, H.J., Kovacs, P., Blakely, L.J., Pajkos, G., Somfay, A., Zatloukal, P., Kazarnowicz, A., Moezi, M.M., Schreeder, M.T., Schnyder, J., Ao-Baslock, A., Pathak, A.K., Berger, M.S., 2014. Randomized phase II study of carboplatin and etoposide with or without obatoclax mesylate in extensive-stage small cell lung cancer. Lung Cancer 85, 420–428.

Lee, T., Christov, P.P., Shaw, S., Tarr, J.C., Zhao, B., Veerasamy, N., Jeon, K.O., Mills, J.J., Bian, Z., Sensintaffar, J.L., Arnold, A.L., Fogarty, S.A., Perry, E., Ramsey, H.E., Cook, R.S., Hollingshead, M., Davis Millin, M., Lee, K.M., Koss, B., Budhraja, A., Opferman, J.T., Kim, K., Arteaga, C.L., Moore, W.J., Olejniczak, E.T., Savona, M.R., Fesik, S.W., 2019. Discovery of potent myeloid cell leukemia-1 (Mcl-1) inhibitors that demonstrate in vivo activity in mouse xenograft models of human cancer. J. Med. Chem. 62, 3971–3988.

Lei, S., Ding, Y., Fu, Y., Wu, S., Xie, X., Wang, C., Liang, H., 2017. The preclinical analysis of TW-37 as a potential anti-colorectal cancer cell agent. PLoS One 12, e0184501.

Letai, A., 2011. The control of mitochondrial apoptosis by the BCL-2 family. In: Reed, J.C., Green, D.R. (Eds.), Apoptosis: Physiology and Pathology. Cambridge University Press, pp. 44–50.

Leverson, J.D., Zhang, H., Chen, J., Tahir, S.K., Phillips, D.C., Xue, J., Nimmer, P., Jin, S., Smith, M., Xiao, Y., Kovar, P., Tanaka, A., Bruncko, M., Sheppard, G.S., Wang, L., Gierke, S., Kategaya, L., Anderson, D.J., Wong, C., Eastham-Anderson, J., Ludlam, M.C., Sampath, D., Fairbrother, W.J., Wertz, I., Rosenberg, S.H., Tse, C., Elmore, S.W., Souers, A.J., 2015. Potent and selective small-molecule MCL-1 inhibitors demonstrate on-target cancer cell killing activity as single agents and in combination with ABT-263 (navitoclax). Cell Death Dis. 6, e1590.

Li, J., Viallet, J., Haura, E.B., 2008. A small molecule pan-Bcl-2 family inhibitor, GX15-070, induces apoptosis and enhances cisplatin-induced apoptosis in non-small cell lung cancer cells. Cancer Chemother. Pharmacol. 61, 525–534.

Li, G., Liu, L., Shan, C., Cheng, Q., Budhraja, A., Zhou, T., Cui, H., Gao, N., 2014. RhoA/ROCK/PTEN signaling is involved in AT-101-mediated apoptosis in human leukemia cells in vitro and in vivo. Cell Death Dis. 5, e998.

Li, L., Pongtornpipat, P., Tiutan, T., Kendrick, S.L., Park, S., Persky, D.O., Rimsza, L.M., Puvvada, S.D., Schatz, J.H., 2015. Synergistic induction of apoptosis in high-risk DLBCL by BCL2 inhibition with ABT-199 combined with pharmacologic loss of MCL1. Leukemia 29, 1702–1712.

Li, H., Liu, L., Chang, H., Zou, Z., Xing, D., 2018. Downregulation of MCL-1 and upregulation of PUMA using mTOR inhibitors enhance antitumor efficacy of BH3 mimetics in triple-negative breast cancer article. Cell Death Dis. 9, 137.

Li, Z., He, S., Look, A.T., 2019. The MCL1-specific inhibitor S63845 acts synergistically with venetoclax/ABT-199 to induce apoptosis in T-cell acute lymphoblastic leukemia cells. Leukemia 33, 262–266.

Liang, L.Z., Ma, B., Liang, Y.J., Liu, H.C., Zhang, T.H., Zheng, G.S., Su, Y.X., Liao, G.Q., 2015. Obatoclax induces Beclin 1- and ATG5-dependent apoptosis and autophagy in adenoid cystic carcinoma cells. Oral Dis. 21, 470–477.

Liang, Y.Y., Deng, X.B., Zeng, L.S., Lin, X.T., Shao, X.F., Wang, B., Mo, Z.W., Yuan, Y.W., 2018. RASSF6-mediated inhibition of Mcl-1 through JNK activation improves the anti-tumor effects of sorafenib in renal cell carcinoma. Cancer Lett. 432, 75–83.

Lin, C.-L., Chen, C.-M., Lin, C.-L., Cheng, C.-W., Lee, C.-H., Hsieh, Y.-H., 2017a. Norcantharidin induces mitochondrial-dependent apoptosis through Mcl-1 inhibition in human prostate cancer cells. Biochim. Biophys. Acta Mol. Cell Res. 1864, 1867–1876.

Lin, J., Fu, D., Dai, Y., Lin, J., Xu, T., 2017b. Mcl-1 inhibitor suppresses tumor growth of esophageal squamous cell carcinoma in a mouse model. Oncotarget 8, 114457–114462.

Lu, Y., Huang, H., Yang, H., Chen, D., Wu, S., Jiang, Z., Wang, R., 2017. Small molecule inhibitor TW-37 is tolerable and synergistic with chemotherapy in nasopharyngeal carcinoma. Cell Cycle 16, 1376–1383.

Lu, X., Liu, Y.-C., Orvig, C., Liang, H., Chen, Z.-F., 2019. Discovery of β-carboline copper(II) complexes as Mcl-1 inhibitor and in vitro and in vivo activity in cancer models. Eur. J. Med. Chem. 181, 111567.

Lucas, K.M., Mohana-Kumaran, N., Lau, D., Zhang, X.D., Hersey, P., Huang, D.C., Weninger, W., Haass, N.K., Allen, J.D., 2012. Modulation of NOXA and MCL-1 as a strategy for sensitizing melanoma cells to the BH3-mimetic ABT-737. Clin. Cancer Res. 18, 783–795.

Mallick, D.J., Soderquist, R.S., Bates, D., Eastman, A., 2019. Confounding off-target effects of BH3 mimetics at commonly used concentrations: MIM1, UMI-77, and A-1210477. Cell Death Dis. 10, e185.

Merino, D., Whittle, J.R., Vaillant, F., Serrano, A., Gong, J.N., Giner, G., Maragno, A.L., Chanrion, M., Schneider, E., Pal, B., Li, X., Dewson, G., Gräsel, J., Liu, K., Lalaoui, N., Segal, D., Herold, M.J., Huang, D.C.S., Smyth, G.K., Geneste, O., Lessene, G.,

Visvader, J.E., Lindeman, G.J., 2017. Synergistic action of the MCL-1 inhibitor S63845 with current therapies in preclinical models of triple-negative and HER2-amplified breast cancer. Sci. Transl. Med. 9, eaam7049.

Milani, M., Byrne, D.P., Greaves, G., Butterworth, M., Cohen, G.M., Eyers, P.A., Varadarajan, S., 2017. DRP-1 is required for BH3 mimetic-mediated mitochondrial fragmentation and apoptosis. Cell Death Dis. 8, e2552.

Mohammad, R.M., Goustin, A.S., Aboukameel, A., Chen, B., Banerjee, S., Wang, G., Nikolovska-Coleska, Z., Wang, S., Al-Katib, A., 2007. Preclinical studies of TW-37, a new nonpeptidic small-molecule inhibitor of Bcl-2, in diffuse large cell lymphoma xenograft model reveal drug action on both Bcl-2 and Mcl-1. Clin. Cancer Res. 13, 2226–2235.

Mojsa, B., Lassot, I., Desagher, S., 2014. Mcl-1 ubiquitination: unique regulation of an essential survival protein. Cell 3, 418–437.

Montero, J., Letai, A., 2016. Dynamic BH3 profiling-poking cancer cells with a stick. Mol. Cell. Oncol. 3, e1040144.

Montero, J., Letai, A., 2018. Why do BCL-2 inhibitors work and where should we use them in the clinic? Cell Death Differ. 25, 56–64.

Moujalled, D.M., Pomilio, G., Ghiurau, C., Ivey, A., Salmon, J., Rijal, S., Macraild, S., Zhang, L., Teh, T.C., Tiong, I.S., Lan, P., Chanrion, M., Claperon, A., Rocchetti, F., Zichi, A., Kraus-Berthier, L., Wang, Y., Halilovic, E., Morris, E., Colland, F., Segal, D., Huang, D., Roberts, A.W., Maragno, A.L., Lessene, G., Geneste, O., Wei, A.H., 2018. Combining BH3-mimetics to target both BCL-2 and MCL1 has potent activity in pre-clinical models of acute myeloid leukemia. Leukemia 33, 905–917.

Muppidi, A., Doi, K., Edwardraja, S., Drake, E.J., Gulick, A.M., Wang, H.G., Lin, Q., 2012. Rational design of proteolytically stable, cell-permeable peptide-based selective Mcl-1 inhibitors. J. Am. Chem. Soc. 134, 14734–14737.

Muppidi, A., Doi, K., Edwardraja, S., Pulavarti, S.V.S.R.K., Szyperski, T., Wang, H.G., Lin, Q., 2014. Targeted delivery of ubiquitin-conjugated BH3 peptide-based Mcl-1 inhibitors into cancer cells. Bioconjug. Chem. 25, 424–432.

Muto, A., Hori, M., Sasaki, Y., Saitoh, A., Yasuda, I., Maekawa, T., Uchida, T., Asakura, K., Nakazato, T., Kaneda, T., Kizaki, M., Ikeda, Y., Yoshida, T., 2007. Emodin has a cytotoxic activity against human multiple myeloma as a Janus-activated kinase 2 inhibitor. Mol. Cancer Ther. 6, 987–994.

Nangia, V., Siddiqui, F.M., Caenepeel, S., Timonina, D., Bilton, S.J., Phan, N., Gomez-Caraballo, M., Archibald, H.L., Li, C., Fraser, C., Rigas, D., Vajda, K., Ferris, L.A., Lanuti, M., Wright, C.D., Raskin, K.A., Cahill, D.P., Shin, J.H., Keyes, C., Sequist, L.V., Piotrowska, Z., Farago, A.F., Azzoli, C.G., Gainor, J.F., Sarosiek, K.A., Brown, S.P., Coxon, A., Benes, C.H., Hughes, P.E., Hata, A.N., 2018. Exploiting MCL1 dependency with combination MEK + MCL1 inhibitors leads to induction of apoptosis and tumor regression in KRAS-mutant non-small cell lung cancer. Cancer Discov. 8, 1598–1613.

Nguyen, M., Marcellus, R.C., Roulston, A., Watson, M., Serfass, L., Murthy Madiraju, S.R., Goulet, D., Viallet, J., Belec, L., Billot, X., Acoca, S., Purisima, E., Wiegmans, A., Cluse, L., Johnstone, R.W., Beauparlant, P., Shore, G.C., 2007. Small molecule obatoclax (GX15-070) antagonizes MCL-1 and overcomes MCL-1-mediated resistance to apoptosis. Proc. Natl. Acad. Sci. U. S. A. 104, 19512–19517.

Opferman, J.T., Iwasaki, H., Ong, C.C., Suh, H., Mizuno, S.I., Akashi, K., Korsmeyer, S.J., 2005. Obligate role of anti-apoptotic MCL-1 in the survival of hematopoietic stem cells. Science 307, 1101–1104.

Paik, P.K., Rudin, C.M., Brown, A., Rizvi, N.A., Takebe, N., Travis, W., James, L., Ginsberg, M.S., Juergens, R., Markus, S., Tyson, L., Subzwari, S., Kris, M.G.,

Krug, L.M., 2010. A phase I study of obatoclax mesylate, a Bcl-2 antagonist, plus topotecan in solid tumor malignancies. Cancer Chemother. Pharmacol. 66, 1079–1085.

Paik, P.K., Rudin, C.M., Pietanza, M.C., Brown, A., Rizvi, N.A., Takebe, N., Travis, W., James, L., Ginsberg, M.S., Juergens, R., Markus, S., Tyson, L., Subzwari, S., Kris, M.G., Krug, L.M., 2011. A phase II study of obatoclax mesylate, a Bcl-2 antagonist, plus topotecan in relapsed small cell lung cancer. Lung Cancer 74, 481–485.

Palve, V., Mallick, N., Ghaisas, G., Kannan, S., Teni, T., 2014. Overexpression of Mcl-1L splice variant is associated with poor and prognosis chemoresistance in oral cancers. PLoS One 9, e111927.

Pandey, M.K., Gowda, K., Doi, K., Sharma, A.K., Wang, H.G., Amin, S., 2013. Proteasomal degradation of Mcl-1 by maritoclax induces apoptosis and enhances the efficacy of ABT-737 in melanoma cells. PLoS One 8, e78570.

Polier, G., Giaisi, M., Köhler, R., Müller, W.W., Lutz, C., Buss, E.C., Krammer, P.H., Li-Weber, M., 2015. Targeting CDK9 by wogonin and related natural flavones potentiates the anti-cancer efficacy of the Bcl-2 family inhibitor ABT-263. Int. J. Cancer 136, 688–698.

Pongrakhananon, V., Stueckle, T.A., Wang, H.Y.L., O'Doherty, G.A., Dinu, C.Z., Chanvorachote, P., Rojanasakul, Y., 2014. Monosaccharide digitoxin derivative sensitize human non-small cell lung cancer cells to anoikis through Mcl-1 proteasomal degradation. Biochem. Pharmacol. 88, 23–35.

Rahmani, M., Davis, E.M., Bauer, C., Dent, P., Grant, S., 2005. Apoptosis induced by the kinase inhibitor BAY 43-9006 in human leukemia cells involves down-regulation of Mcl-1 through inhibition of translation. J. Biol. Chem. 280, 35217–35227.

Ramsey, H.E., Fischer, M.A., Lee, T., Gorska, A.E., Arrate, M.P., Fuller, L., Boyd, K.L., Strickland, S.A., Sensintaffar, J., Hogdal, L.J., Ayers, G.D., Olejniczak, E.T., Fesik, S.W., Savona, M.R., 2018. A novel MCL1 inhibitor combined with venetoclax rescues venetoclax-resistant acute myelogenous Leukemia. Cancer Discov. 8, 1566–1581.

Ready, N., Karaseva, N.A., Orlov, S.V., Luft, A.V., Popovych, O., Holmlund, J.T., Wood, B.A., Leopold, L., 2011. Double-blind, placebo-controlled, randomized phase 2 study of the proapoptotic agent AT-101 plus docetaxel, in second-line non-small cell lung cancer. J. Thorac. Oncol. 6, 781–785.

Rezaei Araghi, R., Ryan, J.A., Letai, A., Keating, A.E., 2016. Rapid optimization of Mcl-1 inhibitors using stapled peptide libraries including non-natural side chains. ACS Chem. Biol. 11, 1238–1244.

Rezaei Araghi, R., Bird, G.H., Ryan, J.A., Jenson, J.M., Godes, M., Pritz, J.R., Grant, R.A., Letai, A., Walensky, L.D., Keating, A.E., 2018. Iterative optimization yields Mcl-1-targeting stapled peptides with selective cytotoxicity to Mcl-1-dependent cancer cells. Proc. Natl. Acad. Sci. U. S. A. 115, E886–E895.

Rinkenberger, J.L., Horning, S., Klocke, B., Roth, K., Korsmeyer, S.J., 2000. Mcl-1 deficiency results in peri-implantation embryonic lethality. Genes Dev. 14, 23–27.

Sadahira, K., Sagawa, M., Nakazato, T., Uchida, H., Ikeda, Y., Okamoto, S., Nakajima, H., Kizaki, M., 2014. Gossypol induces apoptosis in multiple myeloma cells by inhibition of interleukin-6 signaling and Bcl-2/Mcl-1 pathway. Int. J. Oncol. 45, 2278–2286.

Schafmeister, C.E., Po, J., Verdine, G.L., 2000. An all-hydrocarbon cross-linking system for enhancing the helicity and metabolic stability of peptides [8]. J. Am. Chem. Soc. 122, 5891–5892.

Schelman, W.R., Mohammed, T.A., Traynor, A.M., Kolesar, J.M., Marnocha, R.M., Eickhoff, J., Keppen, M., Alberti, D.B., Wilding, G., Takebe, N., Liu, G., 2014. A phase I study of AT-101 with cisplatin and etoposide in patients with advanced solid tumors with an expanded cohort in extensive-stage small cell lung cancer. Invest. New Drugs 32, 295–302.

Senichkin, V.V., Streletskaia, A.Y., Zhivotovsky, B., Kopeina, G.S., 2019. Molecular comprehension of Mcl-1: from gene structure to cancer therapy. Trends Cell Biol. 29, 549–562.
Senichkin, V.V., Streletskaia, A.Y., Gorbunova, A.S., Zhivotovsky, B., Kopeina, G.S., 2020. Saga of Mcl-1: regulation from transcription to degradation. Cell Death Differ. 27, 405–419.
Shi, P., Oh, Y.T., Deng, L., Zhang, G., Qian, G., Zhang, S., Ren, H., Wu, G., Legendre, B., Jr., Anderson, E., Ramalingam, S.S., Owonikoko, T.K., Chen, M., Sun, S.Y., 2017. Overcoming acquired resistance to AZD9291, a third-generation EGFR inhibitor, through modulation of MEK/ERK-dependent Bim and Mcl-1 degradation. Clin. Cancer Res. 23, 6567–6579.
Soderquist, R., Pletnev, A.A., Danilov, A.V., Eastman, A., 2014a. The putative BH3 mimetic S1 sensitizes leukemia to ABT-737 by increasing reactive oxygen species, inducing endoplasmic reticulum stress, and upregulating the BH3-only protein NOXA. Apoptosis 19, 201–209.
Soderquist, R.S., Danilov, A.V., Eastman, A., 2014b. Gossypol increases expression of the pro-apoptotic BH3-only protein NOXA through a novel mechanism involving phospholipase A2, cytoplasmic calcium, and endoplasmic reticulum stress. J. Biol. Chem. 289, 16190–16199.
Song, L., Coppola, D., Livingston, S., Cress, D., Haura, E.B., 2005. Mcl-1 regulates survival and sensitivity to diverse apoptotic stimuli in human non-small cell lung cancer cells. Cancer Biol. Ther. 4, 267–276.
Song, T., Wang, Z., Ji, F., Feng, Y., Fan, Y., Chai, G., Li, X., Li, Z., Zhang, Z., 2016. Deactivation of Mcl-1 by dual-function small-molecule inhibitors targeting the Bcl-2 homology 3 domain and facilitating Mcl-1 Ubiquitination. Angew. Chem. Int. Ed. 55, 14250–14256.
Sonpavde, G., Matveev, V., Burke, J.M., Caton, J.R., Fleming, M.T., Hutson, T.E., Galsky, M.D., Berry, W.R., Karlov, P., Holmlund, J.T., Wood, B.A., Brookes, M., Leopold, L., 2012. Randomized phase II trial of docetaxel plus prednisone in combination with placebo or AT-101, an oral small molecule Bcl-2 family antagonist, as first-line therapy for metastatic castration-resistant prostate cancer. Ann. Oncol. 23, 1803–1808.
Spijkers-Hagelstein, J.A.P., Schneider, P., Pinhanços, S.M., Garrido Castro, P., Pieters, R., Stam, R.W., 2014. Glucocorticoid sensitisation in mixed lineage leukaemia-rearranged acute lymphoblastic leukaemia by the pan-BCL-2 family inhibitors gossypol and AT-101. Eur. J. Cancer 50, 1665–1674.
Stamelos, V.A., Fisher, N., Bamrah, H., Voisey, C., Price, J.C., Farrell, W.E., Redman, C.W., Richardson, A., 2016. The BH3 mimetic obatoclax accumulates in lysosomes and causes their alkalinization. PLoS One 11, e0150696.
Stein, M.N., Goodin, S., Gounder, M., Gibbon, D., Moss, R., Portal, D., Lindquist, D., Zhao, Y., Takebe, N., Tan, A., Aisner, J., Lin, H., Ready, N., Mehnert, J.M., 2019. A phase I study of AT-101, a BH3 mimetic, in combination with paclitaxel and carboplatin in solid tumors. Invest. New Drugs. https://doi.org/10.1007/s10637-019-00807-2. [Epub ahead of print].
Stewart, M.L., Fire, E., Keating, A.E., Walensky, L.D., 2010. The MCL-1 BH3 helix is an exclusive MCL-1 inhibitor and apoptosis sensitizer. Nat. Chem. Biol. 6, 595–601.
Sulkshane, P., Teni, T., 2016. BH3 mimetic Obatoclax (GX15-070) mediates mitochondrial stress predominantly via MCL-1 inhibition and induces autophagy-dependent necroptosis in human oral cancer cells. Oncotarget 8, 60060–60079.
Szlávik, Z., Ondi, L., Csékei, M., Paczal, A., Szabó, Z.B., Radics, G., Murray, J., Davidson, J., Chen, I., Davis, B., Hubbard, R.E., Pedder, C., Dokurno, P., Surgenor, A., Smith, J., Robertson, A., LeToumelin-Braizat, G., Cauquil, N.,

Zarka, M., Demarles, D., Perron-Sierra, F., Claperon, A., Colland, F., Geneste, O., Kotschy, A., 2019. Structure-guided discovery of a selective Mcl-1 inhibitor with cellular activity. J. Med. Chem. 62, 6913–6924.
Thomas, L.W., Lam, C., Edwards, S.W., 2010. Mcl-1; the molecular regulation of protein function. FEBS Lett. 584, 2981–2989.
Timme, C.R., Gruidl, M., Yeatman, T.J., 2013. Gamma-secretase inhibition attenuates oxaliplatin-induced apoptosis through increased Mcl-1 and/or Bcl-xL in human colon cancer cells. Apoptosis 18, 1163–1174.
Tron, A.E., Belmonte, M.A., Adam, A., Aquila, B.M., Boise, L.H., Chiarparin, E., Cidado, J., Embrey, K.J., Gangl, E., Gibbons, F.D., Gregory, G.P., Hargreaves, D., Hendricks, J.A., Johannes, J.W., Johnstone, R.W., Kazmirski, S.L., Kettle, J.G., Lamb, M.L., Matulis, S.M., Nooka, A.K., Packer, M.J., Peng, B., Rawlins, P.B., Robbins, D.W., Schuller, A.G., Su, N., Yang, W., Ye, Q., Zheng, X., Secrist, J.P., Clark, E.A., Wilson, D.M., Fawell, S.E., Hird, A.W., 2018. Discovery of Mcl-1-specific inhibitor AZD5991 and preclinical activity in multiple myeloma and acute myeloid leukemia. Nat. Commun. 9, 5341.
Tse, C., Shoemaker, A.R., Adickes, J., Anderson, M.G., Chen, J., Jin, S., Johnson, E.F., Marsh, K.C., Mitten, M.J., Nimmer, P., Roberts, L., Tahir, S.K., Xiao, Y., Yang, X., Zhang, H., Fesik, S., Rosenberg, S.H., Elmore, S.W., 2008. ABT-263: a potent and orally bioavailable Bcl-2 family inhibitor. Cancer Res. 68, 3421–3428.
Tutusaus, A., Stefanovic, M., Cucarull, B., Mari, M., Boix, L., Reig, M., Bruix, J., Morales, A., 2018. Sorafenib/regorafenib resistance and BH3-mimetics efficacy in hepatocellular carcinoma treatment is determined by mitochondrial changes in the BCL-2 profile. J. Hepatol. 68, S686–S687.
Varadarajan, S., Vogler, M., Butterworth, M., Dinsdale, D., Walensky, L.D., Cohen, G.M., 2013. Evaluation and critical assessment of putative MCL-1 inhibitors. Cell Death Differ. 20, 1475–1484.
Varadarajan, S., Poornima, P., Milani, M., Gowda, K., Amin, S., Wang, H.-G., Cohen, G.M., 2015. Maritoclax and dinaciclib inhibit MCL-1 activity and induce apoptosis in both a MCL-1-dependent and -independent manner. Oncotarget 6, 12668–12681.
Villalobos-Ortiz, M., Ryan, J., Mashaka, T.N., Opferman, J.T., Letai, A., 2019. BH3 profiling discriminates on-target small molecule BH3 mimetics from putative mimetics. Cell Death Differ. https://doi.org/10.1038/s41418-019-0391-9[Epub ahead of print].
Voss, V., Senft, C., Lang, V., Ronellenfitsch, M.W., Steinbach, J.P., Seifert, V., Kogel, D., 2010. The Pan-Bcl-2 inhibitor (−)-gossypol triggers autophagic cell death in malignant glioma. Mol. Cancer Res. 8, 1002–1016.
Waiwut, P., Inujima, A., Inoue, H., Saiki, I., Sakurai, H., 2012. Bufotalin sensitizes death receptor-induced apoptosis via Bid- and STAT1-dependent pathways. Int. J. Oncol. 40, 203–208.
Walensky, L., Kung, A., Escher, I., Malia, T., Barbuto, S., Wright, R., Wagner, G., Verdine, G., Korsmeyer, S., 2004. 464 Regulation of apoptosis by synthetic helices of the BCL-2 family. Eur. J. Cancer Suppl. 2, 141.
Walensky, L.D., Pitter, K., Morash, J., Oh, K.J., Barbuto, S., Fisher, J., Smith, E., Verdine, G.L., Korsmeyer, S.J., 2006. A stapled BID BH3 helix directly binds and activates BAX. Mol. Cell 24, 199–210.
Wang, G., Nikolovska-Coleska, Z., Yang, C.Y., Wang, R., Tang, G., Guo, J., Shangary, S., Qiu, S., Gao, W., Yang, D., Meagher, J., Stuckey, J., Krajewski, K., Jiang, S., Roller, P.P., Abaan, H.O., Tomita, Y., Wang, S., 2006. Structure-based design of potent small-molecule inhibitors of anti-apoptotic Bcl-2 proteins. J. Med. Chem. 49, 6139–6142.

Wang, B., Xie, M., Li, R., Owonikoko, T.K., Ramalingam, S.S., Khuri, F.R., Curran, W.J., Wang, Y., Deng, X., 2014. Role of Ku70 in deubiquitination of Mcl-1 and suppression of apoptosis. Cell Death Differ. 21, 1160–1169.

Wang, R., Xia, L., Gabrilove, J., Waxman, S., Jing, Y., 2016. Sorafenib inhibition of Mcl-1 accelerates ATRA-induced apoptosis in differentiation-responsive AML cells. Clin. Cancer Res. 22, 1211–1221.

Weeden, C.E., Ah-Cann, C., Holik, A.Z., Pasquet, J., Garnier, J.M., Merino, D., Lessene, G., Asselin-Labat, M.L., 2018. Dual inhibition of BCL-XL and MCL-1 is required to induce tumour regression in lung squamous cell carcinomas sensitive to FGFR inhibition. Oncogene 37, 4475–4488.

Wei, G., Margolin, A., Haery, L., Brown, E., Cucolo, L., Julian, B., Shehata, S., Kung, A., Beroukhim, R., Golub, T., 2012. Chemical genomics identifies small-molecule MCL1 repressors and BCL-xL as a predictor of MCL1 dependency. Cancer Cell 21, 547–562.

Wertz, I.E., Kusam, S., Lam, C., Okamoto, T., Sandoval, W., Anderson, D.J., Helgason, E., Ernst, J.A., Eby, M., Liu, J., Belmont, L.D., Kaminker, J.S., O'Rourke, K.M., Pujara, K., Kohli, P.B., Johnson, A.R., Chiu, M.L., Lill, J.R., Jackson, P.K., Fairbrother, W.J., Seshagiri, S., Ludlam, M.J.C., Leong, K.G., Dueber, E.C., Maecker, H., Huang, D.C.S., Dixit, V.M., 2011. Sensitivity to antitubulin chemotherapeutics is regulated by MCL1 and FBW7. Nature 471, 110–114.

Wroblewski, D., Jiang, C.C., Croft, A., Farrelly, M.L., Zhang, X.D., Hersey, P., 2013. OBATOCLAX and ABT-737 induce ER stress responses in human melanoma cells that limit induction of apoptosis. PLoS One 8, e84073.

Wu, X., Luo, Q., Zhao, P., Chang, W., Wang, Y., Shu, T., Ding, F., Li, B., Liu, Z., 2019. MGMT-activated DUB3 stabilizes MCL1 and drives chemoresistance in ovarian cancer. Proc. Natl. Acad. Sci. U. S. A. 116, 2961–2966.

Xie, H., Yin, J., Shah, M.H., Menefee, M.E., Bible, K.C., Reidy-Lagunes, D., Kane, M.A., Quinn, D.I., Gandara, D.R., Erlichman, C., Adjei, A.A., 2019. A phase II study of the orally administered negative enantiomer of gossypol (AT-101), a BH3 mimetic, in patients with advanced adrenal cortical carcinoma. Invest. New Drugs 37, 755–762.

Zeidner, J.F., Karp, J.E., 2015. Clinical activity of alvocidib (flavopiridol) in acute myeloid leukemia. Leuk. Res. 39, 1312–1318.

Zhai, D., Jin, C., Satterthwait, A.C., Reed, J.C., 2006. Comparison of chemical inhibitors of antiapoptotic Bcl-2-family proteins. Cell Death Differ. 13, 1419–1421.

Zhang, Z., Jin, L., Qian, X., Wei, M., Wang, Y., Wang, J., Yang, Y., Xu, Q., Xu, Y., Liu, F., 2007. Novel Bcl-2 inhibitors: discovery and mechanism study of small organic apoptosis-inducing agents. Chembiochem 8, 113–121.

Zhang, Z., Song, T., Zhang, T., Gao, J., Wu, G., An, L., Du, G., 2011. A novel BH3 mimetic S1 potently induces Bax/Bak-dependent apoptosis by targeting both Bcl-2 and Mcl-1. Int. J. Cancer 128, 1724–1735.

Zhong, D., Gu, C., Shi, L., Xun, T., Li, X., Liu, S., Yu, L., 2014. Obatoclax induces G1/G0-phase arrest via p38/p21 $^{waf1/Cip1}$ signaling pathway in human esophageal cancer cells. J. Cell. Biochem. 115, 1624–1635.

CHAPTER THREE

TRAIL receptor signaling: From the basics of canonical signal transduction toward its entanglement with ER stress and the unfolded protein response

Daniela Stöhr[a,b,*], Albert Jeltsch[c], Markus Rehm[a,b,d]
[a]University of Stuttgart, Institute of Cell Biology and Immunology, Stuttgart, Germany
[b]University of Stuttgart, Stuttgart Research Center Systems Biology, Stuttgart, Germany
[c]Department of Biochemistry, University of Stuttgart, Institute of Biochemistry and Technical Biochemistry, Stuttgart, Germany
[d]University of Stuttgart, Stuttgart Centre for Simulation Science, Stuttgart, Germany
*Corresponding author: e-mail address: daniela.stoehr@izi.uni-stuttgart.de

Contents

Abstract

The cytokine tumor necrosis factor (TNF)-related apoptosis-inducing ligand (TRAIL) is a member of the large TNF superfamily that can trigger apoptosis in transformed or infected cells by binding and activating two receptors, TRAIL receptor 1 (TRAILR1) and TRAIL receptor 2 (TRAILR2). Compared to other death ligands of the same family,

International Review of Cell and Molecular Biology, Volume 351
ISSN 1937-6448
https://doi.org/10.1016/bs.ircmb.2020.02.002

TRAIL induces apoptosis preferentially in malignant cells while sparing normal tissue and has therefore been extensively investigated for its suitability as an anti-cancer agent. Recently, it was noticed that TRAIL receptor signaling is also linked to endoplasmic reticulum (ER) stress and the unfolded protein response (UPR). The role of TRAIL receptors in regulating cellular apoptosis susceptibility therefore is broader than previously thought. Here, we provide an overview of TRAIL-induced signaling, covering the core signal transduction during extrinsic apoptosis as well as its link to alternative outcomes, such as necroptosis or NF-κB activation. We discuss how environmental factors, transcriptional regulators, and genetic or epigenetic alterations regulate TRAIL receptors and thus alter cellular TRAIL susceptibility. Finally, we provide insight into the role of TRAIL receptors in signaling scenarios that engage the unfolded protein response and discuss how these findings might be translated into new combination therapies for cancer treatment.

1. Apoptotic cell death: The basics

An estimated 1 million damaged or overaged cells die in the human body every second to ensure tissue homeostasis (Green, 2011). In contrast to necrosis, a form of uncontrolled cell lysis that occurs, for example, in response to physical damage, the process of physiological apoptotic cell death is highly regulated to ensure that selected cells are removed without harming the surrounding tissue by the release of cellular components. The basic concept of programmed cell death has been known since 1842, however, only in 1972 was it referred to for the first time with the word apoptosis by Kerr et al. (Kerr et al., 1972; Vogt, 1842). Gradually it was found that apoptosis is not only involved in tissue homeostasis in the adult human body but that it also plays an important role during the development of all metazoa. Examples are numerous, ranging from the depletion of surplus neurons in brain development, the elimination of connecting tissue between fingers and toes in the formation of extremities, to the removal of the tadpole tail during their metamorphosis to frogs (Kerr et al., 1974; Mori et al., 1995; Yuan and Yanker, 2000). Apoptotic cells morphologically die in a characteristic way, starting with rounding up and simultaneously shrinking, which is followed by condensation of chromatin and fragmentation of DNA. During the whole process, the cell membrane stays intact, largely preventing the release of damage-associated molecular patterns (*DAMPs*) that lead to inflammation as observed after necrotic cell death. Instead, apoptotic cells disintegrate into smaller apoptotic bodies that are engulfed by, e.g., macrophages due to the secretion and presentation of "find me" and "eat me signals" such as phosphatidylserine. During apoptosis, this glycerophospholipid translocates from the inner layer of the cell membrane to the cell surface and, since this can be prevented by apoptosis inhibitors, allows the facile detection of apoptotic cell death prior to cell debris engulfment or secondary rupture of apoptotic cell bodies (Arandjelovic and Ravichandran, 2015; Kerr et al., 1972; Kiraz et al., 2016). Alongside being involved in tissue renewal and embryonic development, apoptosis also represents a defense against the formation of cancer. Normally, DNA damage, withdrawal of survival signals or enhanced expression of oncoproteins is sensed within the cell, resulting in its elimination by apoptosis. Furthermore, leukocytes can also induce apoptosis in tumor cells by activating death receptors on their surfaces. However, during the development and progression of proliferative malignancies, cells often acquire resistance against autonomous or immune-mediated attempts to induce programmed cell death. As such, cancer cells can be considered to be primed for programmed cell death, but loss of function of sensor proteins as well as downregulation of proteins that promote apoptosis and/or the upregulation of cell death inhibitors can prevent cell death execution. This priming flags the need of additional pro-apoptotic stress to eliminate these cells, and this can be achieved during therapy by apoptosis-inducing or apoptosis-sensitizing anti-cancer agents.

2. Apoptosis signal transduction and caspase activation

Apoptotic cell death can be executed via the extrinsic and the intrinsic pathways (illustrated in Fig. 3). External and internal stimuli such as irradiation, genotoxic agents, viral infection, formation of reactive oxygen species, growth factor withdrawal or endoplasmic reticulum (ER) stress can contribute to inducing the intrinsic pathway of apoptosis in which mitochondrial outer membrane permeabilization (MOMP) plays a central role. In addition, ligands that belong to the tumor necrosis factor (TNF) receptor superfamily such TNFα, TNF-related apoptosis-inducing ligand (TRAIL) and Fas ligand (FasL) can bind to their cognate receptors, TNF receptor 1 (TNFR1), TRAIL receptor 1/2 (TRAILR1/2 or DR4/DR5) and Fas, activating the extrinsic pathway of apoptosis (Ichim and Tait, 2016; Redza-Dutordoir and Averill-Bates, 2016; Roos and Kaina, 2006; Takuma et al., 2005; Thomson, 2001). In both signaling cascades, cell death is executed via an interplay between initiator and effector aspartate-specific cysteine proteases (caspases) that activate each other, resulting finally in the proteolytic cleavage of numerous substrates such as anti-apoptotic key players, structural proteins like spectrin, DNA repair poly ADP (adenosine diphosphate)-ribose polymerase (PARP) and different cyclins and kinases such as protein kinase B/Akt. Furthermore, effector caspases also indirectly activate the caspase-activated DNase (CAD) by proteolysis of its inhibitor, resulting in the fragmentation of DNA (Fischer et al., 2003; Ichim and Tait, 2016; Kiraz et al., 2016; Nagata, 2000). All caspases are synthesized as inactive precursors (zymogens) that consist out of an N-terminal prodomain and a C-terminal protease domain. The prodomain of initiator caspase-8 and 10 contains two death effector domains (DEDs) that enable their recruitment to the death-inducing signaling complex (DISC) and the binding of further proteins. In contrast to that, the prodomain of initiator caspase-9 is comprised of a caspase activation and recruitment domain (CARD) which is necessary for its binding to the apoptosome, a complex of different proteins that causes the activation of caspase-9. If initiator caspases come into close proximity to each other within their respective complexes, their protease domains dimerise, resulting in activation and proteolytic cleavage (Creagh, 2014). However, only recently it was shown that pro-caspase-9 can additionally be activated when forming a heterodimer with the apoptosome without the requirement for proteolytic cleavage (Wu et al., 2016). In contrast to initiator caspases, the proforms of effector caspases (caspase-3, -7) preassemble as homodimers in non-apoptotic cells. After cleavage by active initiator caspases, they subsequently exert their function as executioners of apoptosis (Creagh, 2014). Executioner caspase-3 can also activate pro-caspase-6 which, besides cleaving nuclear lamins, can cleave pro-caspase-8 (Cowling and Downward, 2002; Inoue et al., 2009). Even though initially thought to be a positive feedback loop, this rather results in cleaved caspase-8 monomers that do not form active caspase-8 dimers (Oberst et al., 2010; Würstle et al., 2010). Caspase-3 also cleaves pro-caspase-9 and auto-processed caspase-9 (p35/p12) into p37/p10 and p35/p10 forms, respectively, which prevents the association of the X-linked inhibitor of apoptosis protein (XIAP) and thereby contributes to forming a positive feedback loop (Denault et al., 2007; Rehm et al., 2006).

3. Core signal transduction during extrinsic apoptosis

The extrinsic pathway of apoptosis is triggered by binding of apoptosis-inducing ligands to their cognate receptors. In response to stimulation with TRAIL, TRAIL receptors trimerise and form the DISC, with the latter containing the receptors, Fas-associated protein with death domain

(FADD), cellular FLICE-like inhibitory protein (cFlip, FLICE = FADD-like interleukin-1β-converting enzyme, another name for caspase-8), pro-caspase-8 and pro-caspase-10 as key apoptosis regulating components. Of note, while both initiator caspases are expressed in humans, mice lack pro-caspase-10. During DISC formation, FADD is recruited to the receptor trimer by interactions of death domains (DDs) that are present on the intracellular domain of the death receptors and in FADD. Controversy remains as to the exact stoichiometry of FADD to death receptors, with 1 FADD per trimer and 3 FADDs per trimer having been suggested (Dickens et al., 2012; Majkut et al., 2014; Schleich et al., 2016). FADD also contains a DED, allowing it to subsequently recruit pro-caspase-8 via binding to its DED1. Thereafter, further pro-caspase-8 proteins are recruited via their DEDs forming a large filament in which proteolytic cleavage and activation takes place (Dickens et al., 2012; Fu et al., 2016; Schleich et al., 2016). Beside further pro-caspase molecules, pro-caspase-8 can also recruit cFlip, a protein that displays high similarity to caspase-8 but lacks its protease activity. Originally, it was believed that cFlip can compete with pro-caspase-8 for binding to the DED of FADD; however, more recently it has been demonstrated that pro-caspase-8 is required for recruitment of cFlip to the DISC, giving rise to hierarchal binding where pro-caspase-8 is first recruited to FADD to initiate filament extension on the DISC (Hughes et al., 2016). The two short isoforms of cFLIP, cFlip-S and cFlip-R, lack the protease domain and inhibit caspase-8 activation when being part of the DISC. The role of the long isoform (cFlip-L) is more complex. Even though cFlip-L contains an inactive protease domain, it enhances apoptosis when present at physiological concentrations. Heterodimerization with pro-caspase-8 stabilizes the catalytic domain of the caspase, allowing for localized catalytic activity. Higher concentrations of cFlip-L instead inhibit apoptotic signaling, probably due to its excessive presence in the DISC that prevents the formation of interactions with pro-caspase-8 (Fricker et al., 2010; Hughes et al., 2016). It has also been suggested that the heterodimer of cFlip-L and caspase-8 alters substrate specificity of caspase-8, thus modulating downstream signaling (Pop et al., 2011). Beside caspase-8, also the close homolog, caspase-10, can be recruited into the filament. Only recently the assumption of caspase-10 being involved in promoting apoptosis similarly to caspase-8 was challenged by the finding that, in different cell lines as well as in a cell free system, caspase-10 inhibited apoptosis and rather shifted the DISC signaling to the induction of pro-survival genes (Horn et al., 2017). However, whether caspase-10 inhibits apoptosis by preventing filament assembly or forming inactive heterodimers with caspase-8 has yet to be investigated. Taken together, DISC assembly is a highly complex cooperative and hierarchical process that comprises different proteins and can shift signaling between either pro- or anti-apoptotic outcomes. Alongside type I cells, in which direct activation of pro-caspase-3 by caspase-8 is in principle sufficient to induce apoptosis, type II cells need to amplify the apoptotic signal via the intrinsic MOMP-dependent route (Scaffidi et al., 1998). It needs to be noted though, that also type I cells typically would engage this amplification pathway (Aldridge et al., 2011). MOMP requires the formation of pores in the mitochondrial outer membrane, subsequently resulting in the activation of effector caspases. Besides intrinsic apoptosis signaling cascades, also caspase-8 activation can trigger MOMP. Caspase-8 cleaves the B cell lymphoma-2 (Bcl-2) homology (BH3)-only protein Bid, and truncated Bid (tBid) translocates to the mitochondrial outer membrane to initiate the formation of homo-oligomeric pores generated from two pro-apoptotic effector proteins, Bcl-2 associated X protein (Bax) and Bcl-2 antagonist or killer (Bak) (Martinou and Youle, 2011).

Bid, Bax and Bak are members of the large Bcl-2 family that comprises proteins containing one to four BH domains. This protein family is further subdivided into three groups according to protein function. Bax and Bak belong to the first group of pro-apoptotic pore-forming proteins and are functionally essential for MOMP. The second group of the family comprises anti-apoptotic proteins such as Bcl-2, Bcl-xl and myeloid cell leukemia-1 (Mcl-1), all of which can directly bind to Bax and/or Bak and prevent pore formation. These latter proteins are neutralized by members of the third Bcl-2 family group which comprises pro-apoptotic proteins with a single BH3 domain (BH3-only proteins), for example, Bid, Bcl-2 interacting mediator of cell death (Bim), Bcl-2-antagonist of cell death (BAD), p53 upregulated modulator of apoptosis (PUMA) and NOXA. Interestingly, some of these BH3-only proteins, such as Bid and Bim, also directly activate the two pore-forming proteins Bax and Bak. Thus, a complex interplay between pro- and anti-apoptotic Bcl-2 family proteins and their relative abundances dictate the fate of cells (Finlay et al., 2017; Kiraz et al., 2016).

After pore formation, cytochrome *c* is released from the mitochondrial intermembrane space and forms, together with apoptotic protease activating factor-1 (Apaf-1) and deoxyadenosine triphosphate/adenosine triphosphate (dATP/ATP), a heptameric complex that recruits pro-caspase-9 via interaction between the CARD domains present in both, Apaf-1 and pro-caspase-9 (Dorstyn et al., 2018). This structure, called the apoptosome, thus is the activation platform for initiator caspase-9.

Besides anti-apoptotic Bcl-2 family members that control MOMP, another group of proteins repress caspase activation, the so-called inhibitors of apoptosis proteins (IAPs). A common feature of these proteins is the occurrence of one to three baculovirus IAP repeats (BIR) domains, zinc finger folds responsible for various protein-protein interactions. XIAP contains three BIR domains and binds to and inactivates caspases-3, -7 and -9. Furthermore, due to its additional ubiquitin-associated (UBA) domain and its really interesting new gene (RING) domain, XIAP can also target proteins for ubiquitin-dependent degradation. Two other members of the IAP family, cellular IAP1 (cIAP1) and cellular IAP2 (cIAP2), containing three BIR and one RING domain(s), also bind to caspases-3, -7 and -9, but instead of inhibiting their activity likely rather accelerate their degradation. Furthermore, both cIAPs activate nuclear factor "kappa-light-chain-enhancer" of activated B-cells (NF-κB) and mitogen-activated protein kinase (MAPK) signaling, in particular in survival signaling complexes that form after death receptor activation (Kiraz et al., 2016). In this context, the E3-ligase activity of cIAPs prominently contributes to formation and regulation of the linear ubiquitin chain assembly complex ((LUBAC) see later section in this chapter) (Lafont et al., 2017).

XIAP, cIAP1 and cIAP2 can be antagonized by second mitochondria-derived activator of caspases (Smac). Upon MOMP, Smac is released from the mitochondria and binds as a homodimer to IAPs, resulting in the release and de-repression of active caspases (Kiraz et al., 2016). Similarly, Smac and IAP-antagonizing synthetic compounds can suppress the pro-survival function of cIAPs in response to death receptor activation.

Taken together, apoptosis therefore is a complex, highly regulated signaling pathway that can be activated by numerous external and internal stimuli, resulting in the eradication of mutated or damaged cells.

4. TRAIL and TRAIL receptors as members of the TNF superfamily

Among the various ligands that can induce extrinsic apoptosis, the cytokine TRAIL has been extensively studied. Within the group of death receptor ligands it lends itself best for translational and thus clinical application, since primarily cancer cells rather than non-transformed cells respond with apoptosis upon exposure to TRAIL. TRAIL is part of the TNF superfamily that comprises, to date, of 19 ligands and 29 receptors, all of which are involved in pro-inflammatory signaling. However, individual ligands can induce apoptosis as well as promote proliferation and differentiation (Aggarwal et al., 2012). Synthesized as type II transmembrane proteins with an extracellular C-terminal domain and an intracellular N-terminus, members of the TNF superfamily activate their cognate receptors in a homotrimeric assembly mediated by the TNF homology domain (THD) located in the C-terminal end of the proteins (Vanamee and Faustman, 2018). Even though most ligands execute signaling when bound to membranes, some TNF superfamily members, such as TRAIL, can be proteolytically cleaved off the cell surface which, depending on the protein, either activates or inhibits their function (Bodmer et al., 2002).

While Coley understood that bacterial infection-mediated inflammation can induce tumor regression and consequently started to treat cancer patients with a mixture of different bacteria in 1893 (Coley, 1891; McCarthy, 2006), only in 1975 it was found that the factor responsible for tumor shrinkage was not a bacterial derived endotoxin, but a compound produced by the human body in response to infection, namely TNF (Carswell et al., 1975). TNF was structurally described in the mid-1980s by a group around Aggarwal et al. (1985). Since then, further members of the TNF superfamily were identified due to human complementary DNA (cDNA) sequence homology, with TRAIL being discovered in the 1990s by two independent groups (Pitti et al., 1996; Wiley et al., 1995). In contrast to other TNF superfamily ligands, stable trimerization and hence activity of TRAIL monomers requires a zinc ion that connects the single proteins by binding to cysteine residue 230 (Bodmer et al., 2000; Hymowitz et al., 2000). In the human body, TRAIL binds to five TRAIL receptors, TRAILR1, TRAILR2, TRAILR3, TRAILR4 and osteoprotegerin (OPG). Each receptor-binding site is formed between two monomers of the TRAIL trimer, resulting in the trimerization of TRAIL receptors after binding of the trimeric ligand (Hymowitz et al., 1999; Ramamurthy et al., 2015). Mainly produced by cells of the immune system such as natural killer (NK) and T cells, this TNF superfamily member is not only involved in the elimination of infected or malignant cells by inducing apoptosis, but also influences the development of leukocytes (Collison et al., 2009; Danish et al., 2017; Mizrahi and Askenasy, 2014). However, beside its role in the immune system, physiological TRAIL signaling might also regulate germ cell apoptosis and protects spermatozoa in the human testis (Lin and Richburg, 2014; Zauli et al., 2013). Furthermore, TRAIL produced by the medial smooth layer of the aorta and pulmonary arteries has been demonstrated to positively influence endothelial cell proliferation as well as survival (Secchiero et al., 2003). Additionally, exposing endothelial cells to TRAIL was shown to influence the adhesion of leukocytes to the endothelium (Secchiero et al., 2005). Alongside the previously mentioned cell types and organs, TRAIL and its receptors are also expressed by a range of other tissues such as the colon, the lung and the kidney, with the physiological function in these organs needing further investigation (Danish et al., 2017).

Receptors of the TNF superfamily are characterized by the presence of at least one cysteine rich domain (CRD) in their extracellular part which is/are responsible for ligand binding (Bodmer et al., 2002). Searching the human genome database for sequence homologies with TNFR1, Pan et al. discovered the first TRAIL-binding receptor in 1997, TRAILR1 (Pan et al., 1997b). Comprised altogether of 468 amino acids, the extracellular part of TRAILR1 is build up from two complete CRDs that form the ligand-binding site and one partial CRD that harbors a putative pre-ligand binding assembly domain (PLAD), allowing ligand-independent homo- and heterooligomerization of TRAILRs on the cell membrane (Chan, 2007; Clancy et al., 2005; Danish et al., 2017; Neumann et al., 2014). As mentioned above, the intracellular moiety of TRAILR1 contains a death domain that is necessary for the formation of signaling complexes. TRAILR2, the second receptor of the TRAILR family able to induce apoptosis, was described only shortly after the discovery of TRAILR1 by several groups (MacFarlane et al., 1997; Pan et al., 1997a; Schneider et al., 1997; Screaton et al., 1997; Walczak et al., 1997; Wu et al., 1997). Two isoforms of TRAILR2 can be found in the human body. The short isoform (TRAILR2-S) comprises of 411 amino acids that, similar to TRAILR1, form two complete and one partial CRD as well as an intracellular death domain. In contrast to TRAILR2-S, TRAILR2-L contains an additional 29 amino acids close to the cell membrane prior to the first CRD. As this part of the protein is rich in threonine, alanine, proline and glutamine it is referred to as TAPE (Schneider et al., 1997; von Karstedt et al., 2017). In addition to TRAILR1 and TRAILR2, two further TRAIL-binding receptors, TRAILR3 and TRAILR4 were discovered in the late 1990s (Degli-Esposti et al., 1997a,b; Marsters et al., 1997; Mongkolsapaya et al., 1998; Pan et al., 1998). In contrast to the two apoptosis-inducing receptors, the 259 amino acids comprising TRAILR3 lacks an intracellular domain but is bound to the cell membrane via a glycosylphosphatidylinositol (GPI) anchor. Two complete and one partial CRD are connected to the anchor via five TAPE domains. Due to the lack of an intracellular domain, direct signaling via TRAILR3 is highly unlikely, although signal transduction via the GPI anchor cannot be completely excluded (Kimberley and Screaton, 2004). Nevertheless, the main function of TRAILR3 is highly likely that of a regulatory receptor, either inhibiting signaling via TRAILR1 and TRAILR2 by sequestering TRAIL or forming signaling-incompetent heteromeric complexes (Neumann et al., 2014). The conformation of the 386 amino acids comprising TRAILR4 is closely related to those of TRAILR1/R2, containing two complete and one partial CRD. However, in contrast to the other two receptors, TRAILR4 only comprises a truncated DD and is therefore believed to be incapable of recruiting FADD and thus inducing apoptotic signaling. Whether TRAILR4 is able to activate the NF-κB and Akt pathway, as it was suggested by different studies, is under debate and needs further investigation (Degli-Esposti et al., 1997b; Harper et al., 2001; Hu et al., 1999; Lalaoui et al., 2011; Marsters et al., 1997; Neumann et al., 2014). Similarly to TRAILR3, TRAILR4 most likely acts solely as a regulatory receptor. The fifth receptor known to bind TRAIL is OPG. OPG is a soluble protein that is mainly known for its ability to regulate osteoclast differentiation by sequestering receptor activator of NF-κB ligand (RANKL) (Boyle et al., 2003). Even though OPG binds TRAIL only with low affinity, it was shown to inhibit apoptosis induction in vitro (De Toni et al., 2008; Emery et al., 1998; Lane et al., 2012). However, whether this finding is of physiological relevance has yet to be investigated.

5. Mechanistic insight into canonical TRAILR activation

As mentioned above, ligands of the TNF superfamily such as TRAIL form noncovalently-linked homotrimers that attach with their outside part to cognate receptors. Thereby, two TRAIL monomers form an interface at which one receptor can be bound. Currently it is presumed that TRAILRs are pre-arranged on the cell membrane as inactive antiparallel dimers, mediated by the PLAD domain located in the last CRD of the receptors. Binding of the trimeric ligand subsequently leads to the formation of large hexagonal structures of activated receptors, resulting in DISC formation and execution of downstream signaling (Pan et al., 2019; Vanamee and Faustman, 2018). As stated earlier, TRAILR3 and TRAILR4 do not induce apoptosis, but their overexpression can render cells TRAIL-resistant (Degli-Esposti et al., 1997a,b). Furthermore, evidence exists that malignant and normal diploid cells can upregulate TRAILR3 and TRAILR4 expression, thereby acquiring resistance against TRAIL (Bouralexis et al., 2003; Clancy et al., 2005; Davidovich et al., 2004; Riccioni et al., 2005). Initially, it was thought that both receptors simply inhibit signaling by sequestering TRAIL from binding to apoptosis-inducing TRAILR1 and TRAILR2. However, since TRAILR3 and TRAILR4 display lower affinities to the ligand than TRAILR1 and TRAILR2 and the observation that excess TRAIL does not abolish inhibition argues against TRAILR3 and TRAILR4 solely being decoy receptors (Lee et al., 2005). Since at least TRAILR4 can form heteromers with TRAILR1 and TRAILR2, resulting in inhibition of apoptosis induction, a model of mixed receptor complexes that regulate signaling is much more convincing (Clancy et al., 2005; Merino et al., 2006; Neumann et al., 2014). The formation of heterotrimers with TRAILR3 is less likely, due to its preferential localization in lipid rafts and a larger linker between its transmembrane domain and the ligand binding CRDs (Mongkolsapaya et al., 1998; Sharma et al., 2004; Varma and Mayor, 1998). Whether TRAILR3 nevertheless forms heterotrimers with TRAILR1/R2, inhibiting apoptosis, or instead simply serves as a decoy receptor still needs to be investigated. Of note, since TRAILR2-S displays a linker length comparable to TRAILR1 whereas the connecting domain in TRAILR2-L is much longer, it can be speculated that TRAILR2-L could preferentially form homotrimers (Screaton et al., 1997).

6. TRAILR signaling: Additional signaling routes to death or survival

In addition to DISC-dependent apoptosis, activation of TRAILR1 and TRAILR2 can also result in receptor-interacting serine/threonine protein kinase 1 (RIPK1)-dependent apoptosis, necroptotic cell death or pro-survival signaling (illustrated in Fig. 1). The first complex forming after binding of TRAIL always is complex I, comprising the important key players for the execution of apoptosis, namely TRAILRs, FADD, pro-caspase-8 and cFlip-L/S (complex Ia). However, recently it was demonstrated that, dependent on RIPK1 and cIAP1/2, the linear ubiquitin chain assembly complex (LUBAC) can be recruited to complex I, resulting in linear ubiquitination of pro-caspase-8 and RIPK1 (complex Ib). As a consequence, apoptosis is prevented since ubiquitination of pro-caspase-8 inhibits its activation (Lafont et al., 2017). At the same time, linear ubiquitination of RIPK1 allows for the recruitment of the transforming growth factor beta-activated kinase 1

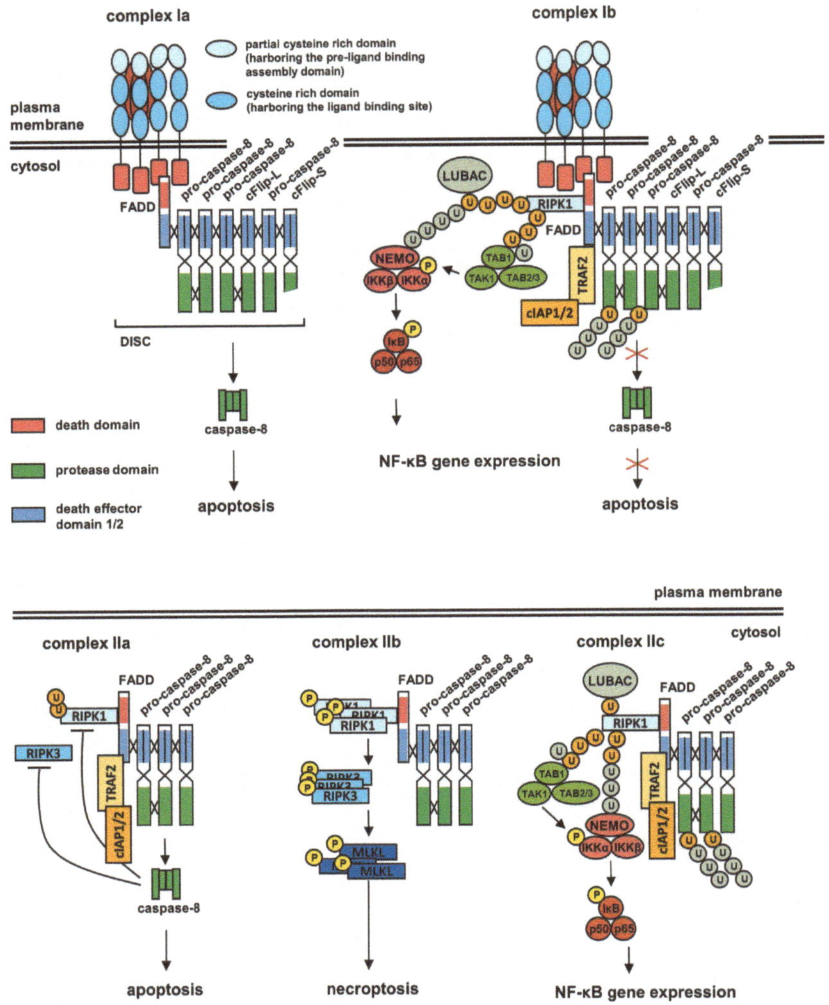

Fig. 1 TRAIL-induced formation of different protein complexes and their signaling outcome. After binding of TRAIL to its cognate receptors, complex Ia is formed that comprises the TRAILRs, FADD, pro-caspase-8 and cFlip-L/S. Within this assembly, pro-caspase-8 is activated and induces apoptosis. In complex Ib, the linear ubiquitin chain assembly complex (LUBAC) is recruited in a cIAP1/2 dependent manner (ubiquitin added by cIAP1/2 = U in an orange circle). LUBAC in turn linearly ubiquitinates pro-caspase-8 and RIPK1 which results in the inhibition of apoptosis, recruitment of the transforming growth factor beta-activated kinase 1 (TAK1) complex as well as the IκB kinase complex and the subsequent activation of NF-κB (ubiquitin added by LUBAC = U in a light green circle). After internalization, a secondary complex is formed which is devoid of TRAILRs. Complex IIa consists of inactive RIPK1 that, together with RIPK3 and FADD, provides a scaffold for pro-caspase-8 activation and induction of apoptosis. Complex IIb is build when apoptosis execution by caspase-8 is blocked. In this case, activated RIPK1 recruits and phosphorylates RIPK3 which in turn phosphorylates MLKL leading to necroptotic cell death by the formation of MLKL pores in the cell membrane. In complex IIc, recruited LUBAC ubiquitinates RIPK1 and pro-caspase-8 in the same way as in complex Ib, resulting in the activation of NF-κB.

(TAK1) complex as well as the IκB kinase complex and the subsequent activation of NF-κB (Henry and Martin, 2017; Lafont et al., 2017; Wajant, 2017).

After internalization of complex I, a secondary complex forms that lacks TRAILRs. Depending on the recruitment and activation of cIAP1/2 and LUBAC, the signaling outcome can either be apoptosis, necroptosis or survival signaling. Similar to the DISC (complex Ia), complex IIa can induce apoptosis with RIPK1, RIPK3 and FADD providing a scaffold for pro-caspase-8 activation (Dondelinger et al., 2013). If caspase-8 activity is blocked, receptor-interacting serine/threonine protein kinase 3 (RIPK3) is recruited and phosphorylated in presence of phosphorylated RIPK1 (complex IIb). Activated RIPK3 can in turn bind to and phosphorylate mixed lineage kinase domain like pseudokinase (MLKL) resulting in its oligomerization and insertion into the cell membrane, where it forms pores. Similarly to necrosis, components of the cell are then released and induce inflammation, thus earning this process the name necroptosis (Chen et al., 2016). Finally, when both, RIPK1 and pro-caspase-8, are retained in their respective inactive states, complex IIc forms that, similarly to complex Ib, activates NF-κB (Lafont et al., 2017).

Beside NF-κB-dependent gene expression, binding of TRAIL to its cognate receptors can also enhance pro-survival signaling by triggering the c-Jun N-terminal kinase (JNK), MAPK and phosphatidylinositol-4,5-bisphosphate 3-kinase (PI3K)/Akt pathways, and increase migration in non-small cell lung cancer (NSCLC) cell lines via the RIPK1/protein kinase SRC/signal transducer and activator of transcription 3 (STAT3) axis (De Miguel et al., 2016; von Karstedt et al., 2017). Of note, the activation of TRAILRs can result either in the induction of cell death or transcription of pro-survival genes, which might explain that TRAILR expression failed as a prognostic marker in cancer treatments (Danish et al., 2017). Furthermore, even if caspases are activated in response to TRAIL, this might not necessarily be linked to a positive outcome for patient treated with TRAIL-based therapeutics in the long-term. For example, the activation of caspases can result in the increased production of arachidonic acid, which is subsequently converted by cyclooxygenase (COX) to prostaglandin E_2 (PGE_2). PGE_2 released into the tissue microenvironment can promote proliferation of the surrounding tissue, including surviving cancer cells that might repopulate the tumor ("phoenix rising" effect) (Li et al., 2010). Furthermore, the "find and eat me" signaling of apoptotic cells can enhance angiogenesis and attract macrophages, responses that were reported to possibly contribute to cancer progression (Tang et al., 2015; Tardáguila and Mañes, 2013). Lastly, it has been shown that cells might survive low amounts of caspase activation, which can cause sublethal DNA damage and genome instability, thereby enhancing tumorigenesis (Ichim and Tait, 2016). Thus, combining apoptosis-inducing agents such as TRAIL-derived biologics with drugs that prevent the unwanted pro-tumorigenic effects might need to be considered for future therapeutic exploitation.

7. Differential roles of TRAILR1 and TRAILR2 in signal transduction?

Interestingly, only humans and chimpanzees are known to possess two apoptosis-inducing TRAILRs while in all other vertebrates studied so far only one receptor is expressed, pointing to the evolution of non-redundant functions of these receptors in these two species (Bridgham et al., 2003; Kelley and Ashkenazi, 2004). Whether ligand-induced apoptosis is equally triggered via both, TRAILR1 and TRAILR2, or mainly executed via one of the two receptors, largely depends on the cell type. In leukemic malignancies apoptosis apparently is preferentially induced by TRAILR1. In solid tumors instead, apoptosis induction can be mediated either via TRAILR2 or TRAILR1, but also equal contributions of both receptors have been observed (van Roosmalen et al., 2014). Even though it was suggested that endoplasmic reticulum (ER) stress-induced apoptosis to be mediated in part by accumulation of intracellular TRAILR2 (see later section in this chapter), evidence also exists for a contribution of TRAILR1, speaking against a unique function of TRAILR2 in the ER stress setting (Li et al., 2015; Lu et al., 2014). Taken together, the two signaling-competent TRAILRs might not necessarily fulfill different functions in apoptosis signaling. More recent findings, however, suggest that TRAILR2 plays a superior role in pro-survival signaling (Haselmann et al., 2014; von Karstedt et al., 2015). Besides being expressed on the cell surface, substantial amounts of TRAILR2 have been reported to reside intracellularly, being localized to the cytoplasm or the nucleus in both, cell lines and tissue samples of patients with advanced or recurrent non-small-cell lung cancer or pancreatic ductal adenocarcinoma (Chen et al., 2012; Haselmann et al., 2014; Kojima et al., 2011; Leithner et al., 2009). Functional studies on nuclear TRAILR2 showed that interaction of the receptor with the core microprocessor components Drosha and DGCR8 results in the decrease of the mature miRNA let-7. Subsequently, this leads to an increased expression of the let-7 targets LIN28B and HMGA2 and enhanced cell proliferation (Haselmann et al., 2014). Furthermore, Karstedt et al. demonstrated that cancer progression, invasion and metastasis was induced by TRAILR2 in complex with TRAIL in KRAS-mutated tumor cells (von Karstedt et al., 2015).

8. Regulation of TRAILR amounts and modulation of TRAILR signaling

8.1 Environmental factors and their influence on TRAILR amounts

Changes in the amounts of TRAILRs can alter a cellular susceptibility to TRAIL. Known factors to interfere with receptor levels are, among other things, adverse environmental conditions such as hypoxia or glucose starvation. These can manifest in solid, rapidly growing tumors where vascularization often cannot develop fast enough to supply sufficient oxygen and nutrients (Forster et al., 2017). Park et al. showed that depriving lung cancer cells of oxygen leads to increased total amounts of TRAILR2 and decreased TRAILR1 expression (Park et al., 2002). However, another study suggested that the regulation of TRAILR expression is cell line-dependent, since only 1 out of 7 investigated cell lines displayed increased amounts of surface TRAILR2 after oxygen deprivation (Gobbi et al., 2010). Regarding nutrient starvation it was shown that glucose withdrawal can upregulate TRAILR1 amounts non-transcriptionally in HeLa cells but failed to do so in human colorectal cancer and melanoma cell lines (Carr et al., 2016; Iurlaro et al., 2017; Liu et al., 2009). Various studies instead demonstrated that glucose withdrawal enhances TRAILR2 transcription and expression in diverse cell lines (Carr et al., 2016; Iurlaro et al., 2017; Liu et al., 2009). Interestingly, Iurlaro et al. also demonstrated that not only surface expression of TRAILR2 increases after glucose deprivation but that the receptor also accumulates intracellularly (Iurlaro et al., 2017). Upregulation of TRAILR2 induced upon glucose deprivation sensitizes cells to TRAIL-induced apoptosis (Carr et al., 2016). Pei et al. demonstrated that hypoxia can upregulate TRAILR4 expression, probably through HIF-1α acting as a transcription factor, in five different colon carcinoma cell lines, and this resulted in attenuated TRAIL-induced cell death (Pei et al., 2010). Taking together, cells within solid tumors that are undersupplied with nutrients and oxygen therefore likely undergo changes in TRAILR expression that influence their TRAIL susceptibility.

Insight into TRAILR expression within tissue or 3D contexture remains limited. In colon carcinoma, the highest expression of TRAILR2 can be found in cells within the invasive front (Sträter et al., 2002). Furthermore, in NSCLCs with squamous cell differentiation, both the expression of TRAILR1 and TRAILR2 seems highest in the basal layer of tumors, whereas cells located in supra basal areas show decreased expression or are completely devoid of the two receptors (Spierings et al., 2003). In multicellular tumor spheroids, an in vitro model mimicking solid avascularized tumors, the overall expression of both, TRAILR1 and TRAILR2, was dramatically reduced in comparison to conventional 2D cultures of the same cell line. As expected, this strong downregulation of TRAILRs correlated with reduced TRAIL susceptibility of spheroids (Chandrasekaran et al., 2014). Nevertheless, a detailed characterization of spatiotemporal TRAILR1 or TRAILR2 expression patterns in 3D grow environments has not been performed for spheroid models yet.

8.2 Regulation of TRAILR expression by transcription factors

TRAILR expression is controlled by several transcription factors that can be activated or induced in adaptation to environmental conditions. TRAILR1 and TRAILR2 expression was shown to be regulated by NF-κB, p53, forkhead-box-protein O3a (FOXO3a), activator protein 1 (AP1) and specificity protein 1 (SP1) in response to stress, while the transcriptional repressor GLI3 was shown to inhibit TRAILR1 expression (van Roosmalen et al., 2014). Regarding the two regulating TRAILRs, TRAILR3 and TRAILR4, it was found that their expression is controlled by p53 (Liu et al., 2005; Toscano et al., 2008). Furthermore, as mentioned before, expression of TRAILR4 is directly inducible by HIF-1α, whereas expression of TRAILR3 can be induced by NF-κB activation (Bernard et al., 2001; Pei et al., 2010). As later described in more detail, adverse conditions such as hypoxia and nutrient starvation can induce ER stress which, if not resolved, triggers apoptosis. Prolonged ER stress was shown to result in the induction of the transcription factor C/EBP homologous protein (CHOP), which can upregulate, with or without the contribution of two further transcription factors, ELK-1 and cyclic adenosine monophosphate (cAMP)-dependent transcription factor 3 (ATF3), the expression of TRAILR2 (Edagawa et al., 2014; Oh et al., 2010; Yamaguchi and Wang, 2004). However, a CHOP-independent TRAILR2 upregulation upon ER stress can be mediated by further proteins involved in the ER stress-induced unfolded protein response (UPR) (Iurlaro et al., 2017; Liu et al., 2009) (illustrated in Fig. 3).

8.3 Genetic and epigenetic regulation of TRAILR expression

The amount of TRAIL receptors is not only controlled by a wide range of transcriptional regulators. Several studies demonstrated that expression of TRAILR1, TRAILR2, TRAILR3 and TRAILR4 can be suppressed by hypermethylation of CpG islands (CGIs) in the corresponding promotor regions (Lee et al., 2009; Shivapurkar et al., 2004; Yoshida and Sakai, 2004). DNA modification is widespread in human DNA where CpG sites are methylated at levels of 70–80% in a cell type specific pattern (Jeltsch et al., 2018; Jones, 2012). CpG islands, regions with a higher density of CpG sites than the bulk genome, which are frequently found in gene promoters, commonly are unmethylated. DNA methylation is connected to gene expression, where hypermethylation of promoter CGIs typically leads to strong gene silencing, while methylation of gene bodies is correlated with active gene expression. The methylome of most cancer cells is heavily deranged, with global DNA methylation in the bulk genome coexisting with hypermethylation of many promotor CGIs, resulting in silencing of the corresponding genes (Bergman and Cedar, 2013). This process is particularly relevant in cases of tumor suppressor proteins, expression of which is silenced by aberrant promoter methylation, as described in the cases of TRAILR1/2.

COSMIC data reveal that downregulation and promoter methylation of TRAILR1/2 is mainly observed in cancers of the large intestine, liver, prostate and breast, but other cancers may show overexpression of these factors (Tate et al., 2019). In addition, genetic alterations of the locus encoding TRAILR1/2 are reported in the cBioportal database (Gao et al., 2014). In particularly, recurrent deep deletions of the TRAILR1/2 locus are found mainly in colon, bladder, prostate, liver and ovarian cancer. These data reflect the general observation in cancer cells, that epigenetic silencing of tumor suppressor genes is roughly as common as their loss-of-function through mutations and deletions, but the exact frequencies depend on the affected gene and tumor type (Flavahan et al., 2018).

DNA methylation abnormalities in cancer cells often are caused by mutations in enzymes directly or indirectly involved in setting or removing DNA methylation or their regulatory factors (Feinberg et al., 2016; Plass et al., 2013). This refers particularly to mutations in the DNMT3A DNA methyltransferase, TET2 methylcytosine dioxygenase and isocitrate dehydrogenases (IDH1 or IHD2). DNMT3A mutations have been associated with reduced activity and altered flanking sequence preferences (Emperle et al., 2018, 2019; Yang et al., 2015). The TET2 enzyme is involved in

DNA demethylation and mutations in TET2 mainly cause loss-of-function. Mutations in IDH enzymes alter the reactivity of the enzyme and lead to the generation of 2-hydroxyglutarate, an abnormal metabolite that is a strong and competitive inhibitor of TET enzymes (Kroeze et al., 2015). How these specifically are linked to modified TRAILR expression, however, currently remains understudied.

Based on the consideration that epigenetic changes are reversible, hypomethylating agents like decitabine are in use for the treatment of some cancers (Baylin and Jones, 2012; Jones et al., 2016). They can lead to global DNA demethylation accompanied by the reactivation of previously silenced genes, including tumor suppressor genes and repeat elements. This effect can trigger apoptosis through the activity of the re-expressed tumor suppressor genes, or lead to an immune attack on the tumor cells caused by the expression of neo-antigens (Jones, 2019; Topper et al., 2019). Indeed, it has been reported that treatment of resistant melanoma cell lines with the hypomethylating agent 5-aza-2′-deoxycytidine restored TRAILR1 expression and rendered cells TRAIL susceptible (Bae et al., 2008). Besides that, a more targeted approach has been proposed in which epigenetic alterations that are causative for the cancerous phenotype could be selectively reverted by Epigenome Editing (Kungulovski and Jeltsch, 2016).

9. TRAIL-based strategies for the development of anti-cancer agents

TRAIL preferential induces apoptosis in cancer cells. Although initial trials with recombinantly produced TRAIL formats also showed toxicity against untransformed cells, namely hepatocytes, this later could be attributed to the formation of higher order aggregates, mediated by FLAG- or hexahistidine (His)-tags meant for purification purposes during production (Ganten et al., 2006; Koschny et al., 2007; Lawrence et al., 2001). Without the expression of such moieties, TRAIL exclusively targets cancer cells. While the underlying reasons for this observation are still not completely resolved, evidence exists that non-transformed cells possibly acquire TRAIL resistance by expressing high amounts of TRAILR3, cFlip, XIAP, Bcl-2 and low expression of TRAILR1 and TRAILR2 (Sheridan et al., 1997; van Dijk et al., 2013; Zhang et al., 2000). Additionally, cancer cells might become "addicted" to elevated TRAIL receptor expression due to these receptors being able to drive cell proliferation (von Karstedt et al., 2015).

The first generation of TRAIL-based anti-cancer therapeutics comprised the soluble, recombinant human TRAIL, dulanermin, and agonistic monoclonal antibodies against TRAILR1 and TRAILR2. Unfortunately, even though well tolerated in clinical trials, the overall results were largely disappointing as it turned out that only very few cancer patients benefited from treatment (Camidge et al., 2010; den Hollander et al., 2013; Herbst et al., 2010; Tolcher et al., 2007). This failure of the first wave of TRAILR activating agents has various reasons. Alongside the problem that many tumors are initially TRAIL-resistant, dulanermin displays an in vivo half-life of just 30 min and fails to efficiently cluster TRAILRs. Furthermore, due to its non-selectivity, it also binds to the regulatory receptors TRAILR3 and TRAILR4. While agonistic antibodies solely bind to the death-inducing TRAILRs and display extended half-lives in comparison to dulanermin, also these failed in clinical trials, likely due to their dimeric format being ineffectively in trimerizing their cognate receptors in vivo. The reason why agonistic TRAILR2 antibodies showed at least some cytotoxic activity in vivo was only recently attributed to their fragment crystallizable (Fc) part. This Fc part binds to Fcγ receptors on tumor-associated leukocytes, which supports their local aggregation and in turn supports more efficient TRAILR2 activation (Wilson et al., 2011). However, in the same way these antibodies can also bind to cells in the blood stream, which diminishes their concentration at the tumor site (von Karstedt et al., 2017).

In order to overcome these drawbacks, efforts in recent years focused on improving and increasing stability, pharmacokinetics, targeting and valency of TRAIL receptor-binding anti-cancer agents. Since agonistic TRAILR-specific antibodies failed to efficiently cluster TRAILRs due to their bivalent format, antibodies with higher valency (nanobodies) were developed (Huet et al., 2014; Papadopoulos et al., 2015). Among these, TAS266 was the first to enter a clinical trial, which was, however, terminated due to strong toxicity of the compound (Papadopoulos et al., 2015). An alternative approach to increase the activity of TRAIL-specific antibodies is the generation of bispecific formats that not only attach to and activate TRAILRs but are also capable of binding to further antigens present on the tumor cells (Kontermann, 2012; Michaelson et al., 2009). It is conceivable that this not only improves targeting of the antibodies but that the simultaneous activation/inhibition of different receptor types might also enhance the bioactivity of the compound. Michaelson et al., for example, developed a bispecific antibody that targets both TRAILR2 and lymphotoxin β receptor, which displayed enhanced activity when compared to the respective antibodies that target only one receptor type (Michaelson et al., 2009).

Similar to the development of more advanced TRAILR-specific antibodies, also soluble recombinant human TRAIL was further developed as a potential anti-cancer agent. To improve its stability, TRAIL monomers were fused to tags such as poly-histidine, FLAG epitope, leucine zipper, isoleucine zipper, the trimerization domain of tenascin-C, human serum albumin or human adenovirus type 5 fiber, all of which enforce homotrimerization (Berg et al., 2007; von Karstedt et al., 2017; Yan et al., 2016). An alternative strategy lies in linking the extracellular domains of three TRAIL monomers by short peptides, resulting in single-chain TRAIL (scTRAIL) formats that can easily be combined with antibody moieties, for example, to further increase stability but also to allow efficient targeting to tumor specific antigens. In 2012, Siegemund et al. developed an antibody TRAIL fusion protein comprised of the single-chain fragment variable (scFv) fragment of the epidermal growth factor receptor (EGFR)-binding antibody Cetuximab and one scTRAIL. By reducing the linker length within the scFv fragment, non-covalent dimerization of two scFV-scTRAIL molecules was achieved, resulting in $Db_{\alpha EGFR}$-scTRAIL (Siegemund et al., 2012). Of note, recently this protein format was further improved by varying the linker lengths between the single TRAIL monomers (Siegemund et al., 2016). In an alternative approach, an scTRAIL moiety was fused to the Fc part of an (immunoglobulin G) IgG antibody. Thereby, formation of disulfide bonds between two Fc parts resulted in a hexavalent TRAILR agonist, Fc-scTRAIL (or Fc-scTRAIL 1551). EGFR targeting was likewise achieved by fusing Cetuximab-derived scFv to the N-terminus of Fc-scTRAIL (Hutt et al., 2017). In subsequent studies it was shown that being able to bind to the EGFR significantly increased the bioactivity of antibody TRAIL fusion proteins in vitro, most likely due to increased clustering of TRAILR. However, surprisingly this positive effect of targeting did not translate to better xenograft tumor control in vivo when compared to hexavalent TRAIL constructs. It can be speculated that high affinity to the EGFR impaired tumor penetration and/or resulted in binding of the protein to tumor distant tissue that expressed low amounts of the receptor (Hutt et al., 2018). Similar to the above-mentioned antibody-TRAIL fusion proteins, two further promising TRAILR agonists, APG350 and the derived ABBV-621, were generated by fusing scTRAIL chains to antibody moieties. Both variants show strong anti-cancer activity in preclinical settings, and ABBV-621 is currently tested in a clinical phase I study (Gieffers et al., 2013; Ratain et al., 2019). Taken together, first generation engineered TRAILR agonists did not meet the expectations as potent novel anti-cancer therapeutics, but there is legitimate hope that latest formats with increased stability, valency and thus potency might prove clinically useful. Since it can be expected that not all patients will be responsive to single agent treatments, it will be important to co-investigate TRAIL sensitization strategies and to develop reliable markers for the pre-selection of patients most likely to respond to treatment.

10. Induction of ER stress: A promising approach to sensitize cancer cells for TRAIL treatment?

Beside the development of improved TRAIL variants, it is equally important to identify treatment strategies that sensitize otherwise resistant cancer cells to TRAIL-based therapy. Taking into account the recent literature on ER stress and its link to apoptosis induction, combining TRAIL with ER stress-inducing agents such as the FDA-approved anti-inflammatory drug celecoxib, might present a promising approach for patients that will not benefit from TRAIL monotherapy. Of note, while literature exists linking ER stress also to necroptosis as a possible outcome, none of these studies are implicating the TRAIL signaling pathway (Kishino et al., 2019; Saveljeva et al., 2015). In eukaryotic cells, the ER regulates Ca^{2+} homeostasis and is a primary site for folding and transport of newly generated proteins. Conditions that disturb the fragile equilibrium cause ER stress, which induces molecular stress response programs (Almanza et al., 2019; Hetz, 2012; Iurlaro and Muñoz-Pinedo, 2016; Lindholm et al., 2017). Common environmental stress factors, in particular in poorly vasculated tumor tissues, are the lack of oxygen and nutrients. Both can contribute to ER stress, for example, by causing ATP depletion, which in turn limits the supply of energy equivalents required for the energy-intensive processes of protein folding. Furthermore, oxygen is also needed for the formation of disulfide bonds while glucose is necessary for the ER/Golgi-based glycosylation of proteins (Kaufman et al., 2002; Schönenberger and Kovacs, 2015; Sies and Bruene, 2007). Alongside hypoxia and starvation, ER stress can be induced by disturbed Ca^{2+} homeostasis, which results in the inhibition of calcium-dependent chaperons, or by viral infections that overload the cell with virus-encoded proteins (Coe and Michalak, 2009; He, 2006; von dem Bussche et al., 2010). Under physiological conditions, ER associated degradation (ERAD) prevents the accumulation of unfolded and misfolded proteins by utilization of the proteasomal degradation machinery. Consequently, inhibition of the proteasome can contribute to ER stress and induction of the unfolded protein response (UPR) (Obeng et al., 2016; Takenokuchi et al., 2015). During stress onset, the UPR pathway attempts to reduce the amount of damaged proteins by breakdown of mRNA, enlargement of the ER membrane and synthesis of protein folding chaperons; however, in unresolved ER stress apoptosis is initiated (illustrated in Fig. 2). Thereby, the accumulation of misfolded proteins first results in the detachment of the chaperon binding immunoglobulin protein (BiP) from different sensor proteins, allowing the autoactivation of the latter. Interestingly, hereby BiP acts both as an ATP cycle-dependent molecular chaperon and at the same time as a sensor protein for ER stress. Very recently it was shown that these functions of BiP are separated, which is achieved the suppression of the BiP ATPase activity upon binding to ER stress sensors (Kopp et al., 2019). In the inositol-requiring protein 1α-X box-binding protein-1 (IRE1α-XBP1) branch of the UPR, sequestering of BiP allows the dimerization and

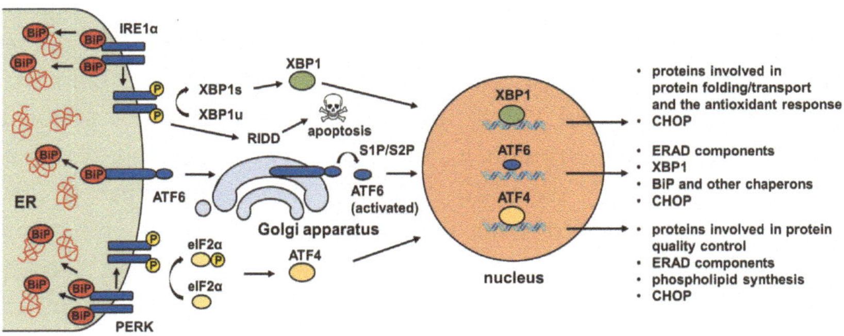

Fig. 2 The unfolded protein response. Induction of ER stress results in the detachment of BiP from the different ER stress sensors (IRE1α, ATF6 and PERK) leading to their subsequent autoactivation. In response to that, the transcriptional regulators XBP1, ATF6 and ATF4 are generated which induce the expression of proteins involved in protein quality control, transport, folding as well as in ER associated degradation (ERAD) and apoptosis. Beside the generation of the transcription factor XBP1, activation of IRE1α also results in IRE1α-dependent decay of mRNAs (RIDD) which first aims to alleviate ER stress by decreasing the amount of generated proteins but later contributes to the induction of apoptosis.

subsequent autophosphorylation of the protein IRE1α. IRE1α possesses both kinase and endoribonuclease activity. Under low ER stress, activated IRE1α cleaves the precursor mRNA XBP1u to XBP1s, which results in the translation of the transcription factor XBP1. XBP1 then induces the expression of the transcription factor CHOP as well as proteins involved in protein quality control and ERAD signaling. Furthermore, XBP1 can promote phospholipid synthesis and thus contribute to the enlargement of the ER membrane. During prolonged ER stress, XBP1 translation is reduced and IRE1α-dependent decay of mRNAs (RIDD) becomes more important (Hetz, 2012; Iurlaro and Muñoz-Pinedo, 2016; Lindholm et al., 2017). RIDD already contributes to ER homeostasis by regulating the translation of many proteins involved in protein folding, in cell cycle progression and apoptosis under basal conditions. Within increasing ER stress, RIDD activity increases and can finally overrun the ER stress resolving functions of XBP1 and promote apoptosis induction by degrading mRNAs that code for anti-apoptotic proteins (Maurel et al., 2014). Alongside its function as a ribonuclease, IRE1α was also shown to recruit the adapter protein TRAF2 which can lead to activation of JNKs and subsequent apoptosis induction (Lee et al., 2011; Urano et al., 2000). Signaling via the IRE1α axes therefore might be pro- or anti-apoptotic in vivo, most likely depending on the strength and duration of the ER stress stimulus. While the regulation of this decision-making has not yet been investigated fully, signaling kinetics evoked by ER stress as well as the interplay with the PERK signaling branch (see below) might contribute to casting death or survival decisions (Walter et al., 2015). Alongside the IRE1α axis, the UPR also comprises the activation of the transcriptional regulator ATF6. During ER stress, inactive ATF6 is transported from the ER to the Golgi apparatus, where it is converted into its active form via two intramembranous proteases. Activated ATF6 then induces the expression of ERAD components and the transcription factors XBP1 and CHOP as well as ER chaperons such as BiP (Baumeister et al., 2005; Lindholm et al., 2017; Yoshida et al., 2000). Finally, ER stress also results in the activation of the protein kinase RNA-like endoplasmic reticulum kinase (PERK) leading to phosphorylation and inactivation of the eukaryotic translation initiation factor 2α (eIF2α). Without active eIF2α, translation of proteins that lack an internal site for ribosomal entry is impaired and thus overall protein translation is attenuated. However, the expression of individual proteins such as transcription factor ATF4 is even enhanced in the presence of phosphorylated eIF2α, which results in the transcription of proteins that are involved in protein folding and transport or take part in the antioxidant response (Harding et al., 2003; by Iurlaro and Muñoz-Pinedo, 2016).

Alongside this, ATF4, similar to XBP1 and ATF6, also induces the expression of the transcription factor CHOP as the major driver of ER stress-induced apoptosis (illustrated in Figs. 3 and 4). Transcribed in response to the activation of all three branches of the UPR, this protein facilitates the execution of apoptosis by enhancing the expression of the pro-apoptotic BH3-only proteins Bim and PUMA and by inhibiting the transcription of the anti-apoptotic Bcl-2 protein (Mc Cullough et al., 2001; Puthalakath et al., 2007; Wali et al., 2014). In addition, CHOP induces the expression of TRAILR2, alone or in combination with the two transcription factors ELK-1 and ATF3 (Edagawa et al., 2014; Oh et al., 2010; Yamaguchi and Wang, 2004).

As mentioned above, the decision between the attempt to resolve ER stress and the induction of apoptosis is complex and not yet completely understood. However, Lin et al. suggested that the switch between cytoprotection and pro-survival outcome depends on the duration of individual UPR branch activities. During persistent ER stress, IRE1α and ATF6 signaling decays while PERK signaling persists. As a consequence, this might lead to the induction of apoptosis, arising from prolonged translational inhibition and insufficient protein neosynthesis to restore cellular homeostasis. Additionally, ongoing PERK signaling can result in the accumulation of high amounts of CHOP, sufficient to overcome the threshold for cell death by induction pro-apoptotic transcriptional responses (Lin et al., 2007, 2009).

Regarding the upregulation of TRAILR2 via CHOP, it was shown that early during ER stress TRAILR2 mRNA is constantly degraded by IRE1α. However, if ER stress persists, enhanced TRAILR2 expression can overcome the degradation process (Lu et al., 2014). Interestingly, evidence grows that both, TRAILR2 (TRAILR1) and Caspase-8, are indispensable for the execution of apoptosis in response to ER stress (Iurlaro et al., 2017; Lu et al., 2014). For example, it was shown that glucose starvation can led to increased protein amounts of both TRAILR1 and TRAILR2, which is followed by induction of apoptosis. Of note, TRAILR2 was found to mainly accumulate in the Golgi apparatus. While knockdown of the receptors, Caspase-8 and FADD resulted in reduced cell death, knockdown of TRAIL failed to protect cells, therefore suggesting a ligand-independent but TRAILR-dependent induction of apoptosis in this scenario (Iurlaro et al., 2017). It can be speculated that high amounts of TRAILR2 could accumulate intracellularly in response to ER stress, providing a platform for pro-caspase-8

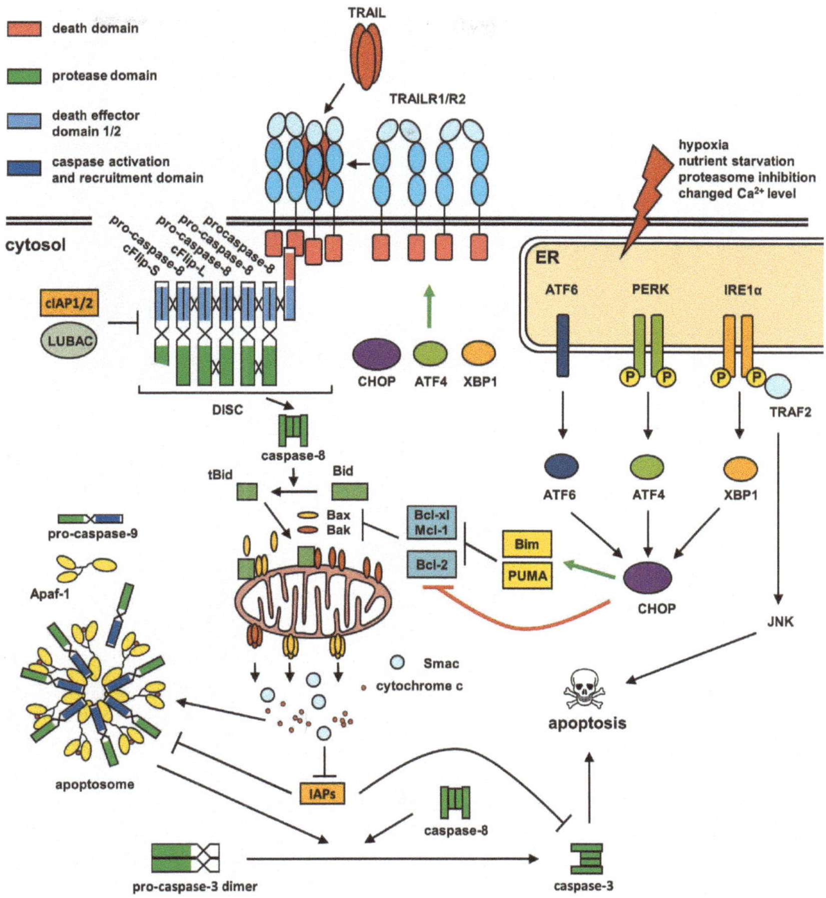

Fig. 3 TRAIL-induced apoptosis and its entanglement with the unfolded protein response. Binding of TRAIL to its cognate receptors results in the formation of the death inducing signaling complex (DISC) comprising the receptors, the adapter protein FADD and a growing filament of cFlipL/S and pro-caspase-8 molecules. Within this complex, pro-caspase-8 is activated and can proteolytically cleave and activate pro-caspase-3 to execute cell death. In type II cells, amplification of the apoptotic signal via the mitochondrion is necessary for the efficient transmission of the apoptotic signal. Thereby Bid, cleaved by caspase-8, facilitates pore formation in the mitochondrial outer membrane resulting in the release of cytochrome *c* and SMAC. Together with Apaf-1 and dATP/ATP, cytochrome *c* forms a platform in which pro-caspase-9 is activated to cleave pro-caspase-3. Apoptosis is further regulated by an interplay of pro- and anti-apoptotic members of the Bcl-2 family as well as IAPs. ER-stress scenarios such as hypoxia, nutrient starvation, proteasome inhibition, or changed amounts of intracellular Ca^{2+} can induce the unfolded protein response (UPR). Activation of its three branches results in the expression of the transcriptional regulators ATF6, ATF4 and XBP1 which are all able to induce the expression of the transcription factor CHOP. CHOP in turn can facilitate apoptosis execution by negatively regulating the amount of anti-apoptotic Bcl-2 and positively regulating the level of the two pro-apoptotic BH3-only proteins, Bim and PUMA as well as the amount of TRAILR1 and TRAILR2. Independently of CHOP, also ATF4 and XBP1 alone were shown to induce TRAILR2 expression.

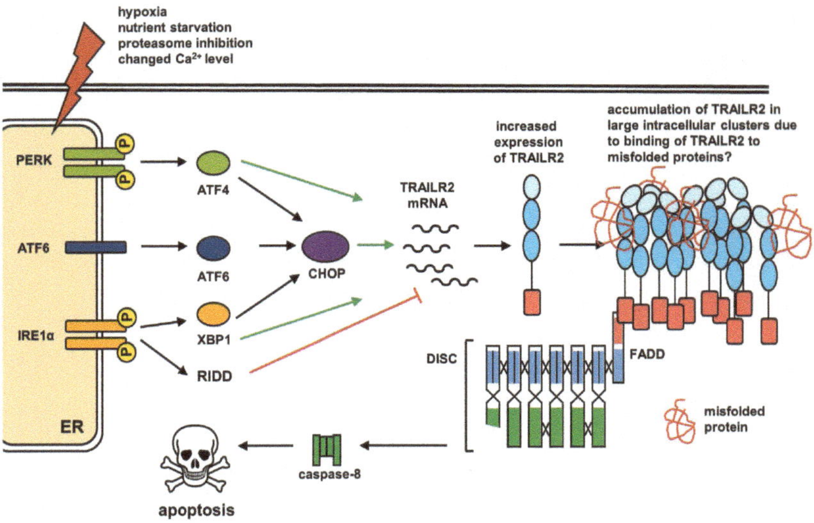

Fig. 4 ER stress-induced intracellular accumulation of TRAILR2 results in activation of pro-caspase-8. Induction of ER stress results in the activation of the ER stress sensors PERK, ATF6 and IRE1α, which in turn leads to the expression of the transcriptional regulators ATF4, ATF6, XBP1 and CHOP. Of these, ATF4, XBP1 and CHOP are known to positively influence TRAILR2 expression (green arrows). Early during ER stress, TRAILR2 mRNA is constantly degraded by IRE1α-dependent decay of mRNAs (RIDD). However, if ER stress persists, induction of TRAILR2 expression exceeds its degradation. During persistent ER stress, high amounts of intracellular TRAILR2 might bind to misfolded proteins, cluster and thereby can act as a scaffold for the activation of pro-caspase-8.

activation (illustrated in Fig. 4). Supporting this hypothesis, it was now shown that upon ER stress, intracellularly accumulating TRAILR2 might bind to misfolded proteins in the ER-Golgi intermediate compartment (ERGIC), leading to TRAILR2 clustering and caspase-8-dependent cell death (Lam et al., 2020). Also proteasome inhibitors can cause a death ligand-independent activation of pro-caspase-8. This activation appears to be more potent than what is observed upon direct ER stress (Laussmann et al., 2011), and whether this response is mechanistically related to ER stress signaling has not yet been studied in detail.

Overall, the ER stress-induced UPR therefore engages a complex interplay of different signaling cascades that regulates ER homeostasis and that also can induce apoptotic cell death. Since tumor cells often grow under ER stress due to nutrient and oxygen deprivation, treatment with drugs that enhance the UPR and thus induce apoptosis might be a promising approach in cancer therapy. Furthermore, since ER-stressed cancer cells are more susceptible to TRAILR activating treatments than unstressed cells, probably due to upregulation of TRAILRs or downregulating anti-apoptotic proteins (Edagawa et al., 2014; Liu et al., 2004; Martín-Pérez et al., 2012; Yamanaka et al., 2006), combinations of TRAILR agonists and ER stress-inducing agents could be promising treatment options.

An FDA-approved drug that was shown to induce ER stress, the accumulation of TRAILR2 and the induction of apoptosis in cancer cells is the Cyclooxygenase II (COX II) inhibitor celecoxib. In the human body, the enzyme COX catalyzes the conversion of arachidonic acid to prostaglandin H_2, which in turn can be converted to various other prostaglandins by prostaglandin synthases. The first isoform, COX I, is constitutively active in most tissues and plays an important role for example in maintaining cell viability in the gastric mucosa and for the aggregation capacity of platelets in response to injury. In contrast to that, the expression of the second isoform, COX II, is induced by various growth factors and cytokines during inflammation (Patrignani and Patrono, 2015; Sobolewski et al., 2010). Taking advantage of the differently regulated expression of the two isoforms, the COX II selective, anti-inflammatory drug celecoxib was developed in the late 1990s and subsequently approved for the treatment of arthritis and pain (Clemett et al., 2000). Interestingly, COX II amounts not only increase in inflamed tissues but also in many tumors of epithelial origin, such as colon, lung, breast and prostate, with COXII amounts correlating with a poor prognosis due to enhanced proliferation, vascularization, invasion and metastasis (Brown and DuBois, 2005; Gupta et al., 2000; Soslow et al., 2000). COX II expression was shown to be regulated by the tumor microenvironment. More precisely, hypoxia enhances the expression of COXII, mechanistically attributable to HIF-1α binding to the promoter region of the COX II gene and driving its transcription (Kaidi et al., 2006). High amounts of COX II were observed to correlate with TRAIL resistance and low expression of TRAILR1 and TRAILR2, both of which could be reversed by knockdown of COXII expression or by treatment with COX II inhibitors (Chandrasekaran et al., 2014; Tang et al., 2002). However, the underlying molecular mechanisms need to be investigated in more details. Interestingly, in many studies, celecoxib was shown to interfere with TRAILR expression independent of its capacity to inhibit COX II. Indeed, it is known that the COX II inhibitors can block the sarco/endoplasmic reticulum Ca^{2+} APTase (SERCA) which in turn leads to a rapid increase of intracellular Ca^{2+} and subsequent induction of ER-stress and the UPR. It therefore appears logical that this route of signaling could be a driver of celecoxib-induced ER stress and induction of CHOP-dependent TRAILR2 expression (Chen et al., 2007; Johnson et al., 2002; Pyrko et al., 2007).

11. Concluding comments

While TRAIL-induced signal transduction toward apoptosis is remarkably well understood at the level of the core signal transduction machinery, in recent years the surprising complexity associated with the modulation and fine-tuning of this signaling cascade emerged. This not only relates to the apoptosis signaling branch but also to alternative signaling outcomes such as necroptosis and cell death independent routes, such as pro-survival and proliferation signaling. Here, we present an overview of the current state of knowledge regarding these processes, but it needs to be noted that our current understanding of the modulation of TRAILR-related signaling, cell intrinsically or through micro-environmental factors, remains still rather descriptive and qualitative. This applies even more to the role TRAILRs appear to play in the nucleus and during ER stress responses, both of which constitute signaling scenarios that do not require the presence of receptor ligands, and the cellular decision-making processes linked to these scenarios. Even though TRAIL and its receptors were discovered over 20 years ago, this limited understanding, together with poor bioactivities and half-lives of recombinant human TRAIL formulations, so far prevented successful clinical exploitation, even though TRAIL-induced signaling preferentially eliminates cancer cells. While latest generation TRAIL-based biologics now demonstrate significantly improved bioactivity and stability and have entered clinical trials (NCT03082209; Gieffers et al., 2013; Hutt et al., 2017), a deeper mechanistic and quantitative understanding of TRAILR-dependent signaling will nevertheless be required to develop knowledge-based biomarkers that allow to reliably predict the responsiveness of individual tumors to TRAILR-targeting treatments and that can assist in devising optimal TRAIL-sensitizing combination treatments. The need for such markers is further highlighted by the finding that TRAIL-based treatments can initiate proliferation and invasion responses, and thus can be pro-oncogenic, in cells incapable to activate apoptosis programs (Ichim and Tait, 2016). This ambitious goal can only be accomplished by building on the improved understanding of TRAILR signaling that was achieved in recent years and by continued interdisciplinary and collaborative research efforts that aim in this direction.

Acknowledgments

We thank Dr. G. Fullstone for help in figure assembly. We gratefully acknowledge funding from the German Research Foundation (FOR2036 (MO 3226/1-1)) and from the German Research Foundation under Germany's Excellence Strategy (EXC 2075—390740016).

References

Aggarwal, B.B., Kohr, W.J., Hass, P.E., Moffat, B., Spencer, S.A., Henzel, W.J., Bringman, T.S., Nedwin, G.E., Goeddel, D.V., Harkins, R.N., 1985. Human tumor necrosis factor. Production, purification, and characterization. J. Biol. Chem. 260, 2345–2354.

Aggarwal, B.B., Gupta, S.C., Kim, J.H., 2012. Historical perspectives on tumor necrosis factor and its superfamily: 25 years later, a golden journey. Blood 119, 651–665. https://doi.org/10.1182/blood-2011-04-325225.

Aldridge, B.B., Gaudet, S., Lauffenburger, D.A., Sorger, P.K., 2011. Lyapunov exponents and phase diagrams reveal multi-factorial control over TRAIL-induced apoptosis. Mol. Syst. Biol. 7, 1–21. https://doi.org/10.1038/msb.2011.85.

Almanza, A., Carlesso, A., Chintha, C., Creedican, S., Doultsinos, D., Leuzzi, B., Lu, A., Mc Carthy, N., Montibeller, L., More, S., Papaioannou, A., Püschel, F., Sassano, M.L., Skoko, J., Agostinis, P., de Belleroche, J., Eriksson, L.A., Fulda, S., Gorman, A.M., Healy, S., Kozlov, A., Muñoz-Pinedo, C., Rehm, M., Chevet, E., Samali, A., 2019. Endoplasmatic reticulum stress signalling—from basic mechanisms to clinical applications. FEBS J. 286, 241–278. https://doi.org/10.1111/febs.

Arandjelovic, S., Ravichandran, K.S., 2015. Phagocytosis of apoptotic cells in homeostasis. Nat. Immunol. 16, 907–917. https://doi.org/10.1038/ni.3253.

Bae, S.I., Cheriyath, V., Jacobs, B.S., Reu, F.J., Borden, E.C., 2008. Reversal of methylation silencing of Apo2L/TRAIL receptor 1 (DR4) expression overcomes resistance of SK-MEL-3 and SK-MEL-28 melanoma cells to interferons (IFNs) or Apo2L/TRAIL. Oncogene 27, 490–498. https://doi.org/10.1038/sj.onc.1210655.

Baumeister, P., Luo, S., Skarnes, W.C., Sui, G., Seto, E., Shi, Y., Lee, A.S., 2005. Endoplasmic reticulum stress induction of the Grp78/BiP promoter: activating mechanisms mediated by YY1 and its interactive chromatin modifiers. Mol. Cell Biol. 25, 4529–4540. https://doi.org/10.1128/MCB.25.11.4529.

Baylin, S.B., Jones, P.A., 2012. A decade of exploring the cancer epigenome-biological and translational implications. Nat. Rev. Cancer 11, 726–734. https://doi.org/10.1038/nrc3130.

Berg, D., Lehne, M., Müller, N., Siegmund, D., Münkel, S., Sebald, W., Pfizenmaier, K., Wajant, H., 2007. Enforced covalent trimerization increases the activity of the TNF ligand family members TRAIL and CD95L. Cell Death Differ. 14, 2021–2034. https://doi.org/10.1038/sj.cdd.4402213.

Bergman, Y., Cedar, H., 2013. DNA methylation dynamics in health and disease. Nat. Publ. Group 20, 274–281. https://doi.org/10.1038/nsmb.2518.

Bernard, D., Quatannens, B., Vandenbunder, B., Abbadie, C., 2001. Rel/NF-κB transcription factors protect against tumor necrosis factor (TNF)-related apoptosis-inducing ligand (TRAIL)-induced apoptosis by up-regulating the TRAIL decoy receptor DcR1. J. Biol. Chem. 276, 27322–27328. https://doi.org/10.1074/jbc.M011183200.

Bodmer, J.L., Meier, P., Tschopp, J., Schneider, P., 2000. Cysteine 230 is essential for the structure and activity of the cytotoxic ligand TRAIL. J. Biol. Chem. 275, 20632–20637. https://doi.org/10.1074/jbc.M909721199.

Bodmer, J.L., Schneider, P., Tschopp, J., 2002. The molecular architecture of the TNF superfamily. Trends Biochem. Sci. 27, 19–26. https://doi.org/10.1016/S0968-0004(01)01995-8.

Bouralexis, S., Findlay, D.M., Atkins, G.J., Labrinidis, A., Hay, S., Evdokiou, A., 2003. Progressive resistance of BTK-143 osteosarcoma cells to Apo2L/TRAIL-induced apoptosis is mediated by acquisition of DcR2/TRAIL-R4 expression: Resensitisation with chemotherapy. Br. J. Cancer 89, 206–214. https://doi.org/10.1038/sj.bjc.6601021.

Boyle, W.J., Simonet, W.S., Lacey, D.L., 2003. Osteoclast differentiation and activation. Nature 423, 337–342. https://doi.org/10.1038/nature01658.

Bridgham, J.T., Wilder, J.A., Hollocher, H., Johnson, A.L., 2003. All in the family: evolutionary and functional relationships among death receptors. Cell Death Differ. 10, 19–25. https://doi.org/10.1038/sj.cdd.4401174.

Brown, J.R., DuBois, R.N., 2005. COX-2: a molecular target for colorectal cancer prevention. J. Clin. Oncol. 23, 2840–2855. https://doi.org/10.1200/JCO.2005.09.051.

Camidge, D.R., Herbst, R.S., Gordon, M.S., Eckhardt, S.G., Kurzrock, R., Durbin, B., Ing, J., Tohnya, T.M., Sager, J., Ashkenazi, A., Bray, G., Mendelson, D., 2010. A phase I safety and pharmacokinetic study of the death receptor 5 agonistic antibody PRO95780 in patients with advanced malignancies. Clin. Cancer Res. 16, 1256–1263. https://doi.org/10.1158/1078-0432.CCR-09-1267.

Carr, R.M., Qiao, G., Qin, J., Jayaraman, S., Prabhakar, B.S., Maker, A.V., 2016. Targeting the metabolic pathway of human colon cancer overcomes resistance to TRAIL-induced apoptosis. Cell Death Dis. 2, 16067. https://doi.org/10.1038/cddiscovery.2016.67.

Carswell, E.A., Old, L.J., Kassel, R.L., Green, S., Fiore, N., Williamson, B., 1975. An endotoxin-induced serum factor that causes necrosis of tumors. Proc. Natl. Acad. Sci. U. S. A. 72, 3666–3670. https://doi.org/10.1073/pnas.72.9.3666.

Chan, F.K.M., 2007. Three is better than one: pre-ligand receptor assembly in the regulation of TNF receptor signaling. Cytokine 37, 101–107. https://doi.org/10.1016/j.cyto.2007.03.005.

Chandrasekaran, S., Marshall, J.R., Messing, J.A., Hsu, J.-W., King, M.R., 2014. TRAIL-mediated apoptosis in breast cancer cells cultured as 3D spheroids. PLoS One 9, e111487. https://doi.org/10.1371/journal.pone.0111487.

Chen, S., Liu, X., Yue, P., Schönthal, A., 2007. CCAAT/Enhancer binding protein homologous protein-dependent death receptor 5 induction and ubiquitin/proteasome-mediated cellular FLICE-inhibitory protein. Mol. Pharmacol. 6, 1269–1279. https://doi.org/10.1124/mol.107.037465.

Chen, J.-J., Shen, H.-C.J., Rivera Rosado, L.A., Zhang, Y., Di, X., Zhang, B., 2012. Mislocalization of death receptors correlates with cellular resistance to their cognate ligands in human breast cancer cells. Oncotarget 3, 833–842. https://doi.org/10.18632/oncotarget.542.

Chen, D., Yu, J., Zhang, L., 2016. Necroptosis: an alternative cell death program defending against cancer. Biochim. Biophys. Acta Rev. Cancer 1865, 228–236. https://doi.org/10.1016/j.bbcan.2016.03.003.

Clancy, L., Mruk, K., Archer, K., Woelfel, M., Mongkolsapaya, J., Screaton, G., Lenardo, M.J., Chan, F.K.-M., 2005. Preligand assembly domain-mediated ligand-independent association between TRAIL receptor 4 (TR4) and TR2 regulates TRAIL-induced apoptosis. Proc. Natl. Acad. Sci. U. S. A. 102, 18099–18104. https://doi.org/10.1073/pnas.0507329102.

Clemett, D., Goa, K.L., Bolten, W., Immunologie, K., Wiesbaden, K.D.L., Ryan, P.J.F., Centre, C.M., Simon, L., 2000. A review of its use in osteoarthritis, rheumatoid arthritis and acute pain. Drugs 59, 957–980.

Coe, H., Michalak, M., 2009. Calcium binding chaperones of the endoplasmic reticulum. Gen. Physiol. Biophys. 28, 96–103. https://doi.org/20093733.

Coley, W.B., 1891. Contribution to the knowledge of sarcoma. Ann. Surg. 14, 199–200. https://doi.org/10.1016/S0140-6736(01)32995-1.
Collison, A., Foster, P.S., Mattes, J., 2009. Emerging role of tumour necrosis factor-related apoptosis-inducing ligand (TRAIL) as a key regulator of inflammatory responses. Clin. Exp. Pharmacol. Physiol. 36, 1049–1053. https://doi.org/10.1111/j.1440-1681.2009.05258.x.
Cowling, V., Downward, J., 2002. Caspase-6 is the direct activator of caspase-8 in the cytochrome c-induced apoptosis pathway: absolute requirement for removal of caspase-6 prodomain. Cell Death Differ. 9, 1046–1056. https://doi.org/10.1038/sj.cdd.4401065.
Creagh, E.M., 2014. Caspase crosstalk: integration of apoptotic and innate immune signalling pathways. Trends Immunol. 35, 631–640. https://doi.org/10.1016/j.it.2014.10.004.
Danish, L., Stöhr, D., Scheurich, P., Pollak, N., 2017. TRAIL-R3/R4 and inhibition of TRAIL signalling in cancer. In: Micheau, O. (Ed.), TRAIL, Fas Ligand, TNF and TLR3 in Cancer. Springer International Publishing, pp. 27–59.
Davidovich, I.A., Levenson, A.S., Levenson, V.V., 2004. Overexpression of DcR1 and survivin in genetically modified cells with pleiotropic drug resistance. Cancer Lett. 211, 189–197. https://doi.org/10.1016/j.canlet.2004.03.026.
De Miguel, D., Lemke, J., Anel, A., Walczak, H., Martinez-Lostao, L., 2016. Onto better TRAILs for cancer treatment. Cell Death Differ. 23, 733–747. https://doi.org/10.1038/cdd.2015.174.
De Toni, E.N., Thieme, S.E., Herbst, A., Behrens, A., Stieber, P., Jung, A., Blum, H., Göke, B., Kolligs, F.T., 2008. OPG is regulated by β-catenin and mediates resistance to TRAIL-induced apoptosis in colon cancer. Clin. Cancer Res. 14, 4713–4718. https://doi.org/10.1158/1078-0432.CCR-07-5019.
Degli-Esposti, M.A., Smolak, P.J., Walczak, H., Waugh, J., Huang, C.-P., DuBose, R.F., Goodwin, R.G., Smith, C.A., 1997a. Cloning and characterization of TRAIL-R3, a novel member of the emerging TRAIL receptor family. J. Exp. Med. 186, 1165–1170. https://doi.org/10.1084/jem.186.7.1165.
Degli-Esposti, M.A., Dougall, W.C., Smolak, P.J., Waugh, J.Y., Smith, C.A., Goodwin, R.G., 1997b. The novel receptor TRAIL-R4 induces NF-kappaB and protects against TRAIL-mediated apoptosis, yet remains an incomplete death domain. Immunity 7, 813–820. https://doi.org/10.1016/S1074-7613(00)80399-4.
den Hollander, M.W., Gietema, J.A., de Jong, S., Walenkamp, A.M.E., Reyners, A.K.L., Oldenhuis, C.N.A.M., De Vries, E.G.E., 2013. Translating TRAIL-receptor targeting agents to the clinic. Cancer Lett. 332, 194–201. https://doi.org/10.1016/j.canlet.2012.04.007.
Denault, J.-B., Eckelman, B.P., Shin, H., Pop, C., Salvesen, G.S., 2007. Caspase-3 attenuates XIAP (X-linked inhibitor of apoptosis protein)-mediated inhibition of caspase-9. Biochem. J. 405, 11–19. https://doi.org/10.1042/BJ20070288.
Dickens, L.S., Boyd, R.S., Jukes-Jones, R., Hughes, M.A., Robinson, G.L., Fairall, L., Schwabe, J.W.R., Cain, K., MacFarlane, M., 2012. A death effector domain chain DISC model reveals a crucial role for caspase-8 chain assembly in mediating apoptotic cell death. Mol. Cell 47, 291–305. https://doi.org/10.1016/j.molcel.2012.05.004.
Dondelinger, Y., Aguileta, M.A., Goossens, V., Dubuisson, C., Grootjans, S., Dejardin, E., Vandenabeele, P., 2013. RIPK3 contributes to TNFR1-mediated RIPK1 kinase-dependent apoptosis in conditions of cIAP1/2 depletion or TAK1 kinase inhibition. Cell Death Differ. 20, 1381–1392. https://doi.org/10.1038/cdd.2013.94.
Dorstyn, L., Akey, C.W., Kumar, S., 2018. New insights into apoptosome structure and function. Cell Death Differ. 25, 1194–1208. https://doi.org/10.1038/s41418-017-0025-z.
Edagawa, M., Kawauchi, J., Hirata, M., Goshima, H., Inoue, M., Okamoto, T., Murakami, A., Maehara, Y., Kitajima, S., 2014. Role of activating transcription factor

3 (ATF3) in endoplasmic reticulum (ER) stress-induced sensitization of p53-deficient human colon cancer cells to tumor necrosis factor (TNF)-related apoptosis-inducing ligand (TRAIL)-mediated apoptosis through up-reulation of death receptor 5 (DR5) by zerumbone and celecoxib. J. Biol. Chem. 289, 21544–21561. https://doi.org/10.1074/jbc.M114.558890.

Emery, J.G., Mcdonnell, P., Burke, M.B., Deen, K.C., Lyn, S., Silverman, C., Dul, E., Appelbaum, E.R., Eichman, C., Diprinzio, R., Dodds, R.A., James, I.E., Rosenberg, M., Lee, J.C., Young, P.R., 1998. Osteoprotegerin is a receptor for the cytotoxic ligand TRAIL. J. Biol. Chem. 273, 14363–14367. https://doi.org/10.1074/jbc.273.23.14363.

Emperle, M., Rajavelu, A., Kunert, S., Arimondo, P.B., Reinhardt, R., Jurkowska, R.Z., Jeltsch, A., 2018. The DNMT3A R882H mutant displays altered flanking sequence preferences. Nucleic Acids Res. 46, 3130–3139. https://doi.org/10.1093/nar/gky168.

Emperle, M., Adam, S., Kunert, S., Dukatz, M., Baude, A., Plass, C., Rathert, P., Bashtrykov, P., Jeltsch, A., 2019. Mutations of R882 change flanking sequence preferences of the DNA methyltransferase DNMT3A and cellular methylation patterns. Nucleic Acids Res. 47, 11355–11367. https://doi.org/10.1093/nar/gkz911.

Feinberg, A.P., Koldobskiy, M.A., Göndör, A., 2016. Epigenetic modulators, modifiers and mediators in cancer aetiology and progression. Nat. Rev. Drug Discov. 17, 284–299. https://doi.org/10.1038/nrg.2016.13.

Finlay, D., Teriete, P., Vamos, M., Cosford, N.D.P., Vuori, K., 2017. Inducing death in tumor cells: roles of the inhibitor of apoptosis proteins. F1000Res. 6, 587. https://doi.org/10.12688/f1000research.10625.1.

Fischer, U., Jänicke, R.U., Schulze-Osthoff, K., 2003. Many cuts to ruin: a comprehensive update of caspase substrates. Cell Death Differ. 10, 76–100. https://doi.org/10.1038/sj.cdd.4401160.

Flavahan, W.A., Gaskell, E., Bernstein, B.E., 2018. Epigenetic plasticity and the hallmarks of cancer. Science 357. https://doi.org/10.1126/science.aal2380, pii: eaal2380.

Forster, J.C., Harriss-Phillips, W.M., Douglass, M.J., Bezak, E., 2017. A review of the development of tumor vasculature and its effects on the tumor microenvironment. Hypoxia (Auckl) 5, 21–32. https://doi.org/10.2147/HP.S133231.

Fricker, N., Beaudouin, J., Richter, P., Eils, R., Krammer, P.H., Lavrik, I.N., 2010. Model-based dissection of CD95 signaling dynamics reveals both a pro- and antiapoptotic role of c-FLIP$_L$. J. Cell Biol. 190, 377–389. https://doi.org/10.1083/jcb.201002060.

Fu, T.M., Li, Y., Lu, A., Li, Z., Vajjhala, P.R., Cruz, A.C., Srivastava, D.B., DiMaio, F., Penczek, P.A., Siegel, R.M., Stacey, K.J., Egelman, E.H., Wu, H., 2016. Cryo-EM structure of caspase-8 tandem DED filament reveals assembly and regulation mechanisms of the death-inducing signaling complex. Mol. Cell 64, 236–250. https://doi.org/10.1016/j.molcel.2016.09.009.

Ganten, T.M., Koschny, R., Sykora, J., Schulze-Bergkamen, H., Büchler, P., Haas, T.L., Schader, M.B., Untergasser, A., Stremmel, W., Walczak, H., 2006. Preclinical differentiation between apparently safe and potentially hepatotoxic applications of TRAIL either alone or in combination with chemotherapeutic drugs. Clin. Cancer Res. 12, 2640–2646. https://doi.org/10.1158/1078-0432.CCR-05-2635.

Gao, J., Aksoy, B.A., Dogrusoz, U., Dresdner, G., Gross, B., Sumer, O., Jacobsen, A., Sinha, R., Larsson, E., Cerami, E., Sander, C., Schultz, N., 2014. Integrative analysis of complex cancer genomics and clinical profiles using the cBioPortal. Sci. Signal. 6, 1–34. https://doi.org/10.1126/scisignal.2004088.

Gieffers, C., Kluge, M., Merz, C., Sykora, J., Thiemann, M., Schaal, R., Fischer, C., Branschadel, M., Abhari, B.A., Hohenberger, P., Fulda, S., Fricke, H., Hill, O., 2013. APG350 induces superior clustering of TRAIL receptors and shows therapeutic

antitumor efficacy independent of cross-linking via Fc receptors. Mol. Cancer Ther. 12, 2735–2747. https://doi.org/10.1158/1535-7163.mct-13-0323.
Gobbi, G., Masselli, E., Micheloni, C., Nouvenne, A., Russo, D., Santi, P., Matteucci, A., Cocco, L., Vitale, M., Mirandola, P., 2010. Hypoxia-induced down-modulation of PKCε promotes TRAIL-mediated apoptosis of tumor cells. Int. J. Oncol. 37, 719–729. https://doi.org/10.3892/ijo.
Green, D.R., 2011. Means to an End: Apoptosis and Other Cell Death Mechanisms. Cold Spring Harbor Laboratory Press, Cold Spring Harbor (N.Y.).
Gupta, S., Srivastava, M., Ahmad, N., Bostwick, D.G., Mukhtar, H., 2000. Over-expression of cyclooxygenase-2 in human prostate adenocarcinoma. Prostate 42, 73–78. https://doi.org/10.1002/(SICI)1097-0045(20000101)42:1<73::AID-PROS9>3.0.CO;2-G.
Harding, H.P., Zhang, Y., Zeng, H., Novoa, I., Lu, P.D., Calfon, M., Sadri, N., Yun, C., Popko, B., Paules, R., Stojdl, D.F., Bell, J.C., Hettmann, T., Leiden, J.M., Ron, D., 2003. An integrated stress response regulates amino acid metabolism and resistance to oxidative stress. Mol. Cell 11, 619–633. https://doi.org/10.1016/S1097-2765(03)00105-9.
Harper, N., Farrow, S.N., Kaptein, A., Cohen, G.M., MacFarlane, M., 2001. Modulation of tumor necrosis factor apoptosis-inducing ligand-induced NF-κB activation by inhibition of apical caspases. J. Biol. Chem. 276, 34743–34752. https://doi.org/10.1074/jbc.M105693200.
Haselmann, V., Kurz, A., Bertsch, U., Hübner, S., Olempska-Müller, M., Fritsch, J., Häsler, R., Pickl, A., Fritsche, H., Annewanter, F., Engler, C., Fleig, B., Bernt, A., Röder, C., Schmidt, H., Gelhaus, C., Hauser, C., Egberts, J.H., Heneweer, C., Rohde, A.M., Böger, C., Knippschild, U., Röcken, C., Adam, D., Walczak, H., Schütze, S., Janssen, O., Wulczyn, F.G., Wajant, H., Kalthoff, H., Trauzold, A., 2014. Nuclear death receptor trail-r2 inhibits maturation of let-7 and promotes proliferation of pancreatic and other tumor cells. Gastroenterology 146, 278–290. https://doi.org/10.1053/j.gastro.2013.10.009.
He, B., 2006. Viruses, endoplasmic reciculum stress, and interferon responses. Cell Death Differ. 13, 393–403. https://doi.org/10.1038/sj.cdd.4401833.
Henry, C.M., Martin, S.J., 2017. Caspase-8 acts in a non-enzymatic role as a scaffold for assembly of a pro-inflammatory "FADDosome" complex upon TRAIL stimulation. Mol. Cell 65, 715–729.e5. https://doi.org/10.1016/j.molcel.2017.01.022.
Herbst, R.S., Eckhardt, S.G., Kurzrock, R., Ebbinghaus, S., O'Dwyer, P.J., Gordon, M.S., Novotny, W., Goldwasser, M.A., Tohnya, T.M., Lum, B.L., Ashkenazi, A., Jubb, A.M., Mendelson, D.S., 2010. Phase I dose-escalation study of recombinant human Apo2L/TRAIL, a dual proapoptotic receptor agonist, in patients with advanced cancer. J. Clin. Oncol. 28, 2839–2846. https://doi.org/10.1200/JCO.2009.25.1991.
Hetz, C., 2012. The unfolded protein response: controlling cell fate decisions under ER stress and beyond. Nat. Rev. Mol. Cell Biol. 13, 89–102. https://doi.org/10.1038/nrm3270.
Horn, S., Hughes, M.A., Schilling, R., Sticht, C., Tenev, T., Ploesser, M., Meier, P., Sprick, M.R., MacFarlane, M., Leverkus, M., 2017. Caspase-10 negatively regulates caspase-8-mediated cell death, switching the response to CD95L in favor of NF-κB activation and cell survival. Cell Rep. 19, 785–797. https://doi.org/10.1016/j.celrep.2017.04.010.
Hu, W.H., Johnson, H., Shu, H.B., 1999. Tumor necrosis factor-related apoptosis-inducing ligand receptors signal NF-kappaB and JNK activation and apoptosis through distinct pathways. J. Biol. Chem. 274, 30603–30610. https://doi.org/10.1074/jbc.274.43.30603.
Huet, H.A., Growney, J.D., Johnson, J.A., Li, J., Bilic, S., Ostrom, L., Zafari, M., Kowal, C., Yang, G., Royo, A., Jensen, M., Dombrecht, B., Meerschaert, K.R.A., Kolkman, J.A., Cromie, K.D., Mosher, R., Gao, H., Schuller, A., Isaacs, R., Sellers, W.R.,

Ettenberg, S.A., 2014. Multivalent nanobodies targeting death receptor 5 elicit superior tumor cell killing through efficient caspase induction. MAbs 6, 1560–1570. https://doi.org/10.4161/19420862.2014.975099.
Hughes, M.A., Powley, I.R., Jukes-Jones, R., Horn, S., Feoktistova, M., Fairall, L., Schwabe, J.W.R., Leverkus, M., Cain, K., MacFarlane, M., 2016. Co-operative and hierarchical binding of c-FLIP and caspase-8: a unified model defines how c-FLIP isoforms differentially control cell fate. Mol. Cell 61, 834–849. https://doi.org/10.1016/j.molcel.2016.02.023.
Hutt, M., Marquardt, L., Seifert, O., Siegemund, M., Müller, I., Kulms, D., Pfizenmaier, K., Kontermann, R.E., 2017. Superior properties of Fc-comprising scTRAIL fusion proteins. Mol. Cancer Ther. 16, 2792–2802. https://doi.org/10.1158/1535-7163.MCT-17-0551.
Hutt, M., Fellermeier-Kopf, S., Seifert, O., Schmitt, L.C., Pfizenmaier, K., Kontermann, R.E., 2018. Targeting scFv-Fc-scTRAIL fusion proteins to tumor cells. Oncotarget 9, 11322–11335. https://doi.org/10.18632/oncotarget.24379.
Hymowitz, S.G., Christinger, H.W., Fuh, G., Ultsch, M., O'Connell, M., Kelley, R.F., Ashkenazi, A., de Vos, A.M., 1999. Triggering cell death: the crystal structure of Apo2L/TRAIL in a complex with death receptor 5. Mol. Cell 4, 563–571. https://doi.org/10.1016/S1097-2765(00)80207-5.
Hymowitz, S.G., Connell, M.P.O., Ultsch, M.H., Hurst, A., Totpal, K., Ashkenazi, A., De Vos, A.M., Kelley, R.F., 2000. A unique zinc-binding site revealed by a high-resolution X-ray structure of homotrimeric Apo2L/TRAIL. Biochemistry 39, 633–640. https://doi.or/10.1021/bi992242l.
Ichim, G., Tait, S.W.G., 2016. A fate worse than death: apoptosis as an oncogenic process. Nat. Rev. Cancer 16, 539–548. https://doi.org/10.1038/nrc.2016.58.
Inoue, S., Browne, G., Melino, G., Cohen, G.M., 2009. Ordering of caspases in cells undergoing apoptosis by the intrinsic pathway. Cell Death Differ. 16, 1053–1061. https://doi.org/10.1038/cdd.2009.29.
Iurlaro, R., Muñoz-Pinedo, C., 2016. Cell death induced by endoplasmic reticulum stress. FEBS J. 283, 2640–2652. https://doi.org/10.1111/febs.13598.
Iurlaro, R., Püschel, F., León-Annicchiarico, C.L., O'Connor, H., Martin, S.J., Palou-Gramón, D., Lucendo, E., Muñoz-Pinedo, C., 2017. Glucose deprivation induces ATF4-mediated apoptosis through TRAIL death receptors. Mol. Cell Biol. 37. https://doi.org/10.1128/MCB.00479-16, pii: e00479-16.
Jeltsch, A., Broche, J., Bashtrykov, P., 2018. Molecular processes connecting DNA methylation patterns with DNA methyltransferases and histone modifications in mammalian genomes. Genes (Basel) 9. https://doi.org/10.3390/genes9110566, pii: E566.
Johnson, A.J., Hsu, A.-L., Lin, H.-P., Song, X., Chen, C.-S., 2002. The cyclo-oxygenase-2 inhibitor celecoxib perturbs intracellular calcium by inhibiting endoplasmic reticulum Ca2+-ATPases: a plausible link with its anti-tumour effect and cardiovascular risks. Biochem. J. 366, 831–837. https://doi.org/10.1042/BJ20020279.
Jones, P.A., 2012. Functions of DNA methylation: islands, start sites, gene bodies and beyond. Nat. Rev. Genet. 13, 484–492. https://doi.org/10.1038/nrg3230.
Jones, P.A., 2019. Epigenetic therapy in immune-oncology. Nat. Rev. Cancer 19, 151–161. https://doi.org/10.1038/s41568-019-0109-9.
Jones, P.A., Issa, J.J., Baylin, S., 2016. Targeting the cancer epigenome for therapy. Nat. Publ. Group 17, 630–641. https://doi.org/10.1038/nrg.2016.93.
Kaidi, A., Qualtrough, D., Williams, A.C., Paraskeva, C., 2006. Direct transcriptional up-regulation of cyclooxygenase-2 by hypoxia-inducible factor (HIF)-1 promotes colorectal tumor cell survival and enhances HIF-1 transcriptional activity during hypoxia. Cancer Res. 66, 6683–6691. https://doi.org/10.1158/0008-5472.CAN-06-0425.

Kaufman, R.J., Scheuner, D., Schröder, M., Shen, X., Lee, K., Liu, C.Y., Arnold, S.M., 2002. The unfolded protein response in nutrient sensing and differentiation. Nat. Rev. Mol. Cell Biol. 3, 411–421. https://doi.org/10.1038/nrm829.
Kelley, S.K., Ashkenazi, A., 2004. Targeting death receptors in cancer with Apo2L/TRAIL. Curr. Opin. Pharmacol. 4, 333–339. https://doi.org/10.1016/j.coph.2004.02.006.
Kerr, J.F.R., Wyllie, A.H., Currie, A.R., 1972. Apoptosis: a basic biological phenomenon with wide-ranging implications in tissue kinetics. J. Intern. Med. 258, 479–517. https://doi.org/10.1038/bjc.1972.33.
Kerr, J.F.R., Harmon, B., Searle, J., 1974. An electron-microscope study of cell deletion in the anuran tadpole tail during spontaneous metamorphosis with special reference to apoptosis of striated muscle fibres. J. Cell Sci. 14, 571–585.
Kimberley, F.C., Screaton, G.R., 2004. Following a TRAIL: update on a ligand and its five receptors. Cell Res. 14, 359–372. https://doi.org/10.1038/sj.cr.7290236.
Kiraz, Y., Adan, A., Kartal Yandim, M., Baran, Y., 2016. Major apoptotic mechanisms and genes involved in apoptosis. Tumor Biol. 37, 8471–8486. https://doi.org/10.1007/s13277-016-5035-9.
Kishino, A., Hayashi, K., Meada, M., Jike, T., Hidai, C., Nomura, Y., Oshima, T., 2019. Caspase-8 regulates endoplasmic reticulum stress-induced necroptosis independent of the apoptosis pathway in auditory cells. Int. J. Mol. Sci. 20, 5896. https://doi.org/10.3390/ijms20235896.
Kojima, Y., Nakayama, M., Nishina, T., Nakano, H., Koyanagi, M., Takeda, K., Okumura, K., Yagita, H., 2011. Importin β1 protein-mediated nuclear localization of death receptor 5 (DR5) limits DR5/tumor necrosis factor (TNF)-related apoptosis-inducing ligand (TRAIL)-induced cell death of human tumor cells. J. Biol. Chem. 286, 43383–43393. https://doi.org/10.1074/jbc.M111.309377.
Kontermann, R.E., 2012. Dual targeting strategies with bispecific antibodies. MAbs 4, 182–197. https://doi.org/10.4161/mabs.4.2.19000.
Kopp, M.C., Larburu, N., Durairaj, V., Adams, C.J., Ali, M.M.U., 2019. UPR proteins IRE1 and PERK switch BiP from chaperone to ER stress sensor. Nat. Struct. Mol. Biol. 1, 1053–1062. https://doi.org/10.1038/s41594-019-0324-9.
Koschny, R., Walczak, H., Ganten, T.M., 2007. The promise of TRAIL—potential and risks of a novel anticancer therapy. J. Mol. Med. 85, 923–935. https://doi.org/10.1007/s00109-007-0194-1.
Kroeze, L.I., Van Der Reijden, B.A., Jansen, J.H., 2015. 5-Hydroxymethylcytosine: an epigenetic mark frequently deregulated in cancer. Biochim. Biophys. Acta Rev. Cancer 1855, 144–154. https://doi.org/10.1016/j.bbcan.2015.01.001.
Kungulovski, G., Jeltsch, A., 2016. Epigenome editing: state of the art, concepts, and perspectives. Trends Genet. 32, 101–113. https://doi.org/10.1016/j.tig.2015.12.001.
Lafont, E., Kantari-Mimoun, C., Draber, P., De Miguel, D., Hartwig, T., Reichert, M., Kupka, S., Shimizu, Y., Taraborrelli, L., Spit, M., Sprick, M.R., Walczak, H., 2017. The linear ubiquitin chain assembly complex regulates TRAIL-induced gene activation and cell death. EMBO J. 36, 1147–1166. https://doi.org/10.15252/embj.201695699.
Lalaoui, N., Morlé, A., Mérino, D., Jacquemin, G., Iessi, E., Morizot, A., Shirley, S., Robert, B., Solary, E., Garrido, C., Micheau, O., 2011. TRAIL-R4 promotes tumor growth and resistance to apoptosis in cervical carcinoma HeLa cells through AKT. PLoS One 6, 2–9. https://doi.org/10.1371/journal.pone.0019679.
Lam, M., Marsters, S.A., Ashkenazi, A., Walter, P., 2020. Misfolded proteins bind and and activate death receptor 5 to induce apoptosis during unresolved endoplasmic reticulum stress. Elife 9. https://doi.org/10.7554/eLife.52291, pii: e52291.
Lane, D., Matte, I., Rancourt, C., Piché, A., 2012. Osteoprotegerin (OPG) protects ovarian cancer cells from TRAIL-induced apoptosis but does not contribute to malignant

ascites-mediated attenuation of TRAIL-induced apoptosis. J. Ovarian Res. 5, 1–9. https://doi.org/10.1186/1757-2215-5-34.

Laussmann, M.A., Passante, E., Rauen, J.A., Würstle, M.L., Delgado, M.E., Devocelle, M., Prehn, J.H.M., Rehm, M., 2011. Proteasome inhibition can induce an autophagy-dependent apical activation of caspase-8. Cell Death Differ. 18, 1584–1597. https://doi.org/10.1038/cdd.2011.27.

Lawrence, D., Shahrokh, Z., Marsters, S., Mounho, B., Hillan, K., Totpal, K., DeForge, L., Schow, P., Hooley, J., Sherwood, S., Pai, R., Leung, S., Khan, L., Gliniak, B., Bussiere, J., Smith, C.A., Strom, S.S., Kelley, S., Fox, J.A., Thomas, D., Ashkenazi, A., 2001. Differential hepatocyte toxicity of recombinant Apo2L/TRAIL versions. Nat. Med. 7, 2000–2002. https://doi.org/10.1038/86397.

Lee, H.W., Lee, S.H., Lee, H.W., Ryu, Y.W., Kwon, M.H., Kim, Y.S., 2005. Homomeric and heteromeric interactions of the extracellular domains of death receptors and death decoy receptors. Biochem. Biophys. Res. Commun. 330, 1205–1212. https://doi.org/10.1016/j.bbrc.2005.03.101.

Lee, K.H., Lim, S.W., Kim, H.G., Kim, D.Y., Ryu, S.Y., Joo, J.K., Kim, J.C., Lee, J.H., 2009. Lack of death receptor 4 (DR4) expression through gene promoter methylation in gastric carcinoma. Langenbecks Arch. Surg. 4, 661–670. https://doi.org/10.1007/s00423-009-0484-x.

Lee, H., Park, M.T., Choi, B.H., Oh, E.T., Song, M.J., Lee, J., Kim, C., Lim, B.U., Park, H.J., 2011. Endoplasmic reticulum stress-induced JNK activation is a critical event leading to mitochondria-mediated cell death caused by β-lapachone treatment. PLoS One 6, 1–12. https://doi.org/10.1371/journal.pone.0021533.

Leithner, K., Stacher, E., Wurm, R., Ploner, F., Quehenberger, F., Wohlkoenig, C., Bálint, Z., Polachova, J., Olschewski, A., Samonigg, H., Popper, H.H., Olschewski, H., 2009. Nuclear and cytopaslmic death receptor 5 as prognostic factors in patients with non-small cell lung cancer treated with chemotherapy. Lung Cancer 65, 98–104. https://doi.org/10.1016/j.lungscan.2008.10.015.

Li, F., Huang, Q., Chen, J., Peng, Y., Roop, D.R., Bedford, J.S., Li, C.-Y., 2010. Apoptotic cells activate the "phoenix rising" pathway to promote wound healing and tissue regeneration. Sci. Signal. 3, 1–10. https://doi.org/10.1126/scisignal.2000634.

Li, T., Su, L., Lei, Y., Liu, X., Zhang, Y., Liu, X., 2015. DDIT3 and KAT2A proteins regulate TNFRSF10A and TNFRSF10B expression in endoplasmic reticulum stress-mediated apoptosis in human lung cancer cells. J. Biol. Chem. 290, 11108–11118. https://doi.org/10.1074/jbc.M115.645333.

Lin, Y., Richburg, J.H., 2014. Characterization of the role of tumor necrosis factor apoptosis inducing ligand (TRAIL) in spermatogenesis through the evaluation of trail gene-deficient mice. PLoS One 9, e93926. https://doi.org/10.1371/journal.pone.0093926.

Lin, J.H., Li, H., Yasumura, D., Cohen, H.R., Zhang, C., Panning, B., Shokat, K.M., Lavail, M.M., Walter, P., 2007. IRE1 signaling effects cell fate during the unfolded protein response. Science 318, 944–949. https://doi.org/10.1126/science.1146361.

Lin, J.H., Li, H., Zhang, Y., Ron, D., Walter, P., 2009. Divergent effects of PERK and IRE1 signaling on cell viability. PLoS One 4, e4170. https://doi.org/10.1371/journal.pone.0004170.

Lindholm, D., Korhonen, L., Eriksson, O., Kõks, S., 2017. Recent insights into the role of unfolded protein response in ER stress in health and disease. Front. Cell Dev. Biol. 5, 1–16. https://doi.org/10.3389/fcell.2017.00048.

Liu, X., Yue, P., Zhou, Z., Khuri, F.R., Sun, S.Y., 2004. Death receptor regulation and celecoxib-induced apoptosis in human lung cancer cells. J. Natl. Cancer Inst. 96, 1769–1780. https://doi.org/10.1093/jnci/djh322.

Liu, X., Yue, P., Khuri, F.R., Sun, S.Y., 2005. Decoy receptor 2 (DcR2) is a p53 target gene and regulates chemosensitivity. Cancer Res. 65, 9169–9175. https://doi.org/10.1158/0008-5472.CAN-05-0939.

Liu, H., Jiang, C.C., Lavis, C.J., Croft, A., Dong, L., Tseng, H.Y., Yang, F., Tay, K.H., Hersey, P., Zhang, X.D., 2009. 2-Deoxy-D-glucose enhances TRAIL-induced apoptosis in human melanoma cells through XBP1-mediated up-regulation of TRAIL-R2. Mol. Cancer 8, 1–17. https://doi.org/10.1186/1476-4598-8-122.
Lu, M., Lawrence, D.A., Marsters, S., Acosta-Alvear, D., Kimmig, P., Mendez, A.S., Paton, A.W., Paton, J.C., Walter, P., Ashkenazi, A., 2014. Opposing unfolded-protein-response signals converge on death receptor 5 to control apoptosis. Science 345, 98–101. https://doi.org/10.1126/science.1254312.
MacFarlane, M., Ahmad, M., Srinivasula, S.M., Fernandes-Alnemri, T., Cohen, G.M., Alnemri, E.S., 1997. Identification and molecular cloning of two novel receptors for the cytotoxic ligand TRAIL. J. Biol. Chem. 272, 25417–25420. https://doi.org/9325248.
Majkut, J., Sgobba, M., Holohan, C., Crawford, N., Logan, A.E., Kerr, E., Higgins, C.A., Redmond, K.L., Riley, J.S., Stasik, I., Fennell, D.A., Van Schaeybroeck, S., Haider, S., Johnston, P.G., Haigh, D., Longley, D.B., 2014. Differential affinity of FLIP and procaspase-8 for FADD's DED binding surfaces regulates DISC assembly. Nat. Commun. 5, 3350. https://doi.org/10.1038/ncomms4350.
Marsters, S.A., Sheridan, J.P., Pitti, R.M., Huang, A., Skubatch, M., Baldwin, D., Yuan, J., Gurney, A., Goddard, A.D., Godowski, P., Ashkenazi, A., 1997. A novel receptor for Apo2L/TRAIL contains a truncated death domain. Curr. Biol. 7, 1003–1006. https://doi.org/10.1016/s0960-9822(06)00422-2.
Martinou, J.C., Youle, R.J., 2011. Mitochondria in apoptosis: Bcl-2 family members and mitochondrial dynamics. Dev. Cell 21, 92–101. https://doi.org/10.1016/j.devcel.2011.06.017.
Martín-Pérez, R., Niwa, M., López-Rivas, A., 2012. ER stress sensitizes cells to TRAIL through down-regulation of FLIP and Mcl-1 and PERK-dependent up-regulation of TRAIL-R2. Apoptosis 17, 349–363. https://doi.org/10.1007/s10495-011-0673-2.
Maurel, M., Chevet, E., Tavernier, J., Gerlo, S., 2014. Getting RIDD of RNA: IRE1 in cell fate regulation. Trends Biochem. Sci. 39, 245–254. https://doi.org/10.1016/j.tibs.2014.02.008.
Mc Cullough, K.D., Martindale, J.L., Aw, T., Holbrook, N.J., Cullough, K.D.M.C., Klotz, L., 2001. Gadd153 sensitizes cells to endoplasmic reticulum stress by down-regulating Bcl2 and perturbing the cellular redox state. Mol. Cell Biol. 21, 1249–1259. https://doi.org/10.1128/MCB.21.4.1249.
McCarthy, E.F., 2006. The toxins of William B. Coley and the treatment of bone and soft-tissue sarcomas. Iowa Orthop. J. 26, 154–158.
Merino, D., Lalaoui, N., Morizot, A., Schneider, P., Solary, E., Micheau, O., 2006. Differential inhibition of TRAIL-mediated DR5-DISC formation by decoy receptors 1 and 2. Mol. Cell Biol. 26, 7046–7055. https://doi.org/10.1128/MCB.00520-06.
Michaelson, J.S., Demarest, S.J., Miller, B., Amatucci, A., Snyder, W.B., Wu, X., Huang, F., Phan, S., Gao, S., Doern, A., Farrington, G.K., Lugovskoy, A., Joseph, I., Bailly, V., Wang, X., Garber, E., Browning, J., Glaser, S.M., 2009. Anti-tumor activity of stability-engineered IgG-like bispecific antibodies targeting TRAIL-R2 and LTβR. MAbs 1, 128–141. https://doi.org/10.4161/mabs.1.2.7631.
Mizrahi, K., Askenasy, N., 2014. Physiological functions of TNF family receptor/ligand interactions in hematopoiesis and transplantation. Blood 124, 176–183. https://doi.org/10.1182/blood-2014-03-559641.
Mongkolsapaya, J., Cowper, A.E., Xu, X.N., Morris, G., McMichael, A.J., Bell, J.I., Screaton, G.R., 1998. Lymphocyte inhibitor of TRAIL (TNF-related apoptosis-inducing ligand): a new receptor protecting lymphocytes from the death ligand TRAIL. J. Immunol. 160, 3–6. https://doi.org/0022-1767/98.
Mori, C., Nakamura, N., Kimura, S., Irie, H., Takigawa, T., Shiota, K., 1995. Programmed cell death in the interdigital tissue of the fetal mouse limb is apoptosis with DNA fragmentation. Anat. Rec. 242, 103–110. https://doi.org/10.1002/ar.1092420114.

Nagata, S., 2000. Apoptotic DNA fragmentation. Exp. Cell Res. 256, 12–18. https://doi.org/10.1006/excr.2000.4834.

Neumann, S., Hasenauer, J., Pollak, N., Scheurich, P., 2014. Dominant negative effects of tumor necrosis factor (tnf)-related apoptosis-inducing ligand (TRAIL) receptor 4 on TRAIL receptor 1 signaling by Formation of heteromeric complexes. J. Biol. Chem. 289, 16576–16587. https://doi.org/10.1074/jbc.M114.559468.

Obeng, E.A., Carlson, L.M., Gutman, D.M., Harrington Jr., W.J., Lee, K.P., Boise, L.H., 2016. Proteasome inhibitors induce a terminal unfolded protein response in multiple myeloma cells. Blood 107, 4907–4917. https://doi.org/10.1182/blood-2005-08-3531.

Oberst, A., Pop, C., Tremblay, A.G., Blais, V., Denault, J., Salvesen, G.S., Green, D.R., 2010. Inducible Dimerization and inducible cleavage reveal a requirement for both processes in caspase-8 activation. J. Biol. Chem. 285, 16632–16642. https://doi.org/10.1074/jbc.M109.095083.

Oh, Y.T., Liu, X., Yue, P., Kang, S., Chen, J., Taunton, J., Khuri, F.R., Sun, S.Y., 2010. ERK/ribosomal S6 kinase (RSK) signaling positively regulates death receptor 5 expression through co-activation of CHOP and Elk1. J. Biol. Chem. 285, 41310–41319. https://doi.org/10.1074/jbc.M110.153775.

Pan, G., Ni, J., Wei, Y.F., Yu, G.I., Gentz, R., Dixit, V.M., 1997a. An antagonist decoy receptor and a death domain-containing receptor for TRAIL. Science 277, 815–818. https://doi.org/10.1126/science.277.5327.815.

Pan, G., O'Rourke, K., Chinnaiyan, A.M., Gentz, R., Ebner, R., Ni, J., Dixit, V.M., 1997b. The receptor for the cytotoxic ligand TRAIL. Science 276, 111–113. https://doi.org/10.1126/science.276.5309.111.

Pan, G., Ni, J., Yu, G.L., Wei, Y.F., Dixit, V.M., 1998. TRUNDD, a new member of the TRAIL receptor family that antagonizes TRAIL signalling. FEBS Lett. 424, 41–45. https://doi.org/10.1016/S0014-5793(98)00135-5.

Pan, L., Fu, T., Zhao, W., Chen, S., Wu, H., Chou, J.J., Pan, L., Fu, T., Zhao, W., Zhao, L., Chen, W., Qiu, C., Liu, W., Liu, Z., 2019. Higher-order clustering of the transmembrane anchor of DR5 drives signaling. Cell 176, 1–13. https://doi.org/10.1016/j.cell.2019.02.001.

Papadopoulos, K.P., Isaacs, R., Bilic, S., Kentsch, K., Huet, H.A., Hofmann, M., Rasco, D., Kundamal, N., Tang, Z., Cooksey, J., Mahipal, A., 2015. Unexpected hepatotoxicity in a phase I study of TAS266, a novel tetravalent agonistic Nanobody® targeting the DR5 receptor. Cancer Chemother. Pharmacol. 75, 887–895. https://doi.org/10.1007/s00280-015-2712-0.

Park, S.Y., Billiar, T.R., Seol, D.W., 2002. Hypoxia inhibition of apoptosis induced by tumor necrosis factor-related apoptosis-inducing ligand (TRAIL). Biochem. Biophys. Res. Commun. 291, 150–153. https://doi.org/10.1006/bbrc.2002.6421.

Patrignani, P., Patrono, C., 2015. Cyclooxygenase inhibitors: from pharmacology to clinical read-outs. Biochim. Biophys. Acta Mol. Cell Biol. Lipids 1851, 422–432. https://doi.org/10.1016/j.bbalip.2014.09.016.

Pei, G.T., Wu, C.W., Lin, W.W., 2010. Hypoxia-induced decoy receptor 2 gene expression is regulated via a hypoxia-inducible factor 1α-mediated mechanism. Biochem. Biophys. Res. Commun. 391, 1274–1279. https://doi.org/10.1016/j.bbrc.2009.12.058.

Pitti, R.M., Marsters, S.A., Ruppert, S., Donahue, C.J., Moore, A., Ashkenazi, A., 1996. Induction of apoptosis by Apo-2 ligand, a new member of the tumor necrosis factor cytokine family. J. Biol. Chem. 271, 12687–12690. https://doi.org/10.1074/jbc.271.22.12687.

Plass, C., Pfister, S.M., Lindroth, A.M., Bogatyrova, O., 2013. Mutations in regulators of the epigenome and their connections to global chromatin patterns in cancer. Nat. Publ. Group 14, 765–780. https://doi.org/10.1038/nrg3554.

Pop, C., Oberst, A., Drag, M., Van Raam, B.J., Riedl, S.J., Green, D.R., Salvesen, G.S., 2011. FLIP(L) induces caspase-8 activity in the absence of interdomain caspase-8 cleavage and alters substrate specificity. Biochem. J. 433, 447–457. https://doi.org/10.1042/BJ20101738.

Puthalakath, H., Reilly, L.A.O., Gunn, P., Lee, L., Kelly, P.N., Huntington, N.D., Hughes, P.D., Michalak, E.M., Mckimm-breschkin, J., Motoyama, N., Gotoh, T., Akira, S., Bouillet, P., Strasser, A., 2007. ER stress triggers apoptosis by activating BH3-only protein bim. Cell 129, 1337–1349. https://doi.org/10.1016/j.cell.2007.04.027.

Pyrko, P., Kardosh, A., Liu, Y.-T., Soriano, N., Xiong, W., Chow, R.H., Uddin, J., Petasis, N.A., Mircheff, A.K., Farley, R.A., Louie, S.G., Chen, T.C., Schonthal, A.H., 2007. Calcium-activated endoplasmic reticulum stress as a major component of tumor cell death induced by 2,5-dimethyl-celecoxib, a non-coxib analogue of celecoxib. Mol. Cancer Ther. 6, 1262–1275. https://doi.org/10.1158/1535-7163.MCT-06-0629.

Ramamurthy, V., Yamniuk, A.P., Lawrence, E.J., Yong, W., Schneeweis, L.A., Cheng, L., Murdock, M., Corbett, M.J., Doyle, M.L., Sheriff, S., 2015. The structure of the death receptor 4-TNF-related apoptosis-inducing ligand (DR4-TRAIL) complex. Acta Crystallogr. Sect. F Struct. Biol. Commun. 71, 1273–1281. https://doi.org/10.1107/S2053230X15016416.

Ratain, M.J., Doi, T., De Jonge, M.J., LoRusso, P., Dunbar, M., Chiney, M., Motwani, M., Glasgow, J., Petrich, A.M., Rasco, D.W., Calvo, E., 2019. Phase 1, first-in-human study of TRAIL receptor agonist fusion protein ABBV-621. J. Clin. Oncol. 15, 3013. https://doi.org/10.1200/JCO.2019.37.15_suppl.3013.

Redza-Dutordoir, M., Averill-Bates, D.A., 2016. Activation of apoptosis signalling pathways by reactive oxygen species. Biochim. Biophys. Acta Mol. Cell Res. 1863, 2977–2992. https://doi.org/10.1016/j.bbamcr.2016.09.012.

Rehm, M., Huber, H.J., Dussmann, H., Prehn, J.H.M., 2006. Systems analysis of effector caspase activation and its control by X-linked inhibitor of apoptosis protein. EMBO J. 25, 4338–4349. https://doi.org/10.1038/sj.emboj.7601295.

Riccioni, R., Pasquini, L., Mariani, G., Saulle, E., Rossini, A., Diverio, D., Pelosi, E., Vitale, A., Chierichini, A., Cedrone, M., Foà, R., Lo Coco, F., Peschle, C., Testa, U., 2005. TRAIL decoy receptors mediate resistance of acute myeloid leukemia cells to TRAIL. Haematologica 90, 612–624.

Roos, W.P., Kaina, B., 2006. DNA damage-induced cell death by apoptosis. Trends Mol. Med. 12, 440–450. https://doi.org/10.1016/j.molmed.2006.07.007.

Saveljeva, S., Mc Laughlin, S.L., Vandenabeele, P., Samali, A., Betrand, M.J., 2015. Endoplasmic reticulum stress induces lignad-independent TNFR1-mediated necroptosis in L929 cells. Cell Death Dis. 8, e1587. https://doi.org/10.1038/cddis.2014.548.

Scaffidi, C., Fulda, S., Srinivasan, A., Friesen, C., Li, F., Tomaselli, K.J., Debatin, K.M., Krammer, P.H., Peter, M.E., 1998. Two CD95 (APO-1/Fas) signaling pathways. EMBO J. 17, 1675–1687. https://doi.org/10.1093/emboj/17.6.1675.

Schleich, K., Buchbinder, J.H., Pietkiewicz, S., Kähne, T., Warnken, U., Öztürk, S., Schnölzer, M., Naumann, M., Krammer, P.H., Lavrik, I.N., 2016. Molecular architecture of the DED chains at the DISC: regulation of pro-caspase-8 activation by short DED proteins c-FLIP and pro-caspase-8 prodomain. Cell Death Differ. 23, 681–694. https://doi.org/10.1038/cdd.2015.137.

Schneider, P., Bodmer, J.L., Thome, M., Hofmann, K., Holler, N., Tschopp, J., 1997. Characterization of two receptors for TRAIL. FEBS Lett. 416, 329–334. https://doi.org/10.1016/S0014-5793(97)01231-3.

Schönenberger, M.J., Kovacs, W.J., 2015. Hypoxia signaling pathways: modulators of oxygen-related organelles. Front. Cell Dev. Biol. 3, 42. https://doi.org/10.3389/fcell.2015.00042.

Screaton, G.R., Mongkolsapaya, J., Xu, X.N., Cowper, A.E., McMichael, A.J., Bell, J.I., 1997. TRICK2, a new alternatively spliced receptor that transduces the cytotoxic signal from TRAIL. Curr. Biol. 7, 693–696. https://doi.org/10.1016/S0960-9822(06)00297-1.

Secchiero, P., Gonelli, A., Carnevale, E., Milani, D., Pandolfi, A., Zella, D., Zauli, G., 2003. TRAIL promotes the survival and proliferation of primary human vascular endothelial cells by activating the Akt and ERK pathways. Circulation 107, 2250–2256. https://doi.org/10.1161/01.CIR.0000062702.60708.C4.

Secchiero, P., Corallini, F., Grazia, M., Gonelli, A., Barbarotto, E., Zauli, G., 2005. TRAIL counteracts the proadhesive activity of inflammatory cytokines in endothelial cells by down-modulating CCL8 and CXCL10 chemokine expression and release. Blood 105, 3413–3419. https://doi.org/10.1182/blood-2004-10-4111.

Sharma, P., Varma, R., Sarasij, R.C., Ira, Gousset, K., Krishnamoorthy, G., Rao, M., Mayor, S., 2004. Nanoscale organization of multiple GPI-anchored proteins in living cell membranes. Cell 116, 577–589. https://doi.org/10.1016/S0092-8674(04)00167-9.

Sheridan, J.P., Marsters, S.A., Pitti, R.M., Gurney, A., Skubatch, M., Baldwin, D., Ramakrishnan, L., Gray, C.L., Baker, K., Wood, W.I., Goddard, A.D., Godowski, P., Ashkenazi, A., 1997. Control of TRAIL-induced apoptosis by a family of signaling and decoy receptors. Science 277, 818–821. https://doi.org/10.1126/science.277.5327.818.

Shivapurkar, N.S., Oyooka, S.T., Oyooka, K.O.T., Eddy, J.R., Iyajima, K.M., Uzuki, M.S., 2004. Aberrant methylation of TRAIL decoy receptor genes is frequent in multiple tumor types. Int. J. Cancer 792, 786–792. https://doi.org/10.1002/ijc.20041.

Siegemund, M., Pollak, N., Seifert, O., Wahl, K., Hanak, K., Vogel, A., Nussler, A.K., Göttsch, D., Münkel, S., Bantel, H., Kontermann, R.E., Pfizenmaier, K., 2012. Superior antitumoral activity of dimerized targeted single-chain TRAIL fusion proteins under retention of tumor selectivity. Cell Death Dis. 3, e295. https://doi.org/10.1038/cddis.2012.29.

Siegemund, M., Seifert, O., Zarani, M., Džinić, T., De Leo, V., Göttsch, D., Münkel, S., Hutt, M., Pfizenmaier, K., Kontermann, R.E., 2016. An optimized antibody-single-chain TRAIL fusion protein for cancer therapy. MAbs 8, 879–891. https://doi.org/10.1080/19420862.2016.1172163.

Sies, H., Bruene, B., 2007. Methods in Enymology-Oxygen Biology and Hypoxia, first ed. Elsevier.

Sobolewski, C., Cerella, C., Dicato, M., Ghibelli, L., Diederich, M., 2010. The role of cyclooxygenase-2 in cell proliferation and cell death in human malignancies. Int. J. Cell Biol. 2010, 1–21. https://doi.org/10.1155/2010/215158.

Soslow, R.A., Dannenberg, A.J., Rush, D., Woerner, B.M., Nasir Khan, K., Masferrer, J., Koki, A.T., 2000. COX-2 is expressed in human pulmonary, colonic, and mammary tumors. Cancer 89, 2637–2645. https://doi.org/10.1002/1097-0142(20001215)89:12<2637::AID-CNCR17>3.0.CO;2-B.

Spierings, D.C.J., de Vries, E.G.E., Timens, W., Groen, H.J.M., Boezen, H.M., de Jong, S., 2003. Expression of TRAIL and TRAIL death receptors in stage III non-small cell lung cancer tumors. Clin. Cancer Res. 9, 3397–3405. https://doi.org/10.1158/1078-0432.ccr-10-0631.

Sträter, J., Hinz, U., Walczak, H., Mechtersheimer, G., Koretz, K., Herfarth, C., Möller, P., Lehnert, T., 2002. Expression of TRAIL and TRAIL receptors in colon carcinoma: TRAIL-R1 is an independent prognostic parameter. Clin. Cancer Res. 8, 3734–3740.

Takenokuchi, M., Miyamoto, K., Saigo, K., Taniguchi, T., Sciences, P., 2015. Bortezomib causes ER stress-related death of acute promyelocytic leukemia cells through excessive accumulation of PML-RARA. Anticancer Res 35, 3307–3316.

Takuma, K., Yan, S.S., Stern, D.M., Yamada, K., 2005. Mitochondrial dysfunction, endoplasmic reticulum stress, and apoptosis in Alzheimer's disease. J. Pharmacol. Sci. 97, 312–316. https://doi.org/10.1254/jphs.CPJ04006X.

Tang, X.M., Sun, Y.J., Half, E., Kuo, M.T., Sinicrope, F., 2002. Cyclooxygenase-2 overexpression inhibits death receptor 5 expression and confers resistance to tumor necrosis factor-related apoptosis-inducing ligand-induced apoptosis in human colon cancer cells. Cancer Res. 62, 4903–4908.

Tang, J., Chen, Y., Cui, R., Li, D., Xiao, L., Lin, P., Du, Y., Sun, H.U.I., Yu, X., Zheng, X., 2015. Upregulation of fractalkine contributes to the proliferative response of prostate cancer cells to hypoxia via promoting the G 1/S phase transition. Mol. Med. Rep. 12, 7907–7914. https://doi.org/10.3892/mmr.2015.4438.

Tardáguila, M., Mañes, S., 2013. Relevance for the progression of ERBB2 + breast carcinoma CX3CL1 at the crossroad of EGF signals. Oncoimmunology 2, 8–10. https://doi.org/10.4161/onci.25669.

Tate, J.G., Bamford, S., Jubb, H.C., Sondka, Z., Beare, D.M., Bindal, N., Boutselakis, H., Cole, C.G., Creatore, C., Dawson, E., Fish, P., Harsha, B., Hathaway, C., Jupe, S.C., Kok, Y., Noble, K., Ponting, L., Ramshaw, C.C., Rye, C.E., Speedy, E., Stefancsik, R., Thompson, S.L., Wang, S., Ward, S., Campbell, J., Forbes, S.A., 2019. COSMIC: the catalogue of somatic mutations in cancer. Nucleic Acids Res. 47, 941–947. https://doi.org/10.1093/nar/gky1015.

Thomson, B.J., 2001. Viruses and apoptosis. Int. J. Exp. Pathol. 82, 65–76. https://doi.org/10.1093/oxfordjournals.bmb.a011627.

Tolcher, A.W., Mita, M., Meropol, N.J., Von Mehren, M., Patnaik, A., Padavic, K., Hill, M., Mays, T., McCoy, T., Fox, N.L., Halpern, W., Corey, A., Cohen, R.B., 2007. Phase I pharmacokinetic and biologic correlative study of mapatumumab, a fully human monoclonal antibody with agonist activity to tumor necrosis factor-related apoptosis-inducing ligand receptor-1. J. Clin. Oncol. 25, 1390–1395. https://doi.org/10.1200/JCO.2006.08.8898.

Topper, M.J., Vaz, M., Marrone, K.A., Brahmer, J.R., Baylin, S.B., 2019. The emerging role of epigenetic therapeutics in immuno-oncology. Nat. Rev. Clin. Oncol. 17, 75–90. https://doi.org/10.1038/s41571-019-0266-5.

Toscano, F., El Fajoui, Z., Gay, F., Lalaoui, N., Parmentier, B., Chayvialle, J.A., Scoazec, J.Y., Micheau, O., Abello, J., Saurin, J.C., 2008. p53-Mediated upregulation of DcR1 impairs oxaliplatin/TRAIL-induced synergistic anti-tumour potential in colon cancer cells. Oncogene 27, 4161–4171. https://doi.org/10.1038/onc.2008.52.

Urano, F., Wang, X.-Z., Bertolotti, A., Zhang, Y., Chung, P., Harding, H.P., Ron, D., 2000. Coupling of stress in the endoplasmic reticulum to activation of JNK protein kinases by transmembrane protein kinase IRE1. Science 287, 664–666. https://doi.org/10.1126/science.287.5453.664.

van Dijk, M., Halpin-McCormick, A., Sessler, T., Samali, A., Szegezdi, E., 2013. Resistance to TRAIL in non-transformed cells is due to multiple redundant pathways. Cell Death Dis. 4, e702–e711. https://doi.org/10.1038/cddis.2013.214.

van Roosmalen, I.A.M., Reis, C.R., Setroikromo, R., Yuvaraj, S., Joseph, J.V., Tepper, P.G., Kruyt, F.A.E., Quax, W.J., 2014. The ER stress inducer DMC enhances TRAIL-induced apoptosis in glioblastoma. Springerplus 3, 1–12. https://doi.org/10.1186/2193-1801-3-495.

Vanamee, É.S., Faustman, D.L., 2018. Structural principles of tumor necrosis factor superfamily signaling. Sci. Signal. 11, 1–12. https://doi.org/10.1126/scisignal.aao4910.

Varma, R., Mayor, S., 1998. GPI-anchored proteins are organized in submicron domains at the cell surface. Nature 394, 798–801. https://doi.org/10.1038/29563.
Vogt, C., 1842. Untersuchungen über die Entwicklungsgeschichte der Geburtshelferkröte. (Alytes obstetricians). Junt und Gassmann, Solothurn.
von dem Bussche, A., MacHida, R., Li, K., Loevinsohn, G., Khander, A., Wang, J., Wakita, T., Wands, J.R., Li, J., 2010. Hepatitis C virus NS2 protein triggers endoplasmic reticulum stress and suppresses its own viral replication. J. Hepatol. 53, 797–804. https://doi.org/10.1016/j.jhep.2010.05.022.
von Karstedt, S., Conti, A., Nobis, M., Montinaro, A., Hartwig, T., Lemke, J., Legler, K., Annewanter, F., Campbell, A.D., Taraborrelli, L., Grosse-Wilde, A., Coy, J.F., El-Bahrawy, M.A., Bergmann, F., Koschny, R., Werner, J., Ganten, T.M., Schweiger, T., Hoetzenecker, K., Kenessey, I., Hegedüs, B., Bergmann, M., Hauser, C., Egberts, J.H., Becker, T., Röcken, C., Kalthoff, H., Trauzold, A., Anderson, K.I., Sansom, O.J., Walczak, H., 2015. Cancer cell-autonomous TRAIL-R signaling promotes KRAS-driven cancer progression, invasion, and metastasis. Cancer Cell 27, 561–573. https://doi.org/10.1016/j.ccell.2015.02.014.
von Karstedt, S., Montinaro, A., Walczak, H., 2017. Exploring the TRAILs less travelled: TRAIL in cancer biology and therapy. Nat. Rev. Cancer 17, 352–366. https://doi.org/10.1038/nrc.2017.28.
Wajant, H., 2017. TRAIL- and TNF-induced signaling complexes—so similar yet so different. EMBO J. 36, 1117–1119. https://doi.org/10.15252/embj.201796997.
Walczak, H., Degli-esposti, M.A., Johnson, R.S., Smolak, P.J., Waugh, J.Y., Boiani, N., Timour, M.S., Gerhart, M.J., Schooley, K.A., Smith, C.A., Goodwin, R.G., Rauch, C.T., 1997. TRAIL-R2: a novel apoptosis-mediating receptor for TRAIL. EMBO J. 16, 5386–5397. https://doi.org/10.1093/emboj/16.17.5386.
Wali, J.A., Rondas, D., Zhao, Y., Elkerbout, L., Fynch, S., Gurzov, E.N., Akira, S., Mathieu, C., Kay, T.W.H., 2014. The proapoptotic BH3-only proteins Bim and Puma are downstream of endoplasmic reticulum and mitochondrial oxidative stress in pancreatic islets in response to glucotoxicity. Cell Death Dis. 5, e1124–e1129. https://doi.org/10.1038/cddis.2014.88.
Walter, F., Schmid, J., Dussmann, H., Concannon, C.G., Prehn, J.H.M., 2015. Imaging of single cell responses to ER stress indicates that the relative dynamics of IRE1/XBP1 and PERK/ATF4 signalling rather than a switch between signalling branches determine cell survival. Cell Death Differ. 22, 1502–1516. https://doi.org/10.1038/cdd.2014.241.
Wiley, S.R., Schooley, K., Smolak, P.J., Din, W.S., Huang, C.P., Nicholl, J.K., Sutherland, G.R., Smith, T.D., Rauch, C., Smith, C.A., Goodwin, R.G., 1995. Identification and characterization of a new member of the TNF family that induces apoptosis. Immunity 3, 673–682. https://doi.org/10.1016/1074-7613(95)90057-8.
Wilson, N.S., Yang, B., Yang, A., Loeser, S., Marsters, S., Lawrence, D., Li, Y., Pitti, R., Totpal, K., Yee, S., Ross, S., Vernes, J., Lu, Y., Adams, C., Offringa, R., Kelley, B., Hymowiz, S., Daniel, D., Meng, G., Ashkenazi, A., 2011. An Fcγ receptor-dependent mechanism drives antibody-mediated tartget-receptor signaling in cancer cells. Cancer Cell 19, 101–113. https://doi.org/10.1016/j.ccr.2010.11.012.
Wu, G.S., Burns, T.F., McDonald, E.R., Jiang, W., Meng, R., Krantz, I.D., Kao, G., Di Gan, D., Zhou, J.Y., Muschel, R., Hamilton, S.R., Spinner, N.B., Markowitz, S., Wu, G., El-Deiry, W.S., 1997. KILLER/DR5 is a DNA damage-inducible p53-regulated death receptor gene. Nat. Genet. 17, 141–143. https://doi.org/10.1038/ng1097-141.
Wu, C.C., Lee, S., Malladi, S., Der Chen, M., Mastrandrea, N.J., Zhang, Z., Bratton, S.B., 2016. The Apaf-1 apoptosome induces formation of caspase-9 homo- and heterodimers with distinct activities. Nat. Commun. 7, 1–14. https://doi.org/10.1038/ncomms13565.

Würstle, M.L., Laussmann, M.A., Rehm, M., 2010. The caspase-8 dimerization/dissociation balance is a highly potent regulator of caspase-8, -3, -6 signaling. J. Biol. Chem. 285, 33209–33218. https://doi.org/10.1074/jbc.M110.113860.

Yamaguchi, H., Wang, H.G., 2004. CHOP is involved in endoplasmic reticulum stress-induced apoptosis by enhancing DR5 expression in human carcinoma cells. J. Biol. Chem. 279, 45495–45502. https://doi.org/10.1074/jbc.M406933200.

Yamanaka, Y., Shiraki, K., Inoue, T., Miyashita, K., Fuke, H., Yamaguchi, Y., Yamamoto, N., Ito, K., Sugimoto, K., Nakano, T., 2006. COX-2 inhibitors sensitize human hepatocellular carcinoma cells to TRAIL-induced apoptosis. Int. J. Mol. Med. 18, 41–47.

Yan, J., Wang, L., Wang, Z., Wang, B., Zhu, R., Bi, J., Wu, J., Zhang, H., Wu, H., Yu, B., Kong, W., Yu, X., 2016. Engineered adenovirus fiber shaft fusion homotrimer of soluble TRAIL with enhanced stability and antitumor activity. Cell Death Dis. 7, e2274. https://doi.org/10.1038/cddis.2016.177.

Yang, L., Rau, R., Goodell, M.A., 2015. DNMT3A in haematological malignancies. Nat. Rev. Cancer 15, 152–165. https://doi.org/10.1038/nrc3895.

Yoshida, T., Sakai, T., 2004. Promoter of TRAIL-R2 gene. Vitam. Horm. 67, 35–49.

Yoshida, H., Okada, T., Haze, K., Yanagi, H., Yura, T., Negishi, M., Mori, K., 2000. ATF6 activated by proteolysis binds in the presence of NF-Y (CBF) directly to the cis-acting element responsible for the mammalian unfolded protein response. Mol. Cell Biol. 20, 6755–6767. https://doi.org/10.1128/MCB.20.18.6755-6767.2000.

Yuan, J., Yanker, B.A., 2000. Apoptosis in the nervous system. Nature 407, 802–809. https://doi.org/10.1038/35037739.

Zauli, G., Celeghini, C., Monasta, L., Martinelli, M., Luppi, S., Gonelli, A., Grill, V., Ricci, G., Secchiero, P., 2013. Soluble TRAIL is present at high concentrations in seminal plasma and promotes spermatozoa survival. Reproduction 148, 191–198. https://doi.org/10.1530/REP-14-0144.

Zhang, X.D., Nguyen, T., Thomas, W.D., Sanders, J.E., Hersey, P., 2000. Mechanisms of resistance of normal cells to TRAIL induced apoptosis vary between different cell types. FEBS Lett. 482, 193–199. https://doi.org/10.1016/S0014-5793(00)02042-1.

CHAPTER FOUR

Type 3 IP_3 receptors: The chameleon in cancer

Nicolas Rosa, Flore Sneyers, Jan B. Parys, Geert Bultynck*
KU Leuven, Laboratory of Molecular and Cellular Signaling, Department of Cellular and Molecular Medicine and Leuven Kanker Instituut (LKI), Leuven, Belgium
*Corresponding author: e-mail address: geert.bultynck@kuleuven.be

Contents

Abstract

Inositol 1,4,5-trisphosphate (IP_3) receptors (IP_3Rs), intracellular calcium (Ca^{2+}) release channels, fulfill key functions in cell death and survival processes, whose dysregulation contributes to oncogenesis. This is essentially due to the presence of IP_3Rs in microdomains of the endoplasmic reticulum (ER) in close proximity to the mitochondria. As such, IP_3Rs enable efficient Ca^{2+} transfers from the ER to the mitochondria, thus regulating metabolism and cell fate. This review focuses on one of the three IP_3R isoforms, the type 3 IP_3R (IP_3R3), which is linked to proapoptotic ER-mitochondrial Ca^{2+} transfers. Alterations in IP_3R3 expression have been highlighted in numerous cancer types, leading to dysregulations of Ca^{2+} signaling and cellular functions. However, the outcome of IP_3R3-mediated Ca^{2+} transfers for mitochondrial function is complex with opposing effects on oncogenesis. IP_3R3 can either suppress cancer by promoting cell death and cellular senescence or support cancer by driving metabolism, anabolic processes, cell cycle progression, proliferation and invasion. The aim of this review is to provide an overview of

International Review of Cell and Molecular Biology, Volume 351
ISSN 1937-6448
https://doi.org/10.1016/bs.ircmb.2020.02.003

IP_3R3 dysregulations in cancer and describe how such dysregulations alter critical cellular processes such as proliferation or cell death and survival. Here, we pose that the IP_3R3 isoform is not only linked to proapoptotic ER-mitochondrial Ca^{2+} transfers but might also be involved in prosurvival signaling.

1. Introduction

Calcium (Ca^{2+}) is a universal second messenger involved in a plethora of physiological processes including neuronal transmission, muscle contraction, gene transcription, secretion, cellular motility, cell proliferation, and apoptosis (Bultynck and Parys, 2018; Parys and Bultynck, 2018a,b; Sammels et al., 2010). Ca^{2+} signals in the cell can arise from the Ca^{2+} release from intracellular stores (i.e. endoplasmic reticulum (ER), mitochondria, Golgi apparatus, etc.) and the entry of Ca^{2+} from the extracellular environment (Bootman and Bultynck, 2019). These mechanisms rely on different kinds of Ca^{2+} channels, pumps, and exchangers and are tightly regulated by a myriad of proteins (Chen et al., 2019; Lewis, 2019; Parys and Vervliet, 2020; Prole and Taylor, 2019).

The inositol 1,4,5-trisphosphate (IP_3) receptor (IP_3R) is a ubiquitous Ca^{2+} channel mediating Ca^{2+} release from the ER, the main intracellular Ca^{2+} storage organelle. IP_3Rs open in response to IP_3, which is produced through hydrolysis of phosphatidyl inositol 4,5-bisphosphate (PIP_2) following phospholipase C (PLC) activation elicited by extracellular stimuli (Berridge, 2009; Berridge et al., 1984). Three isoforms of the IP_3R exist (IP_3R1, IP_3R2 and IP_3R3, respectively encoded by the genes *ITPR1*, *ITPR2*, and *ITPR3* in human) with strong sequence homology (Foskett et al., 2007). Their relative expression levels vary not only according to tissue type, but also to cellular location and stage of development (Ivanova et al., 2014; Taylor et al., 1999; Vermassen et al., 2004). This allows the IP_3R to meet very specific needs of the cell. The channels are assembled in homo- and heterotetramers (Monkawa et al., 1995) and display subtle differences in affinity toward IP_3, Ca^{2+} and adenosine triphosphate (ATP) (Patel et al., 1999). Of note, heterotetramers do not simply exhibit a blend of the behavior of the constitutive subunits (Alzayady et al., 2013; Chandrasekhar et al., 2015, 2016). For example, the activity of channels formed from coupled IP_3R1 and IP_3R2 heterodimers was predominantly dictated by IP_3R2 (Alzayady et al., 2013) while IP_3R2/IP_3R3 complexes exhibited unique properties (Chandrasekhar et al., 2016). At the monomer level, IP_3Rs

display three functional regions: the ligand binding domain, the central, modulatory domain and the C-terminal region. The central, modulatory domain displays more variability among the isoforms while the N-terminal ligand binding domain and the C-terminal region are the most conserved (Prole and Taylor, 2019). The C-terminal region contains the transmembrane channel domain and determinants for tetramer formation (Boehning et al., 2001; Mignery and Sudhof, 1990). All three domains contain interaction sites for an array of modulatory proteins (Ivanova et al., 2014, 2019a; Parys and De Smedt, 2012; Parys and Vervliet, 2020; Prole and Taylor, 2016; Yule et al., 2010).

The IP_3R opens upon binding of IP_3 whereby the isoforms exhibit different binding affinities for IP_3 ($IP_3R2 > IP_3R1 > IP_3R3$) (Hattori et al., 2004; Miyakawa et al., 1999; Tu et al., 2005). Recent work revealed that four IP_3 molecules ought to bind to a tetrameric IP_3R in order to open the channel (Alzayady et al., 2016; Taylor and Konieczny, 2016). It is well known that global Ca^{2+} signals are initiated through Ca^{2+} puffs (Bootman et al., 1997). These puffs arise from the simultaneous opening of a small cluster of IP_3Rs (Thomas et al., 2000). Fascinatingly, not all IP_3Rs within a cell are equally sensitive to IP_3 for generating Ca^{2+} puffs. Recent work indicated that IP_3Rs located close to ER-plasma membrane junctions are "licensed" to respond to IP_3 and generate puffs (Prole and Taylor, 2019; Thillaiappan et al., 2017). The mechanisms that render this subpopulation of IP_3Rs licensed and thus to be responsive to IP_3 for the generation of Ca^{2+} puffs remain to be elucidated.

The regulation of the IP_3Rs is further fine-tuned through Ca^{2+}-dependent modulation of their ability to flux Ca^{2+} in response to IP_3 (Finch et al., 1991; Kaftan et al., 1997). The effect of Ca^{2+} on IP_3R activity is bell-shaped, with a stimulatory effect of cytosolic $[Ca^{2+}]$ in the submicromolar range and an inhibitory effect of high cytosolic $[Ca^{2+}]$ (Bezprozvanny et al., 1991; Finch et al., 1991; Iino, 1990). In addition, the channel is regulated by Ca^{2+} levels in the lumen of the ER, decreasing its affinity toward IP_3 when ER-Ca^{2+} levels are low (Missiaen et al., 1992, 1994; Nunn and Taylor, 1992; Parys et al., 1993a). This mechanism helps to prevent excessive depletion of the ER, protecting the cell from harmful consequences. Moreover, IP_3R activity is regulated by ATP, where its activity is increased by concentrations at the submillimolar level while higher concentrations have an inhibitory effect (Bezprozvanny and Ehrlich, 1993; Iino, 1991). Although all isoforms are modulated by ATP, some isoform-specific characteristics have been observed (Maes et al., 2000; Tu et al., 2005). ATP promotes Ca^{2+} release of IP_3R2 and IP_3R3 by increasing their unitary open probability (the effect on IP_3R2 is observed only at low IP_3 concentrations) while it increases IP_3R1 activity by improving its Ca^{2+} sensitivity (Betzenhauser et al., 2008; Wagner and Yule, 2012).

Furthermore, IP_3Rs are modulated by their redox state. Reactive oxygen species (ROS) cause oxidation of the channel to sensitize Ca^{2+} release and promote mitochondrial Ca^{2+} uptake (Bansaghi et al., 2014). Redox regulation by ROS or thiol-modifying agents occurs via oxidation of cysteine residues in the N-terminal domain of IP_3R (Joseph et al., 2018; Khan et al., 2013). Thiol-modifying reagents, such as thimerosal, modulate the IP_3R in an isoform-specific fashion. While low micromolar concentrations of thimerosal induce IP_3R-medicated Ca^{2+} release, concentrations higher than 10 μM displayed an inhibitory effect (Parys et al., 1993b). Thimerosal acts synergistically with Ca^{2+} to generate its stimulatory effect (Bultynck et al., 2004) on IP_3R1 and IP_3R2, but not IP_3R3 (Bultynck et al., 2004; Khan et al., 2013; Missiaen et al., 1998). This effect is mediated through the stabilization of the channel in an active conformation, characterized by an interaction between the suppressor domain and IP_3-binding domain of the N-terminal region (Bultynck et al., 2004). Redox regulation of IP_3Rs in living cells could be particularly relevant at ER-mitochondria contact sites (Booth et al., 2016).

Finally, an increasing number of proteins interacting with the IP_3R are being identified not only as regulatory, but also as scaffolding proteins, structural proteins and motor proteins (Parys and De Smedt, 2012; Parys and Vervliet, 2020; Prole and Taylor, 2016).

Dysregulation of IP_3Rs and Ca^{2+} signaling has been described to alter cell functions and has been indeed linked to several diseases. Moreover, recent studies also revealed several disease-causing mutations in *ITPR* genes that affected IP_3R-channel function, resulting in abnormal ER Ca^{2+} fluxes (Hisatsune and Mikoshiba, 2017; Kerkhofs et al., 2018; Terry et al., 2018). In cancer, remodeling of the Ca^{2+}-signaling toolkit, including alterations at the level of the IP_3R, impacts several oncogenic hallmarks (Ivanova et al., 2017; Kania et al., 2017; Monteith et al., 2017; Morciano et al., 2018; Roberts-Thomson et al., 2019). Moreover, several oncogenes and tumor suppressors can directly target IP_3Rs to modulate its properties in such a way that oncogenesis, malignant cell survival and chemotherapeutic resistance are favored (Akl and Bultynck, 2013; Bittremieux et al., 2016). Particularly, changes in IP_3R3 function or expression have been associated with oncogenesis and cell death resistance. Downregulation of IP_3R3 has been proposed to be key in oncogenesis as the channel is implicated in transferring proapoptotic Ca^{2+} signals to mitochondria. Yet, it seems that this may not be a universal concept among all cancer types, since IP_3R3 is upregulated in several cancer types and contributes to malignant cell survival and behavior, adding complexity to IP_3R3 in cancer (Sneyers et al., 2019). In this review, we present an up-to-date view on the complex role of IP_3R3 channels in cancer organized in three points (i) the role of IP_3R3-mediated Ca^{2+} signaling in cell death and survival; (ii) the dysregulation of IP_3R3 expression in cancers; and (iii) the consequence of altered IP_3R3-mediated Ca^{2+} release in cancer.

2. Role of IP_3 receptors in cell death and survival

2.1 Ca^{2+} signaling at the mitochondria associated-ER membranes

IP_3Rs play an important role in the Ca^{2+} transfer from the ER to the mitochondria, and subsequent maintenance of mitochondrial bioenergetics (Marchi et al., 2017). The relationship between ER Ca^{2+} and mitochondrial metabolism is highly dependent on IP_3R-mediated Ca^{2+} release. It has been demonstrated that ER Ca^{2+} depletion induced by thapsigargin, an inhibitor of the sarco/endoplasmic reticulum Ca^{2+}-ATPase (SERCA), give rise to Ca^{2+} leakage all over the ER membrane but is unable to increase mitochondrial [Ca^{2+}] (Csordas et al., 1999). It is well established that membranes of the ER and the mitochondria are closely associated, establishing contact sites (Mannella et al., 1998). This close apposition of ER with mitochondria enables "quasi-synaptic" Ca^{2+} transfers from the ER toward the mitochondria (Csordas et al., 1999, 2006; Rizzuto et al., 1998). In the context of these Ca^{2+} transfers, IP_3Rs reside at mitochondria-associated ER membranes (MAM) and link with voltage-dependent anion channels 1 (VDAC1), ion channels that facilitate Ca^{2+} flux across the outer mitochondria membranes (OMM) (Gincel et al., 2001; Rapizzi et al., 2002). At the MAM, ER and mitochondrial membranes are physically tethered by several proteins (Filadi et al., 2018b; Scorrano et al., 2019).

Furthermore, IP_3Rs and VDAC1 were proposed to be directly coupled by the chaperone glucose-regulated protein of 75 kDa (GRP75) (Szabadkai et al., 2006), and this interaction is regulated by the transglutaminase type 2 (D'Eletto et al., 2018) and the protein DJ-1 (Liu et al., 2019). Very recently, the group of Hajnoczky indicated that genetic deletion of all three IP_3R isoforms in DT40 cells (DT40-TKO cells) resulted in decreased

ER-mitochondrial contact formation (Bartok et al., 2019), whereby each IP_3R isoform could rescue contact site formation. Although all three IP_3R isoforms can localize at ER-mitochondrial contact sites and promote such interorganellar contacts, IP_3R2 appeared most efficient in establishing ER-mitochondrial contact sites and mediating ER-mitochondrial Ca^{2+} transfers (Bartok et al., 2019). Very fascinatingly, the close proximity of ER and mitochondria mediated by IP_3R2 channels was independent of its Ca^{2+} flux properties, since a pore-dead mutant of IP_3R2 was equally potent in restoring ER-mitochondrial contacts as wild-type IP_3R2 in the DT40-TKO cells. This indicates that IP_3R channels and in particular IP_3R2 not only fulfill a Ca^{2+} signaling function at ER-mitochondrial contact sites but also a structural role in promoting and sustaining the formation of such contacts (Filadi and Pizzo, 2019).

IP_3R-mediated Ca^{2+} release generates Ca^{2+} microdomains at this confined ER-mitochondrial interface (Rizzuto et al., 1993, 1998). The Ca^{2+} concentration in these microdomains can be 20-fold higher than the bulk cytosolic Ca^{2+} concentration, allowing quasi-synaptic transmission to the mitochondria, even when the global cytosolic Ca^{2+} increase is low (Csordas et al., 1999). It is well documented that the MAM constitute a hub in Ca^{2+} signaling and mitochondrial metabolism, tightly controlled by an increasing list of regulating proteins (reviewed in Bononi et al., 2012; Csordas et al., 2018; Doghman-Bouguerra and Lalli, 2019). The Ca^{2+} that penetrates the mitochondria is shuttled from the intermembrane space to the mitochondrial matrix through the mitochondrial calcium uniporter (MCU) complex (Baughman et al., 2011; De Stefani et al., 2011).

On the one hand, basal mitochondrial Ca^{2+} dynamics sustains cell survival (Wacquier et al., 2019). In the mitochondrial matrix, Ca^{2+} is a cofactor of several enzymes of the citric acid cycle, sustaining oxidative phosphorylation (Denton, 2009; Glancy and Balaban, 2012; McCormack, 1985; Rutter and Denton, 1988; Wan et al., 1989). It was demonstrated long ago that oscillatory Ca^{2+} release is optimal to promote mitochondrial bioenergetics (Hajnoczky et al., 1995; Robb-Gaspers et al., 1998) and, for this aim, it is essential to maintain a constant IP_3R-mediated Ca^{2+} transfer to the mitochondria (Cardenas et al., 2010). A fascinating bidirectional regulation between ER Ca^{2+} signaling and mitochondrial function occurs at the ER-mitochondrial contact sites. Indeed, ER-derived Ca^{2+} oscillations drive mitochondrial H_2O_2 production from the cristae, establishing a local H_2O_2 nanodomain at the ER-mitochondrial interface which in turn promotes IP_3R-driven Ca^{2+} oscillations (Booth et al., 2016). Absence of this ionic transfer would result in activation of prosurvival autophagy (Rimessi et al., 2013) through a mechanism dependent on the AMP-activated protein kinase (AMPK) (Cardenas et al., 2010). In cancer cells, abrogating IP_3R activity and thus ER-mitochondrial Ca^{2+} transfers results in cancer cell death through a mitotic catastrophe (Cardenas et al., 2016), suggesting a unique addiction of cancer cells toward ER-mitochondrial Ca^{2+} signaling (Bultynck, 2016). During metabolic stress, such as nutrient deprivation, the protein Bax inhibitor-1 (BI-1) promotes autophagy by reducing the levels of ER Ca^{2+} release via IP_3R inhibition (Sano et al., 2012). For a review on Ca^{2+} regulation of autophagy, see Bootman et al. (2018) and Kania et al. (2017).

On the other hand, excessive Ca^{2+} uptake in the mitochondria leads to cell death. Increase of mitochondrial matrix $[Ca^{2+}]$ results in apoptosis by triggering the opening of the mitochondrial permeability transition pore (mPTP) and releasing cytochrome *c* and other proapoptotic factors into the cytoplasm (Baumgartner et al., 2009; Schild et al., 2001; Szalai et al., 1999). Several studies have shown that this process is triggered by Ca^{2+} in diverse ways. First, Ca^{2+} binds to the F_1F_0-ATP synthase β subunit and induces a conformational change of this energetic enzyme into an apoptosis-related structure (Giorgio et al., 2017). In fact, dimers of mitochondrial F_1F_0-ATP synthases are required to form the mPTP (Giorgio et al., 2013) and the c-subunit ring of the ATP synthase has been identified as an uncoupling channel that constitutes the pore itself (Alavian et al., 2014; Bonora et al., 2013). Additionally, Ca^{2+} can target a diphosphatidylglycerol lipid of the inner mitochondrial membrane named cardiolipin, eventually disintegrating the respiratory chain complex II, generating excessive reactive oxygen species (ROS) and inducing cell death (Hwang et al., 2014). The interplay between Ca^{2+} signaling and redox regulation has been discussed in several other excellent reviews (Joseph et al., 2018; Yoboue et al., 2018). The pivotal role of IP_3R-mediated Ca^{2+} signaling in apoptosis is further underpinned by the direct binding of cytochrome *c* released from the mitochondria to IP_3Rs, which promotes uncontrolled Ca^{2+}-release activity by preventing the Ca^{2+}-dependent inhibition of IP_3Rs (Boehning et al., 2003).

Of note, Ca^{2+} also promotes proliferation and cell survival in a non-mitochondrial manner (Humeau et al., 2018). Several major transcription factors, including the nuclear factor of activated T-cells (NFAT) and the nuclear factor κB (NF-κB), are stimulated by a rise of cytosolic $[Ca^{2+}]$; and cytosolic high-frequency Ca^{2+} oscillations are even required to activate

NFAT (Dolmetsch et al., 1998). The regulation of the bulk cytosolic Ca^{2+} concentration relies on processes very different from those taking place at the MAM but the spatiotemporal properties of the Ca^{2+} release appear in any case essential for the cellular functions whereby Ca^{2+} oscillations form a hallmark of cell survival.

2.2 Role of the IP_3R3 in cell death and survival

IP_3Rs are ubiquitously expressed in most tissues. Some cell types can express a preferential isoform but many cells possess at least two or even three IP_3R isoforms (Vermassen et al., 2004). The expression pattern of these isoforms is dependent on developmental stages and may stay highly dynamic in differentiated tissues as many factors are described to affect the expression level of IP_3Rs (Taylor et al., 1999). For a review on the pattern of expression of the various IP_3R isoforms, see Ivanova et al., 2014. The three IP_3R isoforms share about 75% similarity in their amino acid sequences. They however are unique proteins and display exclusive features in their activity and regulation. Several groups have characterized the properties of the IP_3R isoforms (Boehning and Joseph, 2000; Hattori et al., 2004; Lee et al., 2016; Maes et al., 2000; Mak et al., 2001, 2003; Missiaen et al., 1998; Miyakawa et al., 1999; Tu et al., 2005; Vanlingen et al., 2000). Of note, the IP_3R3 shows the lowest affinity for IP_3 and ATP.

More interesting are the differences of Ca^{2+} patterns observed upon stimulation. Although recent studies indicated that Ca^{2+} puffs released by the three IP_3R isoforms have very similar characteristics (Lock et al., 2018; Mataragka and Taylor, 2018), it is since long documented that the spatiotemporal Ca^{2+} dynamics can strongly differ from one channel isoform to another. It is assumed that IP_3R1 and IP_3R2 produce regular Ca^{2+} oscillations while IP_3R3 generates monophasic Ca^{2+} transients (Hattori et al., 2004; Miyakawa et al., 1999). These differences are of great importance in the context of cell death and survival. The pattern of IP_3-gated Ca^{2+} release is directly sensed by the mitochondria and defines the variations of the mitochondrial Ca^{2+} concentration (Wacquier et al., 2016). Isoform-specific studies of IP_3R-triggered Ca^{2+} release in general, and IP_3R-mediated apoptosis in particular, have emphasized the role of IP_3R1 and IP_3R3. Due to its expression limited to a few cell types, IP_3R2 is clearly the least documented isoform although it plays a crucial role in these cells (Vervloessem et al., 2015). Moreover, very recent evidence emerged that this isoform is the most effective in delivering

Ca^{2+} to the mitochondria (Bartok et al., 2019). Both IP_3R1 (Boehning et al., 2003; Hirota et al., 1999; Jayaraman and Marks, 1997) and IP_3R3 (Khan et al., 1996; Mendes et al., 2005) can support Ca^{2+}-mediated apoptosis. However, comparative studies suggest a preferential role for IP_3R3 in the transmission of apoptotic Ca^{2+} signals into the mitochondria, even when it is not the predominant isoform (Mendes et al., 2005). This preference is credited to the enrichment of this isoform at the MAM compared to the IP_3R1 (Bartok et al., 2019; Hedskog et al., 2013; Mendes et al., 2005).

For those reasons, even if IP_3R3 is important to promote mitochondrial bioenergetics in healthy conditions (Filadi et al., 2018a), this isoform was commonly associated with apoptosis. Dysregulation of IP_3R3 expression and/or function has been consequently proposed to be key in carcinogenesis. However, the outcome of Ca^{2+} transfers from the ER to the mitochondria is complex, with opposing effects on oncogenesis, either limiting cancer by promoting cell death and cellular senescence or promoting cancer by driving anabolic metabolism, proliferation and tumor invasion. Yet, other studies also provide evidence that IP_3R3 is upregulated in certain cancer types and contributes to malignant cell behavior.

3. Expression of IP_3R3 in cancer cells and tissues

IP_3Rs fulfill key functions in cell death and survival processes. We described in the previous section the pivotal role of IP_3Rs in ER-mitochondria Ca^{2+} transfers, with a marked interest in the third isoform. As the IP_3R3 has been particularly linked to apoptosis, alteration of its expression or function has been suggested to be responsible for tumor development.

3.1 Dysregulation of IP_3R3 in various cancers

The IP_3R3 is expressed in numerous kinds of cell lines and primary cell types and high expression levels have been recognized in proliferative cells (Ivanova et al., 2014). It has been accordingly studied in several categories of cancers. To date, a precise role for the IP_3R3 has been identified in various types of epithelial cancers (carcinoma), including breast (Szatkowski et al., 2010), lung (Iommelli et al., 2018), liver (Guerra et al., 2019), and colorectal (Shibao et al., 2010) cancer as well as glioblastoma (Kang et al., 2010). Moreover, this Ca^{2+} release channel has been highlighted in transcriptomic or predictive genetic studies for supplementary cancer types, such as cervical (Yang et al., 2017), head and neck (Hedberg et al., 2016), and thyroid (Rusinek et al., 2015) carcinomas, as well as gastric cancer (Sakakura et al., 2003) and neuroblastoma (Florea et al., 2017). Most of these studies were performed in human cancer cell lines, with some others extended to patient primary samples. However, protein expression levels can strongly vary between cell lines and primary cells. Indeed, it is important to note that IP_3R3 tends to be expressed much higher in cell lines compared to native tissues (Blondel et al., 1993) and, for this reason, the therapeutical value of studies in cell lines could be limited (Ivanova et al., 2014). Thus, IP_3R3 expression in patient samples should reflect its actual expression in cancers. In any case, even if primary samples are instrumental in the medical context, cell lines remain very useful tools and they permit different research groups to perform experiments on a same cell model.

The expression of IP_3R3 has been measured in different kinds of cancer and compared to nontumorigenic tissues or related cell lines, giving us an overview of IP_3R3 alteration in cancer (Table 1 and Fig. 1).

3.1.1 Brain cancers

In the brain, two types of cancers extensively studied are the glioblastoma and the neuroblastoma. In Kang et al., IP_3R expression was measured by semi-quantitative polymerase chain reaction (PCR) in several glioblastoma cell lines. IP_3R3 levels were increased in the five tested cell lines compared to normal astrocytes in which IP_3R3 was not detected. This difference was also confirmed in multiple patient samples (Kang et al., 2010). Neuroblastoma was also studied in this publication, but the author did not highlight any alteration of IP_3R3 levels. IP_3R3 is neither expressed in normal astrocytes nor in the neuroblastoma SH-SY5Y cell line, however, the latter does show elevated IP_3R1 levels (Kang et al., 2010). Of note, another group showed in the same neuroblastoma cell line that IP_3R3—but also IP_3R1, SERCA1, and SERCA3—are overexpressed upon treatment with chemotherapeutic drugs such as cisplatin or topotecan (Florea et al., 2017).

3.1.2 Head and neck cancers

In the context of head and neck squamous cell carcinoma (HNSCC), a genetic and statistical analysis was performed in order to identify biomarkers for malignancy risk prediction. Forty candidate genes were tested in tumor samples from a cohort of 49 patients by means of the NanoString nCounter gene expression system. High *ITPR3* expression was identified as the strongest risk factor for the development of second primary malignancy.

Table 1 Overview of IP_3R3 expression alterations in cancer.

Organ	Cancer categories	Primary/cell lines	Evidence	Expression	Notes	References
Brain	Glioblastoma	Cell lines (U87MG, U178MG, U373MG, T98G, M059K) and primary tissues	mRNA	↑		Kang et al. (2010)
Brain	Neuroblastoma	Cell line (SH-SY5Y)	mRNA		↑ Upon chemotherapy treatment	Florea et al. (2017)
Head and neck	Squamous cell carcinoma	Primary tissues	mRNA	↑	In second primary malignancy	Bunbanjerdsuk et al. (2019)
Thyroid	Carcinoma	Primary tissues	mRNA	↑	In *BRAF* V600E cancers	Rusinek et al. (2015)
Lung	Non–small cell carcinoma	Cell lines (H1993, H820) and primary tissues	Protein	(↑)	In high K-ras cancers	Iommelli et al. (2018)
Breast	Melanoma	Cell line (MDA-MB-435S)	Protein	↑		Mound et al. (2017)
Breast	Adenocarcinoma	Cell lines (MCF-7, MDA-MB-231) and primary tissues	Protein	↑		Singh et al. (2017a), Singh et al. (2017b), Mound et al. (2017)
Breast	Adenocarcinoma	Cell line (MDA-MB-468)	Protein	(↓)	In high STAT3 cancers	Avalle et al. (2019)
Breast	Adenocarcinoma	Cell line (MCF-7)	Protein	(↑)	In estrogen-induced cancers	Szatkowski et al. (2010)
Breast	Adenocarcinoma	Cell line (MDA-MB-468)	mRNA	(↑)	After EGF-induced EMT	Davis et al. (2013)
Stomach	Carcinoma	Primary-derived cell line	mRNA	↑	Only in metastatic cells	Sakakura et al. (2003)
Liver	Carcinoma	Primary tissues	Protein	↑		Guerra et al. (2019)
Bile duct	Cholangiocarcinoma	Primary tissues	Protein	↑		Ueasilamongkol et al. (2020)
Kidney	Clear cell renal cell carcinoma	Primary tissues	mRNA	↑		Rezuchova et al. (2019)
Colon	Carcinoma	Cell line (HCT116)	Protein	↓	In *KRAS* G13D cancers	Pierro et al. (2014)
Colon	Carcinoma	Cell line (CACO-2) and primary tissues	Protein	↑		Shibao et al. (2010)
Prostate	Adenocarcinoma	Primary tissues	Protein	(↓)	In low PTEN cancers	Kuchay et al. (2017)
Pancreas	Ductal adenocarcinoma	Primary tissues	mRNA	(↑)	Expression positively correlated with IMP2	Dahlem et al. (2019)
Endometrium	Adenocarcinoma	Cell line (MDAH)	mRNA		↓ Upon chemotherapy treatment	Kucukkaya et al. (2019)
BAP1 cancer syndrome		Primary-derived cell line (fibroblasts)	Protein	(↓)	In low BAP1 cancers	Bononi et al. (2017)

The ↑ and ↓ symbols indicate upregulation or downregulation of IP_3R3 expression, respectively. Symbols are in parenthesis when altered expression was deduced from experimental data without a proper comparison between cancerous and noncancerous tissues.

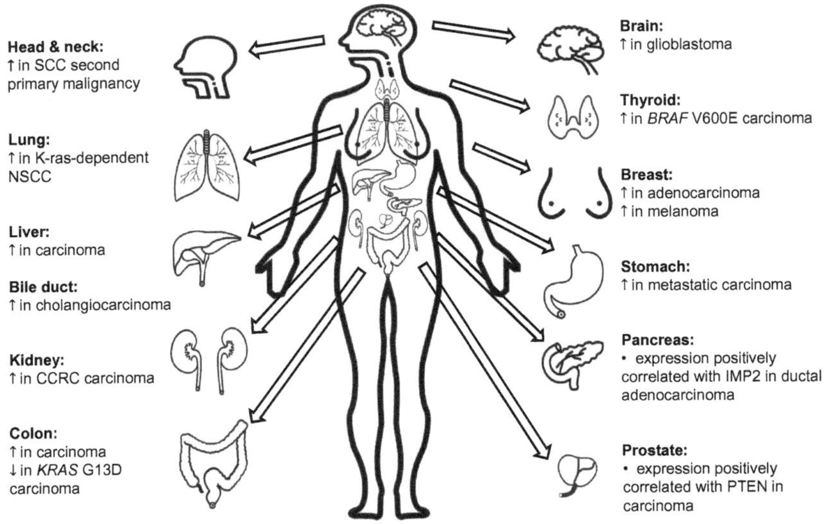

Fig. 1 Overview of dysregulated IP_3R3 expression in various cancers. Organs in which IP_3R3 expression is dysregulated in cancer are represented. The ↑ and ↓ symbols indicate upregulation or downregulation of IP_3R3 expression, respectively. SCC, squamous cell carcinoma; NSCC, non–small cell carcinoma; CCRC, clear cell renal cell. *The figure was created with the following icons under Creative Commons license from the Noun Project website (https://thenounproject.com/): "human body" by Yu-Hsuan Chen, "Brain" by Berkah Icon, "throat" by Maxicons, "Lung", "Liver", "Kidneys", "intestine", "Stomach" by Hermine Blanquart, "breast" by corpus delicti, "Thyroid", "Prostate" by PJ Witt, "Pancreas" by fauzan akbar.*

More precisely, an increase in one standard deviation of *ITPR3* level is associated with three-fold higher risk of this kind of malignancy (Bunbanjerdsuk et al., 2019). In addition, a whole exome sequencing study of 16 HNSCC primary samples revealed that *ITPR3* is specifically mutated in synchronous nodal metastases (Hedberg et al., 2016).

3.1.3 Thyroid cancers

Dysregulation of IP_3R3 expression and an association with the oncogenic Ras-Raf-MEK-ERK pathway might exist in thyroid cancer. The V600E mutation of the serine–threonine kinase B-Raf is well known for its constitutive and oncogenic activation of that pathway. A transcriptomic analysis of papillary thyroid carcinomas from a patient cohort identified the significant dysregulation of 18 genes (of 532 tested) associated with *BRAF* V600E. Among them, *ITPR3* is upregulated (Rusinek et al., 2015).

3.1.4 Lung cancers

In the lung, a relationship between IP_3R3 and non–small cell lung cancer (NSCLC) has been found. In these cancers, aberrant activation of the Ras-Raf-MEK-ERK pathway promotes tumor growth and progression. To investigate this pathway, Iommelli et al. used NSCLC adenocarcinoma cell lines that possess extra copies of the epidermal growth factor receptor (EGFR) and the hepatocyte growth factor receptor (HGFR). The EGFR and HGFR, best known as c-MET, are located upstream of the Ras-Raf-MEK-ERK pathway. Those cells have been implanted in mice and the xenografted-tumors formed were later removed for analysis. Treatment of the animals with EGFR or c-MET tyrosine kinase inhibitors resulted in a significant reduction of IP_3R3 protein level in the xenografts, indicating that abnormal activation of this pathway favors IP_3R3 expression (Iommelli et al., 2018).

3.1.5 Breast cancers

IP_3R3 expression was also widely studied in breast cancer. IP_3R3 overexpression was detected in the MCF-7 adenocarcinoma cell line compared to the nontumorigenic line MCF-10A (Singh et al., 2017a). The same modification was consistently observed by the same group in cancer primary samples, compared to adjacent healthy tissue sections, by quantitative PCR and immunohistochemistry (Singh et al., 2017b). Interestingly, in both studies, IP_3R2 is increased to the same extent as IP_3R3, while IP_3R1 remains unaffected. However, results are more contradictory for the metastatic cell lines MDA-MB-231 (adenocarcinoma) and MDA-MB-435S (melanoma). IP_3R3 protein was found by one group to be more expressed in these cell lines than in MCF-7 (Mound et al., 2017), while another group measured a similar expression level for MDA-MB-231 and nontumorigenic MCF-10A, lower than in MCF-7 (Singh et al., 2017a). Then, in the breast ductal carcinoma line T-47D, visual comparison of IP_3R immunoblots suggest a strong expression of IP_3R3 in those cells when compared to IP_3R1 (Soghoian et al., 2005).

Oestrogens were identified long ago for their role in breast cancer carcinogenesis (Yager and Davidson, 2006) and, very fascinatingly, 7β-estradiol (E2) increases IP_3R3 protein expression in breast cancer cells (Szatkowski et al., 2010). Different amounts of oestrogens in culture medium could explain divergent IP_3R3 protein expression in breast cancer lines. Furthermore, EGF-induced epithelial-mesenchymal transition, a hallmark of metastasis, dysregulates gene expression of several Ca^{2+} signaling-related genes, with a significant increase in mRNA levels of IP_3R3 (Davis et al., 2013).

Lastly, downregulation of IP_3R3 was also observed in breast cancer. In the MDA-MB-468 adenocarcinoma cell line, but not in the MDA-MB-453 metastatic carcinoma line, the signal transducer and activator of transcription 3 (STAT3) negatively regulates the expression of IP_3R3 (Avalle et al., 2019). The implication of STAT3, a prosurvival transcription factor with oncogenic properties, in IP_3R3 stability in cancer will be discussed in Section 3.2.1.

3.1.6 Gastric cancers

Historically, the first dysregulation of the IP_3R3 receptor was discovered in gastric cancer (Sakakura et al., 2003). Overexpression of the channel was detected by microarrays in gastric cancer cell lines (established from malignant tissues) compared to healthy gastric epithelial cells. IP_3R1 and IP_3R2 are weakly or not expressed in these cells. As another gastric cancer cell line (created from primary tumor) does not exhibit any IP_3R3 dysregulation, this study also shows that IP_3R3 expression can vary greatly from one cancer type to another, even in a same organ (Sakakura et al., 2003).

3.1.7 Liver cancers

Recent studies proposed an upregulation of IP_3R3 in hepatic cancer (Guerra et al., 2019; Mangla et al., 2019). Expression of the IP_3Rs has been studied in primary hepatic tissues by real-time PCR and immunohistochemistry. IP_3R1 and IP_3R2 are expressed in normal hepatocytes but IP_3R3 is absent or only expressed in very low amounts. However, IP_3R3 is upregulated in tissues from cohorts of patients with three independent chronic liver diseases, all leading to hepatocellular carcinoma, while IP_3R1 and IP_3R2 remained unaffected. DNA methylation profile analysis has revealed that IP_3R3 upregulation is triggered by an important demethylation of the promoter of *ITPR3* in cancer cells compared to the nontumorous ones. As a confirmation for this mechanism, *ITPR1* and *ITPR2* promoters have been found demethylated in both cancer and noncancer cells. Very fascinatingly, administration of a demethylating agent to mice resulted in *Itpr3* expression in the liver (Guerra et al., 2019). Finally, IP_3R3 expression has been examined in cholangiocytes, epithelial cells of the bile duct, and in cholangiocarcinomas (CCA) (Ueasilamongkol et al., 2020). IP_3R3 proteins are overexpressed in the two CCA cell lines tested in comparison to normal cholangiocytes. The intensity of IP_3R3 staining measured by immunofluorescence and immunochemistry is higher in both hilar CCA and intrahepatic CCA primary tissues compared to healthy liver biopsy.

3.1.8 Kidney cancers

In kidney cancer, IP_3R expression was quantified by microarrays in clear cell renal cell carcinoma primary tumors and compared to healthy parts of the same organs. A strong overexpression of IP_3R3 was found in almost 80% of the tested samples, while IP_3R1 and IP_3R2 are downregulated (Rezuchova et al., 2019).

3.1.9 Colorectal cancers

In the colorectal carcinoma cell line HCT116, tumorigenesis is associated with the G13D mutation of *KRAS*, an oncogene of the Ras-Raf-MEK-ERK pathway. An isogenic derivative cell line of HCT116—HKH2—has been engineered with a deletion of the mutated *RAS* allele by homologous recombination and does not exhibit most of the tumorigenic phenotype of the HCT116 line. Comparison of these cell lines reveals a downregulation of IP_3R3, along with SERCA2b, and an upregulation of IP_3R1 in HCT116 cancer cells compared to HKH2 cells (Pierro et al., 2014). Contrariwise, an overexpression of *ITPR3* has also been reported for colorectal adenocarcinoma, compared to normal colon, in two large transcriptomic screenings (Su et al., 2004). This overexpression of IP_3R3 is rather contradictory with the work of Pierro et al. but another study revealed that IP_3R3 expression in different colorectal adenocarcinoma lines is rather unequal with strong variations among the cell lines (Katona et al., 2019). Finally, IP_3R expression was analyzed in primary cancerous colorectal tissues by immunochemistry (Shibao et al., 2010). It revealed a very disparate pattern of expression, certainly dependent on the original localization of the biopsy in the colorectal mucosa. IP_3R1 and IP_3R2 are uniformly distributed in normal mucosa but some amount is lost in colorectal carcinoma. Nevertheless, IP_3R3 is expressed only in cancerous tissues (Shibao et al., 2010). These authors also show that the colorectal adenocarcinoma cell line CACO-2 expresses IP_3R3 but not IP_3R1 or IP_3R2. In any case, more work will be necessary to identify the IP_3R3 expression pattern in colorectal cancer.

3.1.10 Prostate cancers

Alteration of IP_3R3 expression may be present in prostate cancer. Measurement of IP_3R3 levels in human prostate adenocarcinomas by microarrays reveals a direct correlation between IP_3R3 levels and the expression of the phosphatase and tensin homologue (PTEN), a tumor suppressor (Kuchay et al., 2017). The degradation of IP_3R3 upon loss of PTEN relies on the F-box/LRR-repeat protein 2 (FBXL2) and will be detailed in Section 3.2.1.

3.1.11 Other cancers

To conclude this extensive overview of IP_3R3 expression in cancer, a few other studies may also give some clues about possible dysregulation of its expression in other cancer types.

(i) An analysis of public transcriptomic databases was performed to identify putative partners for the insulin-like growth factor 2 mRNA binding protein (IMP2), an oncogene overexpressed in numerous cancers, including pancreatic cancer. This study reveals a positive correlation between the expression levels of IP_3R3 and IMP2 in pancreatic ductal adenocarcinoma (Dahlem et al., 2019).

(ii) Quantitative RT-PCR revealed a downregulation of IP_3R3 and a dozen other Ca^{2+} channels, pumps, and exchangers in a human endometrioid adenocarcinoma cell line after treatment with the chemotherapeutic drug cisplatin, an inhibitor of DNA replication (Kucukkaya et al., 2019).

(iii) A large transcriptomic screening of countless sorts of cancerous and normal cells by microarrays did not show any alteration of *ITPR3* expression in blood cancers such as acute promyelocytic leukemia, chronic myelogenous leukemia, T-cell acute lymphoblastic leukemia, and B-cell Burkitt's lymphoma (Su et al., 2004).

3.2 Mechanisms that affect IP_3R3 expression in cancers

The previous discussion on IP_3R3 in cancer depicts a rather complex pattern of IP_3R3 expression in cancers, with studies reporting IP_3R3 upregulation versus downregulation even among similar cancer types. Therefore, we will now delve deeper into the mechanisms controlling the expression of IP_3R3 channels in cancers. First, it has become clear that tumor suppressors and oncogenes may control IP_3R3 turnover and thus alterations in these proteins in cancers could affect IP_3R3-protein levels (Fig. 2). Second, mutations in the *ITPR3* gene itself could also account for alterations in IP_3R3-protein levels but until now this topic has been only modestly explored.

3.2.1 Alteration of IP_3R3 stability

Degradation of IP_3Rs is a natural desensitization process required to limit excessive Ca^{2+} signaling after channel activation (Wojcikiewicz et al., 1994b). IP_3R degradation is mediated through the ER-associated

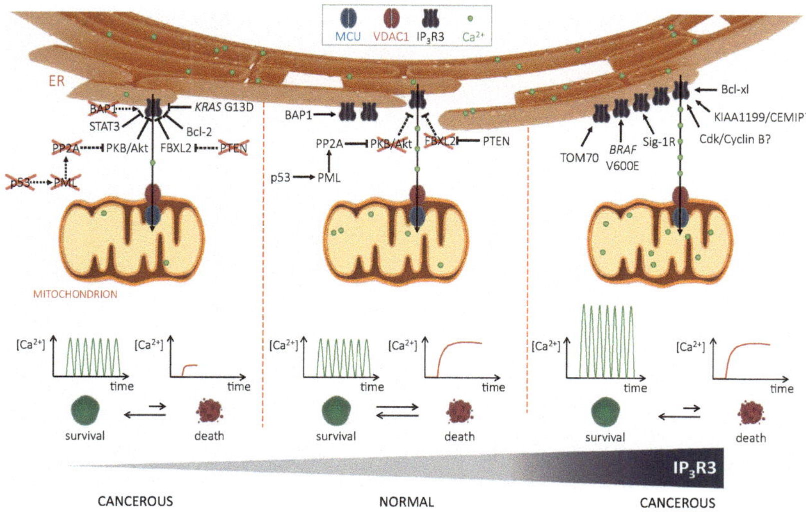

Fig. 2 Regulation of IP_3R3 activity and/or expression in cancer. *Middle*: Normal cells express stable IP_3R3 levels and activity, maintaining the balance between prosurvival (oscillatory) and prodeath (sustained) Ca^{2+} signaling. (i) BAP1, a deubiquitination enzyme, stabilizes IP_3R3 by protecting it from proteasomal degradation. (ii) p53 modulates PML effector functions, allowing recruitment of PP2A to inhibit PKB/Akt-mediated dampening of proapoptotic Ca^{2+} signals. (iii) PTEN competes with FBXL2, an E3 ubiquitin ligase, for binding to IP_3R3, whereby PTEN prevents FBXL2-mediated degradation of the IP_3R3. *Left*: IP_3R3 activity and/or expression can be decreased via several mechanisms to promote cell survival, causing an imbalance in favor of prosurvival Ca^{2+} signaling. (i) In several cancer types, BAP1, which deubiquitinates IP_3Rs, is lost, thus enhancing IP_3R3 degradation. (ii) STAT3 promotes IP_3R3 degradation. (iii) Due to the absence of p53, PML and PP2A in cancer cells, PKB/Akt is able to dampen proapoptotic Ca^{2+} signals. (iv) Upon loss of PTEN, a feature of several cancers, FBXL2 binding to IP_3R3 will be favored, thereby stimulating its degradation through ubiquitination. (v) Bcl-2 is found to be overexpressed in many cancer types and can suppress proapoptotic Ca^{2+} signaling. (vi) The oncogenic G13D mutation of *KRAS* is associated with lower IP_3R3 expression levels. *Right:* IP_3R3 activity and/or expression can be increased via several mechanisms to promote cell survival, favoring prosurvival Ca^{2+} signaling. (i) TOM70 recruits IP3R3 close to the mitochondria, thereby promoting IP_3R3-mediated prosurvival Ca^{2+} signaling which sustains mitochondrial bioenergetics. (ii) The oncogenic *BRAF* V600E mutation underlies IP_3R3 upregulation. (iii) Sig-1R enhances prosurvival Ca^{2+} oscillations from the IP_3R3. (iv) It is proposed that the Cdk/Cyclin B complex phosphorylates IP_3R3 to favor prosurvival Ca^{2+} signaling. (v) KIAA1199/CEMIP interacts with IP_3R3, increasing ER Ca^{2+} release. (vi) Bcl-xl is found to be overexpressed in many cancer types and has been proposed to stimulate prosurvival Ca^{2+} oscillations.

degradation (ERAD) system, as part of the ubiquitin–proteasome pathway. This degradation has been widely studied and we have a good comprehension of the organization of the protein complex involved in this process. The canonical mechanism of IP_3R degradation includes the Erlin1/2 (also known as SPFH1 and SPFH2) complex that mediates the interaction between the ubiquitin ligase RNF170 and activated IP_3Rs. Ubiquitinated IP_3Rs are then targeted by the polyubiquitin-binding cofactors Ufd1 and Npl4 and the ATPase p97 (VCP/Cdc48) that uses ATP to generate energy to extract the channels from the ER membrane (Wright and Wojcikiewicz, 2016). This mechanism induces a > 50% loss of the total IP_3R pool with half-maximal effect approximately 1 h after stimulation (Wojcikiewicz et al., 1994a). However, beyond this classical model, other actors of IP_3R degradation continue to be discovered (Prole and Taylor, 2016), such as the ubiquitin ligase glycoprotein 78 (Gp78) (Pearce et al., 2007) or the ubiquitin fusion degradation 1 (UFD1) protein (Alzayady et al., 2005). Interestingly, some of them could be specific for one of the IP_3R isoforms and differences were thus observed in the characteristics of degradation of the various isoforms (Wojcikiewicz, 1995). In any case, SPFH2 (Pearce et al., 2009), UFD1 (Alzayady et al., 2005) and Gp78 (Pearce et al., 2007) can interact with IP_3R3 and are described to be altered and/or involved in several different cancers (Huiting et al., 2018; Nabi et al., 1992; Wang et al., 2012). All of them could contribute to the dysregulation of IP_3R3 expression in cancer.

More recently, the ubiquitin ligase responsible for IP_3R3 degradation in cancer cells has been identified to be FBXL2, which was linked to prooncogenic IP_3R3 downregulation (Kuchay et al., 2017). FBXL2 belongs to the F-box proteins, one of the three subunits of the Skp, Cullin, F-box (SCF) ubiquitin ligase complex. Fascinatingly, the tumor suppressor PTEN competes with FBXL2 for IP_3R3 binding, preventing the SCF-mediated ubiquitination and proteasomal degradation of the channel (Kuchay et al., 2017). Consistently, *PTEN* knock out in nontumorigenic cells recapitulates the *PTEN*-null cancer phenotype—FBXL2-mediated IP_3R3 degradation—whereas reconstitution of *PTEN*-null cells with wild type PTEN protects IP_3R3 from FBXL2-dependent ubiquitination. At the molecular level, it was shown that FBXL2, through its F-box domain, binds the cytosolic N-terminal region of IP_3R3, a region known to change its conformation upon IP_3 binding. As protein degradation increases with IP_3 production, the authors suggest that binding of IP_3 and activation of the channel unveils its binding site for FBXL2, eventually leading to IP_3R3 ubiquitination and proteasomal degradation (Kuchay et al., 2017). Interestingly, FBXL2 activity itself is Ca^{2+} dependent, but antagonized by calmodulin, a ubiquitous Ca^{2+}-sensing protein (Chen et al., 2011).

This suggests an even more complex system of FBXL2 regulation and channel desensitization. IP_3R3-mediated Ca^{2+} release may be directly sensed by FBXL2 at the MAM and induce channel degradation, depending on the amount of Ca^{2+} released, since calmodulin activation requires a specific Ca^{2+} concentration. *PTEN* is known to be frequently mutated or lost in cancer. The expression levels of PTEN and IP_3R3 directly correlate in primary prostate adenocarcinomas samples (Kuchay et al., 2017). Besides prostate cancer, PTEN is lost in about 30–55% of breast, colon, lung, endometrial, and ovarian cancers. This proportion even reaches 65% of glioblastomas (Alvarez-Garcia et al., 2019). This suggests a possible FBXL2-dependent IP_3R3 degradation in many other cancer types.

One of them could be breast cancer and could involve the prosurvival and oncogenic protein STAT3. This protein is a transcription factor constitutively activated in a high percentage of tumors, often correlating with aggressiveness and poor prognosis. STAT3 aberrant activity drives resistance to apoptosis and chemotherapy, and stimulates migration and invasion, epithelial to mesenchymal transition (EMT) and/or immune evasion of cancerous cells (Avalle et al., 2012; Parys, 2018). In a breast adenocarcinoma cell line, STAT3 was coimmunoprecipitated at the MAM with IP_3R3. There, it negatively regulates the expression of the channel by facilitating its degradation through the proteasome (Avalle et al., 2019). This mechanism is even more interesting as it relies on STAT3 ER localization and is not dependent on its nuclear localization (through phosphorylated tyr^{705}) and thus STAT3 transcriptional activity. The localization of STAT3 at the ER likely relies on ser^{727} phosphorylation (Avalle et al., 2019). The importance of this phosphorylation site in oncogenesis is emphasized by the existence of two splice isoforms of STAT3—STAT3α and STAT3β—whereby ser^{727} is only present in the oncogenic STAT3α (Aigner et al., 2019). This new role of STAT3 is particularly fascinating because it is one of the two transcription factors—with p53 (Giorgi et al., 2015)—recognized for regulating Ca^{2+} signaling independently of its transcriptional activity (Parys, 2018). Moreover, serum starvation/refeeding in the same breast cancer cells resulted in a STAT3-dependent degradation of IP_3R3 (Avalle et al., 2019). The authors link this phenomenon to their precedent work on FBXL2-dependent IP_3R3 degradation since they have established that IP_3R3 is targeted by FBXL2 upon serum starvation/refeeding (Kuchay et al., 2017). Thus, STAT3 could activate directly or indirectly FBXL2 in order to trigger IP_3R3 degradation in cancer.

Another protein that impacts IP_3R3 stability in cancer is the BRCA1-associated protein 1 (BAP1). BAP1 is a deubiquitinating enzyme that acts as a tumor suppressor to regulate environmental carcinogenesis. Mutation or loss of *BAP1* has been reported to generate a BAP1 cancer syndrome, characterized mostly by malignant mesothelioma or uveal melanoma (Testa et al., 2011). The tumor suppressor activity of this protein has been first attributed to its nuclear localization but it has been discovered recently that BAP1 can also localize at the ER where it binds IP_3R3 (Hedgepeth et al., 2015) to protect it from ubiquitination (Bononi et al., 2017). IP_3R3 expression is thus strongly dampened in $BAP1^{+/-}$ and $BAP1^{-/-}$ cells and we can assume this is valid in BAP1 cancers as well.

Finally, we wish to highlight the potential contribution of the Ras-Raf-MEK-ERK pathway in IP_3R3 dysregulation. This ubiquitous pathway is initiated in response to the activation of several cell membrane receptors, such as the EGFR and HGFR, and is implicated in many cellular processes, in particular proliferation. Most of the proteins of the Ras-Raf-MEK-ERK pathway are coded by recognized proto-oncogenes. The most significant are the Ras family proteins, small GTPases located upstream of the pathway. The *RAS* genes are mutated in about one third of human cancers and oncogenic Ras proteins are known to be able to promote tumor initiation in normal cells (Li et al., 2018). Involvement of the Ras-Raf-MEK-ERK pathway in IP_3R3 altered expression has been identified in colorectal, thyroid and lung cancers. As indicated in Section 3.1, the G13D tumorigenic mutation of *KRAS* decreases IP_3R3 levels (Pierro et al., 2014) while the V600E oncogenic mutation of *BRAF* is involved in upregulation of IP_3R3 (Rusinek et al., 2015). The critical role of this pathway has also been highlighted by the use of EGFR and HGFR inhibitors that cause a reduction of IP_3R3 expression in lung cancer. This study is interesting because it also reveals that this decrease of IP_3R3 expression is associated with a loss of interaction between K-Ras and IP_3R3 proteins, mainly because of the dephosphorylation of K-Ras (Iommelli et al., 2018). This study suggests that phospho-K-Ras stabilizes IP_3R3 levels at the MAM. We can thus suppose that oncogenic proteins of the Ras-Raf-MEK-ERK pathway, beyond any transcriptional effect, could interact with IP_3R3 at the ER to regulate the stability of the channel. Of note, direct interaction and modulation of IP_3R3 by another K-Ras variant was previously reported (Sung et al., 2013).

Other oncogenes and tumor suppressors that do not modify IP_3R3 expression but are essential for regulation of its function in cancer will be discussed in the next section.

3.2.2 *Mutations of* ITPR3

Mutations in the *ITPR3* gene itself could explain various expression/stability alterations. However, this approach was very poorly investigated until now. A large amount of mutations of *ITPR1* has been listed and associated to diseases—mainly spinocerebellar ataxia and Gillespie syndrome—but this was not the case for *ITPR3* (Hisatsune and Mikoshiba, 2017; Kerkhofs et al., 2018; Mangla et al., 2019; Terry et al., 2018).

In a first publication, mutations of *ITPR3* have been discovered in steatohepatitis-related hepatocellular carcinoma in a mouse model (Liang et al., 2018). However, pathophysiological consequences of this mutation are unknown.

Then, in cervical squamous cell carcinoma, the association of two single nucleotide polymorphisms (SNP) was found to be significantly associated with cancer in a large cohort of 462 patients (versus 921 healthy controls), while each SNP alone did not seem relevant by itself (Yang et al., 2017). The first (*ITPR3 rs3748079 A*) is interesting because it modifies the 5′ untranslated region of the *ITPR3* promoter and can account for an alteration in its expression. The second (*rs2229634 T*), however, only generates a synonymous variation in the coding sequence (Sherry et al., 2001). This particular association is rather unexpected and more work is necessary to apprehend the consequences of those SNP on protein expression and function.

Lastly, a whole exome sequencing study from a human cohort affected by head and neck squamous cell carcinoma revealed two patients (of 13) with unique *ITPR3* mutations, specifically in synchronous nodal metastases (Hedberg et al., 2016). Nevertheless, these mutations are located in the coding sequence and have thus a greater probability of modifying protein function than expression. More specifically, the mutations generate amino acid modifications—R149L and R64H—in the β-trefoil structures of the ligand-binding domain, in the N-terminal cytosolic region of the protein (Fan et al., 2015; Paknejad and Hite, 2018).

4. The role of IP_3R3 in cancer: Tumor suppression versus oncogenesis

Although IP_3R3 has been reported to be involved in cancer, opposing roles for the channel have surfaced over the years. A vast literature indicates a tumor suppressor role for IP_3R3 by promoting proapoptotic ER-mitochondrial Ca^{2+} signaling, whereby IP_3R3 downregulation or inhibition promotes oncogenesis and cell death resistance. However, several studies reported a dependence of cancer cells toward IP_3R3 for their survival, whereby IP_3R3 upregulation could account for uncontrolled cell growth by promoting cellular metabolism. We wish to converge both seemingly opposing mechanisms whereby both IP_3R3 downregulation as well as upregulation possibly sustain cancer cell survival depending on the cancer stage. It is interesting to note that a similar phenomenon may occur at the level of MCU, potentially counteracting cancer cell survival in an early stage (Marchi et al., 2013) while promoting cancer cell survival in late, metastatic stages (Tosatto et al., 2016).

4.1 IP_3R3 channels acting as tumor suppressors by mediating apoptosis sensitivity

4.1.1 IP_3R3 degradation as an oncogenic feature in apoptosis resistance

An anticancerous role of the IP_3R3 has been elegantly shown in a colorectal carcinoma model in which deletion of the oncogene K-Ras^{G13D} was associated with an upregulation of IP_3R3 expression. This resulted in an increase in agonist-mediated ER Ca^{2+} release and mitochondrial Ca^{2+}, as well as increased susceptibility to apoptosis (Pierro et al., 2014, 2019).

We described in Section 3.2.1 the mechanisms by which the tumor suppressor BAP1 stabilizes IP_3R3 expression. Accordingly, the loss of BAP1 (a feature of the BAP1 cancer syndrome) results in diminished ER Ca^{2+} release and mitochondrial $[Ca^{2+}]$ associated with a decrease of apoptosis (Bononi et al., 2017). This study further advocates for a proapoptotic role for IP_3R3-mediated Ca^{2+} release in cancer.

We also discussed in Section 3.2.1 the implication of the ubiquitin ligase-associated protein FBXL2 and the transcription factor STAT3 in IP_3R3 degradation. In breast cancer cells, STAT3 promotes IP_3R3 degradation. Consistently, knock down of STAT3 restores IP_3R3 levels, promoting Ca^{2+} delivery into the mitochondria and apoptosis susceptibility (Avalle et al., 2019). The effect of STAT3 on IP_3R3 stability could be mediated through FBXL2 but, for now, it is clearly demonstrated that FBXL2 activity is elicited by the loss of the tumor suppressor *PTEN*. In this context, FBXL2 silencing, proteasome inhibition, or PTEN reconstitution restored ER Ca^{2+} release and Ca^{2+}-dependent apoptosis in cancer cells (Kuchay et al., 2017). Consistently, disruption of FBXL2 localization at the ER membrane with a geranylgeranyl transferase inhibitor sensitizes xeno-tumors transplanted into mice to photodynamic therapy, a localized activation of cytotoxic drugs (Kuchay et al., 2017). Moreover, even the hepatitis C virus appears to exploit FBXL2-mediated IP_3R3 downregulation in hepatocytes. The viral protein NS5A has been shown to form a trimeric complex with IP_3R3 and FBXL2, unmasking the IP_3R3 degron for FBXL2 recognition. Degradation of IP_3R3 consequently limits ER Ca^{2+}-mediated cell death and permit the virus to escape host cell apoptosis and establish chronic infection (Kuchay et al., 2018). Those studies demonstrate that Ca^{2+} transfer from the ER to the mitochondria through the IP_3R3 constitutes an apoptotic signal.

4.1.2 Inhibition of IP_3R3-mediated Ca^{2+} release as an oncogenic feature in apoptosis resistance

IP_3R3-mediated Ca^{2+} signaling and apoptosis are controlled by oncogenes such as Akt and tumor suppressors such as the promyelocytic leukemia protein (PML) (Fig. 2). The proto-oncogene Akt (also known as protein kinase B) is a serine/threonine kinase and a major downstream effector of the phosphoinositide 3-kinase (PI3K) pathway. It is involved in several cellular processes such as transcription, cell survival, proliferation, migration, and glucose metabolism. Alterations in the Akt pathway are identified to promote oncogenesis and Akt is found overexpressed in 12% of ovarian carcinomas, 15% of pancreatic carcinomas, 40% of hepatocellular carcinomas and about 57% of colorectal cancer (Altomare and Testa, 2005). All three IP_3R isoforms present a highly conserved phosphorylation motif for Akt in their cytosolic C-terminal tail (Khan et al., 2006; Szado et al., 2008). Comparative experiments in cells expressing only IP_3R1 or IP_3R3 have revealed that phosphorylation of IP_3R3 by Akt impairs ER Ca^{2+} flux to mitochondria, maintains the mitochondrial integrity, and delays the trigger of apoptosis (Marchi et al., 2012). These data propose that Akt could be a putative oncogenic regulator of IP_3R3. We mentioned above the role of the tumor suppressor protein PTEN in the regulation of FBXL2-mediated IP_3R3 degradation (Section 3.2.1). That mechanism is independent of PTEN phosphatase activity (Kuchay et al., 2017) but PTEN can also act as a protein phosphatase at the MAM to counteract Akt regulation of Ca^{2+} apoptotic fluxes (Bononi et al., 2013). Furthermore, the tumor suppressor protein PML seems to be involved in Akt-mediated IP_3R3 regulation (Giorgi et al., 2010). PML was originally identified at the breakpoint of chromosomal translocation of acute promyelocytic leukemia (APL) but it is now recognized to be frequently mutated or lost in several solid and hematopoietic cancers (Missiroli et al., 2017). In normal cells, PML interacts specifically with IP_3R3 at the MAM and promotes IP_3R3-mediated Ca^{2+} release and apoptosis (Giorgi et al., 2010). As PML was found to be enriched in complex with IP_3R3, Akt and the protein phosphatase 2A (PP2A) at the MAM, we can assume that PML recruits PP2A in order to inhibit Akt. In cancer cells, PML and PP2A should be absent, allowing Akt to dampen proapoptotic Ca^{2+} signaling. Fascinatingly, PML is also known to interact with p53 (Pearson et al., 2000), a tumor suppressor and probably the most studied protein in cancer research. p53 directly modulates PML effector functions (de Stanchina et al., 2004) and can localize at the MAM (Giorgi et al., 2015). There, p53 directly regulates the SERCA Ca^{2+} pump but it also acts as an anchor to maintain PML localization in the ER-mitochondria contact sites (Giorgi et al., 2015; Missiroli et al., 2017). One can therefore suggest that p53 could indirectly regulate the IP_3R3/PML complex at the MAM and consequently control IP_3R3-mediated Ca^{2+} release.

Besides these mechanisms, it is important to note that the Bcl-2-family proteins are critical functional modulators of IP_3Rs. Anti- (Bcl-2, Bcl-xl, Mcl-1) and proapoptotic proteins (Bax, Bak, Bim, Bid, etc.) of the Bcl-2 family were described originally to regulate the formation of oligomeric pores in the OMM and the release of apoptotic factors into the cytosol. However, despite this canonical role, it is clear that part of the function of Bcl-2-family proteins is executed through modulation of intracellular Ca^{2+} homeostasis and dynamics (Vervliet et al., 2016). Several Bcl-2-family members directly associate with IP_3R channels, thereby exerting direct control over their Ca^{2+}-flux properties (reviewed in Ivanova et al., 2019a). Overexpression of Bcl-2 was first discovered in B-cell lymphoma (Vaux et al., 1988) but dysregulation of anti- and proapoptotic members of the Bcl-2 family was then identified in many other cancer types, such as prostate (Karnak and Xu, 2010), breast (Hellemans et al., 1995) and lung cancers (Jiang et al., 1995). In particular, many lymphoma and leukemia models—categorized as primed to death—are known to be critically dependent on antiapoptotic Bcl-2 family proteins for their survival (Montero and Letai, 2018). Here we will focus on the role of Bcl-2 and Bcl-xl on IP_3R regulation, although other antiapoptotic Bcl-2-family members can also be involved (Ivanova et al., 2019a). While Bcl-2 inhibits aberrant proapoptotic Ca^{2+} release (Chen et al., 2004; Hanson et al., 2008; Rong et al., 2008), Bcl-xl appears to promote spontaneous prosurvival Ca^{2+} oscillations (Li et al., 2007; White et al., 2005; Yang et al., 2016), sustaining mitochondrial metabolism (Williams et al., 2016). Bcl-2 modulates IP_3R through its binding to several domains of the channel (Ivanova et al., 2016, 2019b; Monaco et al., 2012, 2013; Rong et al., 2009). A peptide derived from the Bcl-2-interacting region in the central, modulatory domain of IP_3R1 could abolish the Bcl-2/IP_3R interaction, rendering T-cell models more sensitive to proapoptotic T-cell receptor stimulation (Rong et al., 2008, 2009). Furthermore, a stabilized cell-permeable version of this peptide was shown to kill diffuse large B-cell lymphoma cancer cells without any additional stimulus (Akl et al., 2013; Bittremieux et al., 2019; Zhong et al., 2011). Thus, small molecules that mimic that peptide and antagonize Bcl-2/IP_3R interaction could function as promising anti-cancer tools (Distelhorst, 2018; Distelhorst and Bootman, 2019; Kerkhofs et al., 2019b). Conversely, peptides corresponding to the helixes 1 and 4 of the carboxyl terminus of IP_3R3 prevented Bcl-xl association with IP_3R and rendered DT40 cells expressing only IP_3R3 more sensitive to staurosporine-induced apoptosis. They could also provoke cancer cell death by themselves in MCF-7 breast cancer and Raji B-cell cancer cells (Yang et al., 2016). Although all three IP_3R isoforms are targets for Bcl-2 (Chen et al., 2004) and Bcl-xl (White et al., 2005), the relevance for IP_3R3/Bcl-2 and IP_3R3/Bcl-xl complexes for cancer cell survival is not fully clarified yet. A study in lung cancer revealed that IP_3R3 can associate with Bcl-2 and Bcl-xl. In this kind of cancer, EGFR tyrosine kinase inhibitor treatment induces cell death but can ultimately confer apoptosis resistance. It was demonstrated that this cell death resistance could be mediated through an increase of Bcl-2 and Bcl-xl interaction with IP_3R3, associated with an increase of cytosolic and mitochondrial [Ca^{2+}] (Zannetti et al., 2008). Indeed, it has been reported that, similarly to Bcl-xl, Bcl-2 can generate prosurvival Ca^{2+} fluxes by binding to the C-terminal domain of IP_3R3 (Eckenrode et al., 2010), even if most studies report inhibition of IP_3Rs by Bcl-2 (Chen et al., 2004; Hanson et al., 2008; Ivanova et al., 2019b; Monaco et al., 2012; Rong et al., 2008). Nevertheless, more work is necessary to decipher the role of the association of IP_3R3 and Bcl-2 family proteins in cancers.

A tumor suppressor related to Bcl-2 is Beclin 1, encoded by the *BECN1* gene (Decuypere et al., 2011). Beclin 1, also known as Atg6, promotes autophagy, thereby inhibiting cancer growth (Liang et al., 1999). *BECN1* is monoallelically deleted in 40–75% of ovarian, breast and prostate cancers (Cao and Klionsky, 2007). Beclin 1 is neutralized by Bcl-2, thus acting not only as an antiapoptotic but also an antiautophagy protein (Pattingre et al., 2005). More recently, our group demonstrated that Beclin 1-dependent autophagy is dependent on IP_3R1 and IP_3R3-mediated Ca^{2+} release. Moreover, upon starvation, a potent trigger of autophagy, Beclin 1 is recruited by IP_3Rs in a Bcl-2-independent manner, resulting in IP_3R sensitization.

We proposed that upon starvation Beclin 1 shuttles from Bcl-2 to IP_3Rs in order to stimulate autophagy (Decuypere et al., 2011).

Finally, it was recently demonstrated that IP_3R3-mediated ER Ca^{2+} release could be deleterious for cancer cells in a mitochondria-independent way. In colorectal cancer cell lines, synergistic combination of EGFR inhibitors and menin (a nuclear epigenetic regulator) inhibitors blocks proliferation and triggers apoptosis through IP_3R3-mediated Ca^{2+} release. In this pathway, Ca^{2+} binds and inhibits menin, altering the methylation status and expression of *SKP2*, a prooncogenic gene (Katona et al., 2019). This is supported by the fact that other tyrosine kinase inhibitors of the EGFR pathway lead to apoptosis of breast cancer cells in an ER Ca^{2+}-dependent but mitochondria-independent manner (Carloni et al., 2010).

4.2 IP_3R3 channels acting as oncogenic drivers by promoting cell survival

4.2.1 *The addiction of cancer cells toward IP_3R3-mediated Ca^{2+} signaling for proliferation and survival*

Upregulation of the IP_3R3 was observed in many cancer types (see Section 3), which seems at odds with IP_3R3 being implicated in proapoptotic ER-mitochondrial Ca^{2+} fluxes. Previous studies indicate that IP_3R3 upregulation could be associated with aggressiveness (Shibao et al., 2010), metastasis (Bunbanjerdsuk et al., 2019; Sakakura et al., 2003), chemotherapy resistance (Florea et al., 2017), and poor prognosis (Guerra et al., 2019). In colorectal (Shibao et al., 2010) and hepatocellular carcinoma (Guerra et al., 2019) in particular, the increase of IP_3R3 expression even correlates with the stage of the disease. Hence, the dichotomy of IP_3R3 being implicated in both pro- and antiapoptotic mechanisms could implicate that IP_3R3 can exert many other cellular functions besides apoptosis sensitivity or that IP_3R3 function diverges between cancer cells with high IP_3R3 expression versus low IP_3R3 expression levels.

In cancers that display a high IP_3R3 expression level, inhibition of IP_3R3-mediated Ca^{2+} release dampens proliferation and generates apoptotic signaling. In gastric epithelial cancer cells, which mainly express IP_3R3, 2-aminoethoxydiphenyl borate (2-APB), a compound with IP_3R-inhibitory properties, provoked apoptosis and impaired proliferation (Sakakura et al., 2003). However, 2-ABP can also inhibit other Ca^{2+}-transport systems including SERCA pumps and store-operated Ca^{2+}-influx channels (Bilmen et al., 2002; Iwasaki et al., 2001; Missiaen et al., 2001; Prakriya and Lewis, 2001). In certain breast cancer cells that express all IP_3R isoforms, but overexpress IP_3R3 upon E2 exposure, inhibition of the channel with caffeine dampened proliferation and cell survival (Szatkowski et al., 2010). A more specific attribution of IP_3R3 to these processes was found in studies using *ITPR3* knock down or knock out. This leads to decreased proliferation in breast adenocarcinoma (Cardenas et al., 2016; Szatkowski et al., 2010) and in cholangiocarcinoma (Ueasilamongkol et al., 2020) and increased cell death in colorectal carcinoma (Shibao et al., 2010), breast adenocarcinoma (Cardenas et al., 2016), cholangiocarcinoma (Ueasilamongkol et al., 2020) and in colorectal carcinoma (Rezuchova et al., 2019). Similarly, colony formation, a feature of cancer cells that reflects their capacity to proliferate indefinitely, is impaired by pharmacological inhibition (caffeine, xestospongin B, etc.) of the IP_3R in breast adenocarcinoma (Cardenas et al., 2016) and glioblastoma (Kang et al., 2010). Furthermore, gene silencing of IP_3R3 compromises cell migration, a hallmark of metastatic cells, in glioblastoma (Kang et al., 2010), melanoma (Cardenas et al., 2016), breast carcinoma (Mound et al., 2017), colorectal adenocarcinoma (Rezuchova et al., 2019) and cholangiocarcinoma (Ueasilamongkol et al., 2020). Fascinatingly, the impairment of migration in breast cancer cells upon *ITPR3* silencing is caused by an alteration of the global morphology of the cell to a more rounded shape via a process involving actin cytoskeletal reorganization through the ARHGAP18/RhoA/mDia1/FAK pathway (Vautrin-Glabik et al., 2018).

At the intracellular level, silencing or inhibition of IP_3R3 leads to a defect in ER Ca^{2+} release (Kang et al., 2010; Szatkowski et al., 2010; Ueasilamongkol et al., 2020) and should result in reduced mitochondrial $[Ca^{2+}]$, as established in cholangiocarcinoma (Ueasilamongkol et al., 2020). The consequence is a dysregulation of mitochondrial metabolism. In breast cancer, it is characterized by a reduction of oxygen consumption (Cardenas et al., 2016), generation of ROS, and a global decrease of the major metabolites such as ATP, glucose, lactate and nicotinamide adenine dinucleotide (NAD^+) (Singh et al., 2017a). Conversely, breast cancer cells with high levels of IP_3R3 exhibit higher concentrations of these metabolites than cancer cells with low levels of IP_3R3 (Singh et al., 2017b). Furthermore, analysis of cell metabolism and signaling pathways indicates that cancer cells enter autophagy upon IP_3R disruption. This is pointed out by the activation of the AMPK enzyme, a cytosolic sensor that detects increases in the AMP/ATP ratio, as well as the expression of autophagic markers such as LC3 and Atg proteins (Cardenas et al., 2016; Singh et al., 2017a). In breast cancer, this autophagy induction is ultimately followed by nonapoptotic cell death, such as dysregulated autophagic cell death (Singh et al., 2017a) or necrosis (Cardenas et al., 2016). The necrosis observed in this model is caused by a mitotic catastrophe, probably due to the absence of cell cycle halt at the G1/S checkpoint. This should account for a requirement of cancer cells for Ca^{2+} to pursue their division (Cardenas et al., 2016). Until now, a clear relationship between ER-Ca^{2+} release and cell cycle has not been demonstrated, although it is well documented that Ca^{2+} is necessary for DNA replication, cytoskeleton reorganization, cell division and cycle progression (Humeau et al., 2018). Futhermore, it is acknowledged that cell cycle progression is strongly dependent on mitochondrial metabolism and ATP production (Mitra et al., 2009). Consistently, *ITPR3* knock out in cholangiocarcinoma impairs cell cycle progression (Ueasilamongkol et al., 2020).

Taken together, those data draw a general cellular pathway in which IP_3R3 provide Ca^{2+} transfer from the ER to the mitochondria to sustain mitochondrial energetics and favor cancer features. Additionally, data in breast cancer suggest that high IP_3R3-expressing cancers could be in an autophagy-related metabolic state in which only IP_3R3-mediated Ca^{2+} flux maintains proliferation and survival. In any case, this mechanism likely relies on VDAC1 (Arif et al., 2014; Zhang et al., 2016) and MCU (Cardenas et al., 2016; Tosatto et al., 2016) to sustain prosurvival Ca^{2+} transfer in the mitochondria. These studies demonstrate that VDAC1 or MCU silencing impairs cancer progression, underlying cancer cells' addiction to ER-mitochondria Ca^{2+} transfers (Bultynck, 2016; Lovy et al., 2016).

Finally, a set of studies performed in breast cancer revealed high IP_3R3 expression levels upon E2 exposure (Szatkowski et al., 2010). In those cells, pharmacological inhibition of all IP_3Rs resulted as expected in an impaired proliferation and an increased cell death. However, specific knock down of *ITPR3* did not modify the global amount of ER Ca^{2+} release but revealed an oscillatory Ca^{2+} pattern, in contrast with the plateau-type pattern of control cancer cells (Mound et al., 2017; Szatkowski et al., 2010; Vautrin-Glabik et al., 2018). This is rather surprising since Ca^{2+} oscillations have been associated with cancer cell survival. Those cells also express IP_3R1 and IP_3R2, which likely drive these Ca^{2+} oscillations. The authors propose that alteration of the ER Ca^{2+} pattern is determinant in the impairment of cancer features but the exact role of those patterns in cellular functions are still unknown. In any case, measurement of mitochondrial $[Ca^{2+}]$ upon IP_3R3 inhibition would be necessary to decipher the role of IP_3R3-mediated Ca^{2+} release on cellular functions and cancer hallmarks.

4.2.2 Proteins favoring IP_3R3-mediated Ca^{2+} fluxes to sustain cancer cell survival

To complete this section, we will discuss several proteins, including oncogenic proteins and tumor suppressors, whose alterations could favor IP_3R3-dependent cancers. We presented in Section 4.1.2 the consequences of dysregulation of several oncogenes and tumor suppressors that lean toward impairment of proapoptotic IP_3R3-mediated Ca^{2+} signaling in cancer. However, dysregulation of other proteins can account for the model of cancers—with high IP_3R3 levels—we described in Section 4.2.1.

The sigma-1 receptor (Sig-1R)—a receptor that binds steroids, neuroleptics, and psychotropic drugs— was found in complexes at the MAM with IP_3R3 and ankyrin B, a cytoskeletal adaptor protein (Hayashi and Su, 2001). Sig-1R enhances constitutive IP_3R3 Ca^{2+} signaling (Hayashi and Su, 2007), and its overexpression under cell stress prevents apoptosis (Wu and Bowen, 2008). Accordingly, Sig-1R activity is clearly associated with cancer proliferation and survival (Spruce et al., 2004).

Furthermore, another interesting protein is the cell migration-inducing and hyaluronan-binding protein (CEMIP), formerly known as KIAA1199. In the SW480 colorectal adenocarcinoma cell line, putative interacting partners for KIAA1199 were identified and endogenous IP_3R3 was successfully co-immunoprecipitated with KIAA1199. Of note, KIAA1199 does not alter the expression of the channel (Tiwari et al., 2013). The exact function of CEMIP/KIAA1199 is unknown but it was recently identified as a marker of poor prognosis in cancer. Actually, this protein promotes EMT, migration and invasion through the enhancement of the Wnt/β-catenin signaling pathway and activation of matrix metalloproteinases (Jia et al., 2017). Fascinatingly, another study described KIAA1199 as a protein anchored to the ER membrane due to the presence of an ER-retention motif and an interaction with the ER-binding immunoglobulin protein (BiP). There, KIAA1199 enhances ER Ca^{2+} leakage that leads to activation of oncogenic PKCα and Ras signaling (Evensen et al., 2013). As IP_3R3 interacts with CEMIP/KIAA1199, we can wonder whether the Ca^{2+} leakage mentioned just above could be mediated by IP_3R3.

We may also include in this list the Heat shock protein 90 (Hsp90). Hsp90 is a ubiquitous molecular chaperone, essential for conformational maturation of proteins. Although Hsp90 is highly expressed in most cell types, it is strongly associated with cancer. This chaperone plays a key role in the maturation of oncogenic proteins and Hsp90 inhibitors can selectively kill some cancer cells but not normal cells. It was discovered that in those cancers, Hsp90 is in a "hyperactive" state, with high ATPase activity and complexed with other chaperones, while Hsp90 from normal cells is in a latent and uncomplexed state (Kamal et al., 2003). Hsp90 has been identified as another binding partner of IP_3R3 (Nguyen et al., 2009). In HEK293 cells, induction of high Hsp90 expression results in a robust IP_3R3/Hsp90 interaction, associated with an oscillatory ER Ca^{2+} release (Nguyen et al., 2009). As we mentioned above, oscillatory calcium patterns are often associated with cell survival and proliferation. Of course, it will be interesting to determine whether "hyperactive" Hsp90 in cancer cells can generate prosurvival oscillatory calcium fluxes.

Moreover, the mitochondrial translocase of the outer membrane 70 (TOM70) makes clusters at the MAM and interacts with IP_3R3 to recruit it close to the mitochondria. This association promotes IP_3R3-dependent Ca^{2+} release to mitochondria and sustains bioenergetics (Filadi et al., 2018a). As TOM70 and TOM20 protein levels correlate in cancer (Bauckman et al., 2015) and as TOM20 is upregulated in breast cancers (Sotgia et al., 2012), we could assume these cancers rely on Ca^{2+}-stimulated mitochondrial metabolism.

Finally, some other proto-oncogenes possibly involved are cyclins and cyclin-dependent kinases (Cdk), some major effectors of the cell cycle. The Cdk1/cyclin B complex interaction and subsequent phosphorylation of IP_3R1 was validated to favor Ca^{2+} release (Malathi et al., 2005). Even though only one of the phosphorylation motives of IP_3R1 is conserved in IP_3R3 (Malathi et al., 2005), IP_3R3 was pulled down with cyclins A and B in breast cancer cells (Soghoian et al., 2005).

4.2.3 Reorganization of IP_3R3 localization in cancer

Finally, some studies point out an alteration of IP_3R3 subcellular localization in high IP_3R3 levels cancers (Mound et al., 2013; Ueasilamongkol et al., 2020). Usually, ER proteins can diffuse toward the nucleus (perinuclear) or toward the cell membrane (periphery). IP_3R3 translational mobility in the ER was previously reported and analyzed in Chinese hamster ovary and COS-7 cells (Ferreri-Jacobia et al., 2005). This study reveals that the large majority of IP_3R3 are potentially free to diffuse throughout the ER. These movements are independent of actin cytoskeleton but are facilitated by ATP and may involve microtubules. Beyond IP_3R3, IP_3R1 was clearly shown to redistribute in a microtubule-dependent way in smooth muscle cells (Vermassen et al., 2003). More recently, high-resolution microscopy demonstrated that IP_3R1 and IP_3R3 move freely within the ER of COS-7 cells and are present in equal amounts in the perinuclear region and periphery. On the contrary, IP_3R2 is less mobile and concentrated in the periphery (Pantazaka and Taylor, 2011).

In breast cancer cells, IP_3R3s are relocated toward the cell membrane, compared to related normal breast cell, where they interact with BK_{Ca} channels, probably in the lipid raft microdomains. BK_{Ca} channels are Ca^{2+} -activated potassium channels of the plasma membrane. In this cancer model, IP_3R3-mediated Ca^{2+} release activates BK_{Ca} channels that contribute to cancer features (Mound et al., 2013). However, the exact consequence of this channel activity on ionic transport in the cytoplasm and Ca^{2+} transfer to the mitochondria is not known. On the contrary, super resolution fluorescence microscopy revealed IP_3R3 localization at the ER-mitochondria contact sites in cholangiocarcinoma and thus outside the regions where IP_3R3 channels reside in normal cholangiocytes, which is in the subapical region of the cell (Ueasilamongkol et al., 2020). This "mislocalization" of oncogenic IP_3R3 channels at the MAM should illuminate functional implications on mitochondrial $[Ca^{2+}]$ and mitochondrial metabolism.

Interestingly, subcellular localization of IP_3R channels is dependent on the K-ras-induced actin-interacting protein (KRAP) (Dingli et al., 2012; Fujimoto et al., 2011). As the oncogene K-ras is highly and heterogeneously mutated and expressed in cancer (Li et al., 2018), we can expect some consequences on KRAP activity in cancer. Anyhow, it would be very fascinating to decipher whether overexpressed IP_3R3s in high IP_3R3 levels cancers are usually mislocalized and whether this mislocalization accounts for different behavior of high and low IP_3R3 levels cancers.

5. Conclusions

We aimed to provide a balanced view on the roles and implication of IP_3R3 in cancer, whereby IP3R3 is associated with both anti- and procancer features. Indeed, in some cancers, loss of IP_3R3 results in apoptosis resistance enabling the survival of malignant cells despite cellular damage and genomic aberrations. Here, loss of IP_3R3 provides cancer cells with a growth advantage and restoring IP_3R3 expression could promote cancer cell death by enhancing apoptotic sensitivity. In other cancers however, IP_3R3 is upregulated and seems to be critical to sustain their survival, e.g. by driving mitochondrial metabolism. In this case, IP_3R3 upregulation provides cancer cells with a growth advantage and suppressing IP_3R3 expression could abrogate cancer cell growth by impairing their mitochondrial bio-energetic fitness. The dual role of IP_3R3 in cancer is puzzling (Sneyers et al., 2019). Yet, it is possible that endogenous IP_3R3 channels have opposing roles in cancer cells depending on the stage of development. In this scenario, IP_3R3 signaling would prevent neoplasia of healthy cells and thus counteract the early stages of cancer by promoting apoptosis. Conversely, it would promote the survival of late-stage cancers by sustaining the mitochondrial metabolism. As such, chronic IP_3R3 upregulation could impose a stress on premalignant cells that requires such Ca^{2+} signaling to rewire their metabolic and proliferative networks. Ultimately, it would not only create a dependence on sustained IP_3R3-mediated signaling for their survival, it would also enable the manifestation of hallmarks of aggressive, late-stage cancers, such as cell motility, migration and invasion. In particular, cells that survive high IP_3R3 expression appear to be resistant to proapoptotic stimuli, likely becoming resistant to genomic stress and further boosting the survival of malignant cells through oncogenic mutations. As such, late-stage cancer cells might become addicted to high levels of IP_3R3 for their survival, which can be exploited by anticancer agents. Indeed, certain chemotherapeutics also trigger Ca^{2+} release from the ER (Doghman-Bouguerra et al., 2016; Kerkhofs et al., 2019a; Pinton et al., 2001), including via IP_3R3 channels that could reside at ER-mitochondrial contact site. In any case, more work will be necessary to decipher the exact role of IP_3R3 in distinct cancer types. In a broader perspective, the search for the precise role of these channels is put to the challenge by the co-expression of several IP_3R isoforms. This further complicates our understanding of IP_3R's control on survival processes in cancers, and how these mechanisms are related to the sensitivity of certain cancer types toward chemotherapeutic drugs.

Acknowledgments

The work was supported by Grants from the Research Foundation—Flanders (FWO) (Grants 6.057.12, G.0819.13, G.0C91.14, G.0A34.16, and G.0901.18) and by the Research Council of the KU Leuven (Grants 14/101 and C14/19/099). N.R. is recipient of a postdoctoral fellowship of the FWO and F.S. is holder of a Ph.D. fellowship of the FWO. G.B. and J.B.P. are part of the FWO Scientific Research Network CaSign (W0.019.17N).

References

Aigner, P., Just, V., Stoiber, D., 2019. STAT3 isoforms: alternative fates in cancer? Cytokine 118, 27–34.

Akl, H., Bultynck, G., 2013. Altered Ca(2+) signaling in cancer cells: proto-oncogenes and tumor suppressors targeting IP3 receptors. Biochim. Biophys. Acta 1835, 180–193.

Akl, H., Monaco, G., La Rovere, R., Welkenhuyzen, K., Kiviluoto, S., Vervliet, T., Molgo, J., Distelhorst, C.W., Missiaen, L., Mikoshiba, K., Parys, J.B., De Smedt, H., Bultynck, G., 2013. IP3R2 levels dictate the apoptotic sensitivity of diffuse large B-cell lymphoma cells to an IP3R-derived peptide targeting the BH4 domain of Bcl-2. Cell Death Dis. 4, e632.

Alavian, K.N., Beutner, G., Lazrove, E., Sacchetti, S., Park, H.-A., Licznerski, P., Li, H., Nabili, P., Hockensmith, K., Graham, M., Porter Jr., G.A., Jonas, E.A., 2014. An uncoupling channel within the c-subunit ring of the F1FO ATP synthase is the mitochondrial permeability transition pore. Proc. Natl. Acad. Sci. U. S. A. 111, 10580–10585.

Altomare, D.A., Testa, J.R., 2005. Perturbations of the AKT signaling pathway in human cancer. Oncogene 24, 7455–7464.

Alvarez-Garcia, V., Tawil, Y., Wise, H.M., Leslie, N.R., 2019. Mechanisms of PTEN loss in cancer: It's all about diversity. Semin. Cancer Biol. 59, 66–79.

Alzayady, K.J., Panning, M.M., Kelley, G.G., Wojcikiewicz, R.J., 2005. Involvement of the p97-Ufd1-Npl4 complex in the regulated endoplasmic reticulum-associated degradation of inositol 1,4,5-trisphosphate receptors. J. Biol. Chem. 280, 34530–34537.

Alzayady, K.J., Wagner 2nd, L.E., Chandrasekhar, R., Monteagudo, A., Godiska, R., Tall, G.G., Joseph, S.K., Yule, D.I., 2013. Functional inositol 1,4,5-trisphosphate receptors assembled from concatenated homo- and heteromeric subunits. J. Biol. Chem. 288, 29772–29784.

Alzayady, K.J., Wang, L., Chandrasekhar, R., Wagner 2nd, L.E., Van Petegem, F., Yule, D.I., 2016. Defining the stoichiometry of inositol 1,4,5-trisphosphate binding required to initiate Ca2+ release. Sci. Signal. 9, ra35.

Arif, T., Vasilkovsky, L., Refaely, Y., Konson, A., Shoshan-Barmatz, V., 2014. Silencing VDAC1 expression by siRNA inhibits cancer cell proliferation and tumor growth in vivo. Mol. Ther. Nucleic Acids 3, e159.

Avalle, L., Regis, G., Poli, V., 2012. Universal and specific functions of STAT3 in solid tumours. In: Decker, T., Müller, M. (Eds.), Jak-Stat Signaling : From Basics to Disease. Springer, Vienna, pp. 305–333.

Avalle, L., Camporeale, A., Morciano, G., Caroccia, N., Ghetti, E., Orecchia, V., Viavattene, D., Giorgi, C., Pinton, P., Poli, V., 2019. STAT3 localizes to the ER, acting as a gatekeeper for ER-mitochondrion Ca(2+) fluxes and apoptotic responses. Cell Death Differ. 26, 932–942.

Bansaghi, S., Golenar, T., Madesh, M., Csordas, G., Ramachandra Rao, S., Sharma, K., Yule, D.I., Joseph, S.K., Hajnoczky, G., 2014. Isoform- and species-specific control of inositol 1,4,5-trisphosphate (IP3) receptors by reactive oxygen species. J. Biol. Chem. 289, 8170–8181.

Bartok, A., Weaver, D., Golenar, T., Nichtova, Z., Katona, M., Bansaghi, S., Alzayady, K.J., Thomas, V.K., Ando, H., Mikoshiba, K., Joseph, S.K., Yule, D.I., Csordas, G., Hajnoczky, G., 2019. IP3 receptor isoforms differently regulate ER-mitochondrial contacts and local calcium transfer. Nat. Commun. 10, 3726.

Bauckman, K., Haller, E., Taran, N., Rockfield, S., Ruiz-Rivera, A., Nanjundan, M., 2015. Iron alters cell survival in a mitochondria-dependent pathway in ovarian cancer cells. Biochem. J. 466, 401–413.

Baughman, J.M., Perocchi, F., Girgis, H.S., Plovanich, M., Belcher-Timme, C.A., Sancak, Y., Bao, X.R., Strittmatter, L., Goldberger, O., Bogorad, R.L., Koteliansky, V., Mootha, V.K., 2011. Integrative genomics identifies MCU as an essential component of the mitochondrial calcium uniporter. Nature 476, 341–345.

Baumgartner, H.K., Gerasimenko, J.V., Thorne, C., Ferdek, P., Pozzan, T., Tepikin, A.V., Petersen, O.H., Sutton, R., Watson, A.J.M., Gerasimenko, O.V., 2009. Calcium elevation in mitochondria is the main Ca2+ requirement for mitochondrial permeability transition pore (mPTP) opening. J. Biol. Chem. 284, 20796–20803.

Berridge, M.J., 2009. Inositol trisphosphate and calcium signalling mechanisms. Biochim. Biophys. Acta 1793, 933–940.

Berridge, M.J., Heslop, J.P., Irvine, R.F., Brown, K.D., 1984. Inositol trisphosphate formation and calcium mobilization in Swiss 3T3 cells in response to platelet-derived growth factor. Biochem. J. 222, 195–201.

Betzenhauser, M.J., Wagner 2nd, L.E., Iwai, M., Michikawa, T., Mikoshiba, K., Yule, D.I., 2008. ATP modulation of Ca2+ release by type-2 and type-3 inositol (1, 4, 5) -triphosphate receptors. Differing ATP sensitivities and molecular determinants of action. J. Biol. Chem. 283, 21579–21587.
Bezprozvanny, I., Ehrlich, B.E., 1993. ATP modulates the function of inositol 1,4, 5-trisphosphate-gated channels at two sites. Neuron 10, 1175–1184.
Bezprozvanny, I., Watras, J., Ehrlich, B.E., 1991. Bell-shaped calcium-response curves of Ins(1,4,5)P3- and calcium-gated channels from endoplasmic reticulum of cerebellum. Nature 351, 751–754.
Bilmen, J.G., Wootton, L.L., Godfrey, R.E., Smart, O.S., Michelangeli, F., 2002. Inhibition of SERCA Ca2+ pumps by 2-aminoethoxydiphenyl borate (2-APB). 2-APB reduces both Ca2+ binding and phosphoryl transfer from ATP, by interfering with the pathway leading to the Ca2+-binding sites. Eur. J. Biochem. 269, 3678–3687.
Bittremieux, M., Parys, J.B., Pinton, P., Bultynck, G., 2016. ER functions of oncogenes and tumor suppressors: modulators of intracellular Ca(2+) signaling. Biochim. Biophys. Acta 1863, 1364–1378.
Bittremieux, M., La Rovere, R.M., Akl, H., Martines, C., Welkenhuyzen, K., Dubron, K., Baes, M., Janssens, A., Vandenberghe, P., Laurenti, L., Rietdorf, K., Morciano, G., Pinton, P., Mikoshiba, K., Bootman, M.D., Efremov, D.G., De Smedt, H., Parys, J.B., Bultynck, G., 2019. Constitutive IP3 signaling underlies the sensitivity of B-cell cancers to the Bcl-2/IP3 receptor disruptor BIRD-2. Cell Death Differ. 26, 531–547.
Blondel, O., Takeda, J., Janssen, H., Seino, S., Bell, G.I., 1993. Sequence and functional characterization of a third inositol trisphosphate receptor subtype, IP3R-3, expressed in pancreatic islets, kidney, gastrointestinal tract, and other tissues. J. Biol. Chem. 268, 11356–11363.
Boehning, D., Joseph, S.K., 2000. Functional properties of recombinant type I and type III inositol 1, 4,5-trisphosphate receptor isoforms expressed in COS-7 cells. J. Biol. Chem. 275, 21492–21499.
Boehning, D., Mak, D.O., Foskett, J.K., Joseph, S.K., 2001. Molecular determinants of ion permeation and selectivity in inositol 1,4,5-trisphosphate receptor Ca2+ channels. J. Biol. Chem. 276, 13509–13512.
Boehning, D., Patterson, R.L., Sedaghat, L., Glebova, N.O., Kurosaki, T., Snyder, S.H., 2003. Cytochrome c binds to inositol (1,4,5) trisphosphate receptors, amplifying calcium-dependent apoptosis. Nat. Cell Biol. 5, 1051–1061.
Bononi, A., Missiroli, S., Poletti, F., Suski, J.M., Agnoletto, C., Bonora, M., De Marchi, E., Giorgi, C., Marchi, S., Patergnani, S., Rimessi, A., Wieckowski, M.R., Pinton, P., 2012. Mitochondria-associated membranes (MAMs) as hotspot Ca(2+) signaling units. Adv. Exp. Med. Biol. 740, 411–437.
Bononi, A., Bonora, M., Marchi, S., Missiroli, S., Poletti, F., Giorgi, C., Pandolfi, P.P., Pinton, P., 2013. Identification of PTEN at the ER and MAMs and its regulation of Ca(2+) signaling and apoptosis in a protein phosphatase-dependent manner. Cell Death Differ. 20, 1631–1643.
Bononi, A., Giorgi, C., Patergnani, S., Larson, D., Verbruggen, K., Tanji, M., Pellegrini, L., Signorato, V., Olivetto, F., Pastorino, S., Nasu, M., Napolitano, A., Gaudino, G., Morris, P., Sakamoto, G., Ferris, L.K., Danese, A., Raimondi, A., Tacchetti, C., Kuchay, S., Pass, H.I., Affar, E.B., Yang, H., Pinton, P., Carbone, M., 2017. BAP1 regulates IP3R3-mediated Ca(2+) flux to mitochondria suppressing cell transformation. Nature 546, 549–553.
Bonora, M., Bononi, A., De Marchi, E., Giorgi, C., Lebiedzinska, M., Marchi, S., Patergnani, S., Rimessi, A., Suski, J.M., Wojtala, A., Wieckowski, M.R., Kroemer, G., Galluzzi, L., Pinton, P., 2013. Role of the c subunit of the FO ATP synthase in mitochondrial permeability transition. Cell Cycle 12, 674–683.

Booth, D.M., Enyedi, B., Geiszt, M., Varnai, P., Hajnoczky, G., 2016. Redox nanodomains are induced by and control calcium signaling at the ER-mitochondrial interface. Mol. Cell 63, 240–248.
Bootman, M.D., Bultynck, G., 2019. Fundamentals of cellular calcium signaling: a primer. In: Bultynck, G., Bootman, M.D., Berridge, M.J., Stutzmann, G.E. (Eds.), Calcium Signaling, second ed. Cold Spring Harbor Laboratory Press, New York, pp. 1–16.
Bootman, M.D., Berridge, M.J., Lipp, P., 1997. Cooking with calcium: the recipes for composing global signals from elementary events. Cell 91, 367–373.
Bootman, M.D., Chehab, T., Bultynck, G., Parys, J.B., Rietdorf, K., 2018. The regulation of autophagy by calcium signals: do we have a consensus? Cell Calcium 70, 32–46.
Bultynck, G., 2016. Onco-IP3Rs feed cancerous cravings for mitochondrial Ca(2+). Trends Biochem. Sci. 41, 390–393.
Bultynck, G., Parys, J.B., 2018. Ca(2+) signaling and cell death: focus on Ca(2+)-transport systems and their implication in cell death and survival. Cell Calcium 69, 1–3.
Bultynck, G., Szlufcik, K., Kasri, N.N., Assefa, Z., Callewaert, G., Missiaen, L., Parys, J.B., De Smedt, H., 2004. Thimerosal stimulates Ca2+ flux through inositol 1,4,5-trisphosphate receptor type 1, but not type 3, via modulation of an isoform-specific Ca2+-dependent intramolecular interaction. Biochem. J. 381, 87–96.
Bunbanjerdsuk, S., Vorasan, N., Saethang, T., Pongrujikorn, T., Pangpunyakulchai, D., Mongkonsiri, N., Arsa, L., Thokanit, N., Pongsapich, W., Anekpuritanang, T., Ngamphaiboon, N., Jinawath, A., Sunpaweravong, S., Pisitkun, T., Suktitipat, B., Jinawath, N., 2019. Oncoproteomic and gene expression analyses identify prognostic biomarkers for second primary malignancy in patients with head and neck squamous cell carcinoma. Mod. Pathol. 32, 943–956.
Cao, Y., Klionsky, D.J., 2007. Physiological functions of Atg6/Beclin 1: a unique autophagy-related protein. Cell Res. 17, 839–849.
Cardenas, C., Miller, R.A., Smith, I., Bui, T., Molgo, J., Muller, M., Vais, H., Cheung, K.-H., Yang, J., Parker, I., Thompson, C.B., Birnbaum, M.J., Hallows, K.R., Foskett, J.K., 2010. Essential regulation of cell bioenergetics by constitutive InsP3 receptor Ca2+ transfer to mitochondria. Cell 142, 270–283.
Cardenas, C., Muller, M., McNeal, A., Lovy, A., Jana, F., Bustos, G., Urra, F., Smith, N., Molgo, J., Diehl, J.A., Ridky, T.W., Foskett, J.K., 2016. Selective vulnerability of cancer cells by inhibition of Ca(2+) transfer from endoplasmic reticulum to mitochondria. Cell Rep. 14, 2313–2324.
Carloni, S., Fabbri, F., Brigliadori, G., Ulivi, P., Silvestrini, R., Amadori, D., Zoli, W., 2010. Tyrosine kinase inhibitors gefitinib, lapatinib and sorafenib induce rapid functional alterations in breast cancer cells. Curr. Cancer Drug Targets 10, 422–431.
Chandrasekhar, R., Alzayady, K.J., Yule, D.I., 2015. Using concatenated subunits to investigate the functional consequences of heterotetrameric inositol 1,4,5-trisphosphate receptors. Biochem. Soc. Trans. 43, 364–370.
Chandrasekhar, R., Alzayady, K.J., Wagner 2nd, L.E., Yule, D.I., 2016. Unique regulatory properties of heterotetrameric inositol 1,4,5-trisphosphate receptors revealed by studying concatenated receptor constructs. J. Biol. Chem. 291, 4846–4860.
Chen, R., Valencia, I., Zhong, F., McColl, K.S., Roderick, H.L., Bootman, M.D., Berridge, M.J., Conway, S.J., Holmes, A.B., Mignery, G.A., Velez, P., Distelhorst, C.W., 2004. Bcl-2 functionally interacts with inositol 1,4,5-trisphosphate receptors to regulate calcium release from the ER in response to inositol 1,4,5-trisphosphate. J. Cell Biol. 166, 193–203.
Chen, B.B., Coon, T.A., Glasser, J.R., Mallampalli, R.K., 2011. Calmodulin antagonizes a calcium-activated SCF ubiquitin E3 ligase subunit, FBXL2, to regulate surfactant homeostasis. Mol. Cell. Biol. 31, 1905–1920.

Chen, J., Sitsel, A., Benoy, V., Sepulveda, M.R., Vangheluwe, P., 2019. Primary active Ca(2+) transport systems in health and disease. In: Bultynck, G., Bootman, M.D., Berridge, M.J., Stutzmann, G.E. (Eds.), Calcium Signaling, second ed. Cold Spring Harbor Laboratory Press, New York, pp. 155–182.

Csordas, G., Thomas, A.P., Hajnoczky, G., 1999. Quasi-synaptic calcium signal transmission between endoplasmic reticulum and mitochondria. EMBO J. 18, 96–108.

Csordas, G., Renken, C., Varnai, P., Walter, L., Weaver, D., Buttle, K.F., Balla, T., Mannella, C.A., Hajnoczky, G., 2006. Structural and functional features and significance of the physical linkage between ER and mitochondria. J. Cell Biol. 174, 915–921.

Csordas, G., Weaver, D., Hajnoczky, G., 2018. Endoplasmic reticulum-mitochondrial contactology: structure and signaling functions. Trends Cell Biol. 28, 523–540.

Dahlem, C., Barghash, A., Puchas, P., Haybaeck, J., Kessler, S.M., 2019. The insulin-like growth factor 2 mRNA binding protein IMP2/IGF2BP2 is overexpressed and correlates with poor survival in pancreatic cancer. Int. J. Mol. Sci. 20, 3204.

Davis, F.M., Parsonage, M.T., Cabot, P.J., Parat, M.O., Thompson, E.W., Roberts-Thomson, S.J., Monteith, G.R., 2013. Assessment of gene expression of intracellular calcium channels, pumps and exchangers with epidermal growth factor-induced epithelial-mesenchymal transition in a breast cancer cell line. Cancer Cell Int. 13, 76.

de Stanchina, E., Querido, E., Narita, M., Davuluri, R.V., Pandolfi, P.P., Ferbeyre, G., Lowe, S.W., 2004. PML is a direct p53 target that modulates p53 effector functions. Mol. Cell 13, 523–535.

De Stefani, D., Raffaello, A., Teardo, E., Szabo, I., Rizzuto, R., 2011. A forty-kilodalton protein of the inner membrane is the mitochondrial calcium uniporter. Nature 476, 336–340.

Decuypere, J.P., Welkenhuyzen, K., Luyten, T., Ponsaerts, R., Dewaele, M., Molgo, J., Agostinis, P., Missiaen, L., De Smedt, H., Parys, J.B., Bultynck, G., 2011. Ins(1,4,5) P3 receptor-mediated Ca2+ signaling and autophagy induction are interrelated. Autophagy 7, 1472–1489.

D'Eletto, M., Rossin, F., Occhigrossi, L., Farrace, M.G., Faccenda, D., Desai, R., Marchi, S., Refolo, G., Falasca, L., Antonioli, M., Ciccosanti, F., Fimia, G.M., Pinton, P., Campanella, M., Piacentini, M., 2018. Transglutaminase type 2 regulates ER-mitochondria contact sites by interacting with GRP75. Cell Rep. 25. (3573–3581 e4).

Denton, R.M., 2009. Regulation of mitochondrial dehydrogenases by calcium ions. Biochim. Biophys. Acta 1787, 1309–1316.

Dingli, F., Parys, J.B., Loew, D., Saule, S., Mery, L., 2012. Vimentin and the K-Ras-induced actin-binding protein control inositol-(1,4,5)-trisphosphate receptor redistribution during MDCK cell differentiation. J. Cell Sci. 125, 5428–5440.

Distelhorst, C.W., 2018. Targeting Bcl-2-IP3 receptor interaction to treat cancer: a novel approach inspired by nearly a century treating cancer with adrenal corticosteroid hormones. Biochim. Biophys. Acta Mol. Cell Res. 1865, 1795–1804.

Distelhorst, C.W., Bootman, M.D., 2019. Creating a new cancer therapeutic agent by targeting the interaction between Bcl-2 and IP3 receptors. In: Bultynck, G., Bootman, M.D., Berridge, M.J., Stutzmann, G.E. (Eds.), Calcium Signaling, second ed. Cold Spring Harbor Laboratory Press, New York, pp. 463–478.

Doghman-Bouguerra, M., Lalli, E., 2019. ER-mitochondria interactions: both strength and weakness within cancer cells. Biochim. Biophys. Acta Mol. Cell Res. 1866, 650–662.

Doghman-Bouguerra, M., Granatiero, V., Sbiera, S., Sbiera, I., Lacas-Gervais, S., Brau, F., Fassnacht, M., Rizzuto, R., Lalli, E., 2016. FATE1 antagonizes calcium- and drug-induced apoptosis by uncoupling ER and mitochondria. EMBO Rep. 17, 1264–1280.

Dolmetsch, R.E., Xu, K., Lewis, R.S., 1998. Calcium oscillations increase the efficiency and specificity of gene expression. Nature 392, 933–936.

Eckenrode, E.F., Yang, J., Velmurugan, G.V., Foskett, J.K., White, C., 2010. Apoptosis protection by Mcl-1 and Bcl-2 modulation of inositol 1,4,5-trisphosphate receptor-dependent Ca2+ signaling. J. Biol. Chem. 285, 13678–13684.

Evensen, N.A., Kuscu, C., Nguyen, H.L., Zarrabi, K., Dufour, A., Kadam, P., Hu, Y.J., Pulkoski-Gross, A., Bahou, W.F., Zucker, S., Cao, J., 2013. Unraveling the role of KIAA1199, a novel endoplasmic reticulum protein, in cancer cell migration. J. Natl. Cancer Inst. 105, 1402–1416.

Fan, G., Baker, M.L., Wang, Z., Baker, M.R., Sinyagovskiy, P.A., Chiu, W., Ludtke, S.J., Serysheva, I.I., 2015. Gating machinery of InsP3R channels revealed by electron cryomicroscopy. Nature 527, 336–341.

Ferreri-Jacobia, M., Mak, D.O., Foskett, J.K., 2005. Translational mobility of the type 3 inositol 1,4,5-trisphosphate receptor Ca2+ release channel in endoplasmic reticulum membrane. J. Biol. Chem. 280, 3824–3831.

Filadi, R., Pizzo, P., 2019. ER-mitochondria tethering and Ca(2+) crosstalk: the IP3R team takes the field. Cell Calcium 84, 102101.

Filadi, R., Leal, N.S., Schreiner, B., Rossi, A., Dentoni, G., Pinho, C.M., Wiehager, B., Cieri, D., Cali, T., Pizzo, P., Ankarcrona, M., 2018a. TOM70 sustains cell bioenergetics by promoting IP3R3-mediated ER to mitochondria Ca(2+) transfer. Curr. Biol. 28 (369–382), e6.

Filadi, R., Pendin, D., Pizzo, P., 2018b. Mitofusin 2: from functions to disease. Cell Death Dis. 9, 330.

Finch, E.A., Turner, T.J., Goldin, S.M., 1991. Calcium as a coagonist of inositol 1,4,5-trisphosphate-induced calcium release. Science (New York, N Y) 252, 443–446.

Florea, A.-M., Varghese, E., McCallum, J.E., Mahgoub, S., Helmy, I., Varghese, S., Gopinath, N., Sass, S., Theis, F.J., Reifenberger, G., Busselberg, D., 2017. Calcium-regulatory proteins as modulators of chemotherapy in human neuroblastoma. Oncotarget 8, 22876–22893.

Foskett, J.K., White, C., Cheung, K.H., Mak, D.O., 2007. Inositol trisphosphate receptor Ca2+ release channels. Physiol. Rev. 87, 593–658.

Fujimoto, T., Machida, T., Tanaka, Y., Tsunoda, T., Doi, K., Ota, T., Okamura, T., Kuroki, M., Shirasawa, S., 2011. KRAS-induced actin-interacting protein is required for the proper localization of inositol 1,4,5-trisphosphate receptor in the epithelial cells. Biochem. Biophys. Res. Commun. 407, 438–443.

Gincel, D., Zaid, H., Shoshan-Barmatz, V., 2001. Calcium binding and translocation by the voltage-dependent anion channel: a possible regulatory mechanism in mitochondrial function. Biochem. J. 358, 147–155.

Giorgi, C., Ito, K., Lin, H.K., Santangelo, C., Wieckowski, M.R., Lebiedzinska, M., Bononi, A., Bonora, M., Duszynski, J., Bernardi, R., Rizzuto, R., Tacchetti, C., Pinton, P., Pandolfi, P.P., 2010. PML regulates apoptosis at endoplasmic reticulum by modulating calcium release. Science 330, 1247–1251.

Giorgi, C., Bonora, M., Sorrentino, G., Missiroli, S., Poletti, F., Suski, J.M., Galindo Ramirez, F., Rizzuto, R., Di Virgilio, F., Zito, E., Pandolfi, P.P., Wieckowski, M.R., Mammano, F., Del Sal, G., Pinton, P., 2015. p53 at the endoplasmic reticulum regulates apoptosis in a Ca2+-dependent manner. Proc. Natl. Acad. Sci. U. S. A. 112, 1779–1784.

Giorgio, V., von Stockum, S., Antoniel, M., Fabbro, A., Fogolari, F., Forte, M., Glick, G.D., Petronilli, V., Zoratti, M., Szabo, I., Lippe, G., Bernardi, P., 2013. Dimers of mitochondrial ATP synthase form the permeability transition pore. Proc. Natl. Acad. Sci. U. S. A. 110, 5887–5892.

Giorgio, V., Burchell, V., Schiavone, M., Bassot, C., Minervini, G., Petronilli, V., Argenton, F., Forte, M., Tosatto, S., Lippe, G., Bernardi, P., 2017. Ca(2+) binding to F-ATP synthase beta subunit triggers the mitochondrial permeability transition. EMBO Rep. 18, 1065–1076.

Glancy, B., Balaban, R.S., 2012. Role of mitochondrial Ca2+ in the regulation of cellular energetics. Biochemistry 51, 2959–2973.
Guerra, M.T., Florentino, R.M., Franca, A., Lima Filho, A.C., Dos Santos, M.L., Fonseca, R.C., Lemos, F.O., Fonseca, M.C., Kruglov, E., Mennone, A., Njei, B., Gibson, J., Guan, F., Cheng, Y.C., Ananthanarayanam, M., Gu, J., Jiang, J., Zhao, H., Lima, C.X., Vidigal, P.T., Oliveira, A.G., Nathanson, M.H., Leite, M.F., 2019. Expression of the type 3 InsP3 receptor is a final common event in the development of hepatocellular carcinoma. Gut 68, 1676–1687.
Hajnoczky, G., Robb-Gaspers, L.D., Seitz, M.B., Thomas, A.P., 1995. Decoding of cytosolic calcium oscillations in the mitochondria. Cell 82, 415–424.
Hanson, C.J., Bootman, M.D., Distelhorst, C.W., Wojcikiewicz, R.J., Roderick, H.L., 2008. Bcl-2 suppresses Ca2+ release through inositol 1,4,5-trisphosphate receptors and inhibits Ca2+ uptake by mitochondria without affecting ER calcium store content. Cell Calcium 44, 324–338.
Hattori, M., Suzuki, A.Z., Higo, T., Miyauchi, H., Michikawa, T., Nakamura, T., Inoue, T., Mikoshiba, K., 2004. Distinct roles of inositol 1,4,5-trisphosphate receptor types 1 and 3 in Ca2+ signaling. J. Biol. Chem. 279, 11967–11975.
Hayashi, T., Su, T.P., 2001. Regulating ankyrin dynamics: roles of sigma-1 receptors. Proc. Natl. Acad. Sci. U. S. A. 98, 491–496.
Hayashi, T., Su, T.-P., 2007. Sigma-1 receptor chaperones at the ER-mitochondrion interface regulate Ca(2+) signaling and cell survival. Cell 131, 596–610.
Hedberg, M.L., Goh, G., Chiosea, S.I., Bauman, J.E., Freilino, M.L., Zeng, Y., Wang, L., Diergaarde, B.B., Gooding, W.E., Lui, V.W., Herbst, R.S., Lifton, R.P., Grandis, J.R., 2016. Genetic landscape of metastatic and recurrent head and neck squamous cell carcinoma. J. Clin. Invest. 126, 169–180.
Hedgepeth, S.C., Garcia, M.I., Wagner 2nd, L.E., Rodriguez, A.M., Chintapalli, S.V., Snyder, R.R., Hankins, G.D., Henderson, B.R., Brodie, K.M., Yule, D.I., van Rossum, D.B., Boehning, D., 2015. The BRCA1 tumor suppressor binds to inositol 1,4,5-trisphosphate receptors to stimulate apoptotic calcium release. J. Biol. Chem. 290, 7304–7313.
Hedskog, L., Pinho, C.M., Filadi, R., Ronnback, A., Hertwig, L., Wiehager, B., Larssen, P., Gellhaar, S., Sandebring, A., Westerlund, M., Graff, C., Winblad, B., Galter, D., Behbahani, H., Pizzo, P., Glaser, E., Ankarcrona, M., 2013. Modulation of the endoplasmic reticulum-mitochondria interface in Alzheimer's disease and related models. Proc. Natl. Acad. Sci. U. S. A. 110, 7916–7921.
Hellemans, P., van Dam, P.A., Weyler, J., van Oosterom, A.T., Buytaert, P., Van Marck, E., 1995. Prognostic value of bcl-2 expression in invasive breast cancer. Br. J. Cancer 72, 354–360.
Hirota, J., Furuichi, T., Mikoshiba, K., 1999. Inositol 1,4,5-trisphosphate receptor type 1 is a substrate for caspase-3 and is cleaved during apoptosis in a caspase-3-dependent manner. J. Biol. Chem. 274, 34433–34437.
Hisatsune, C., Mikoshiba, K., 2017. IP3 receptor mutations and brain diseases in human and rodents. J. Neurochem. 141, 790–807.
Huiting, L.N., Samaha, Y., Zhang, G.L., Roderick, J.E., Li, B., Anderson, N.M., Wang, Y.W., Wang, L., Laroche, F., Choi, J.W., Liu, C.T., Kelliher, M.A., Feng, H., 2018. UFD1 contributes to MYC-mediated leukemia aggressiveness through suppression of the proapoptotic unfolded protein response. Leukemia 32, 2339–2351.
Humeau, J., Bravo-San Pedro, J.M., Vitale, I., Nunez, L., Villalobos, C., Kroemer, G., Senovilla, L., 2018. Calcium signaling and cell cycle: progression or death. Cell Calcium 70, 3–15.

Hwang, M.S., Schwall, C.T., Pazarentzos, E., Datler, C., Alder, N.N., Grimm, S., 2014. Mitochondrial Ca(2+) influx targets cardiolipin to disintegrate respiratory chain complex II for cell death induction. Cell Death Differ. 21, 1733–1745.

Iino, M., 1990. Biphasic Ca^{2+} dependence of inositol 1,4,5-trisphosphate-induced Ca release in smooth muscle cells of the guinea pig taenia caeci. J. Gen. Physiol. 95 (6), 1103–1122.

Iino, M., 1991. Effects of adenine nucleotides on inositol 1,4,5-trisphosphate-induced calcium release in vascular smooth muscle cells. J. Gen. Physiol. 98, 681–698.

Iommelli, F., De Rosa, V., Terlizzi, C., Monti, M., Panico, M., Fonti, R., Del Vecchio, S., 2018. Inositol trisphosphate receptor type 3-mediated enhancement of EGFR and MET cotargeting efficacy in non-small cell lung cancer detected by (18)F-fluorothymidine. Clin. Cancer Res. 24, 3126–3136.

Ivanova, H., Vervliet, T., Missiaen, L., Parys, J.B., De Smedt, H., Bultynck, G., 2014. Inositol 1,4,5-trisphosphate receptor-isoform diversity in cell death and survival. Biochim. Biophys. Acta 1843, 2164–2183.

Ivanova, H., Ritaine, A., Wagner, L., Luyten, T., Shapovalov, G., Welkenhuyzen, K., Seitaj, B., Monaco, G., De Smedt, H., Prevarskaya, N., Yule, D.I., Parys, J.B., Bultynck, G., 2016. The trans-membrane domain of Bcl-2alpha, but not its hydrophobic cleft, is a critical determinant for efficient IP3 receptor inhibition. Oncotarget 7, 55704–55720.

Ivanova, H., Kerkhofs, M., La Rovere, R.M., Bultynck, G., 2017. Endoplasmic reticulum-mitochondrial Ca(2+) fluxes underlying cancer cell survival. Front. Oncol. 7, 70.

Ivanova, H., Vervliet, T., Monaco, G., Terry, L.E., Rosa, N., Baker, M.R., Parys, J.B., Serysheva, I.I., Yule, D.I., Bultynck, G., 2019a. Bcl-2-protein family as modulators of IP3 receptors and other organellar Ca(2+) channels. In: Bultynck, G., Bootman, M.D., Berridge, M.J., Stutzmann, G.E. (Eds.), Calcium Signaling, second ed. Cold Spring Harbor Laboratory Press, New York, pp. 127–154.

Ivanova, H., Wagner 2nd, L.E., Tanimura, A., Vandermarliere, E., Luyten, T., Welkenhuyzen, K., Alzayady, K.J., Wang, L., Hamada, K., Mikoshiba, K., De Smedt, H., Martens, L., Yule, D.I., Parys, J.B., Bultynck, G., 2019b. Bcl-2 and IP3 compete for the ligand-binding domain of IP3Rs modulating Ca(2+) signaling output. Cell. Mol. Life Sci. 76 (19), 3843–3859.

Iwasaki, H., Mori, Y., Hara, Y., Uchida, K., Zhou, H., Mikoshiba, K., 2001. 2-Aminoethoxydiphenyl borate (2-APB) inhibits capacitative calcium entry independently of the function of inositol 1,4,5-trisphosphate receptors. Receptors Channels 7, 429–439.

Jayaraman, T., Marks, A.R., 1997. T cells deficient in inositol 1,4,5-trisphosphate receptor are resistant to apoptosis. Mol. Cell. Biol. 17, 3005–3012.

Jia, S., Qu, T., Wang, X., Feng, M., Yang, Y., Feng, X., Ma, R., Li, W., Hu, Y., Feng, Y., Ji, K., Li, Z., Jiang, W., Ji, J., 2017. KIAA1199 promotes migration and invasion by Wnt/beta-catenin pathway and MMPs mediated EMT progression and serves as a poor prognosis marker in gastric cancer. PLoS One 12, e0175058.

Jiang, S.X., Sato, Y., Kuwao, S., Kameya, T., 1995. Expression of bcl-2 oncogene protein is prevalent in small cell lung carcinomas. J. Pathol. 177, 135–138.

Joseph, S.K., Young, M.P., Alzayady, K., Yule, D.I., Ali, M., Booth, D.M., Hajnoczky, G., 2018. Redox regulation of type-I inositol trisphosphate receptors in intact mammalian cells. J. Biol. Chem. 293, 17464–17476.

Kaftan, E.J., Ehrlich, B.E., Watras, J., 1997. Inositol 1,4,5-trisphosphate (InsP3) and calcium interact to increase the dynamic range of InsP3 receptor-dependent calcium signaling. J. Gen. Physiol. 110, 529–538.

Kamal, A., Thao, L., Sensintaffar, J., Zhang, L., Boehm, M.F., Fritz, L.C., Burrows, F.J., 2003. A high-affinity conformation of Hsp90 confers tumour selectivity on Hsp90 inhibitors. Nature 425, 407–410.

Kang, S.S., Han, K.S., Ku, B.M., Lee, Y.K., Hong, J., Shin, H.Y., Almonte, A.G., Woo, D.H., Brat, D.J., Hwang, E.M., Yoo, S.H., Chung, C.K., Park, S.H., Paek, S.H., Roh, E.J., Lee, S.J., Park, J.Y., Traynelis, S.F., Lee, C.J., 2010. Caffeine-mediated inhibition of calcium release channel inositol 1,4,5-trisphosphate receptor subtype 3 blocks glioblastoma invasion and extends survival. Cancer Res. 70, 1173–1183.

Kania, E., Roest, G., Vervliet, T., Parys, J.B., Bultynck, G., 2017. IP3 receptor-mediated calcium signaling and its role in autophagy in cancer. Front. Oncol. 7, 140.

Karnak, D., Xu, L., 2010. Chemosensitization of prostate cancer by modulating Bcl-2 family proteins. Curr. Drug Targets 11, 699–707.

Katona, B.W., Glynn, R.A., Paulosky, K.E., Feng, Z., Davis, C.I., Ma, J., Berry, C.T., Szigety, K.M., Matkar, S., Liu, Y., Wang, H., Wu, Y., He, X., Freedman, B.D., Brady, D.C., Hua, X., 2019. Combined menin and EGFR inhibitors synergize to suppress colorectal cancer via EGFR-independent and calcium-mediated repression of SKP2 transcription. Cancer Res. 79, 2195–2207.

Kerkhofs, M., Seitaj, B., Ivanova, H., Monaco, G., Bultynck, G., Parys, J.B., 2018. Pathophysiological consequences of isoform-specific IP3 receptor mutations. Biochim. Biophys. Acta Mol. Cell Res. 1865, 1707–1717.

Kerkhofs, M., Bultynck, G., Vervliet, T., Monaco, G., 2019a. Therapeutic implications of novel peptides targeting ER-mitochondria Ca(2+)-flux systems. Drug Discov. Today 24, 1092–1103.

Kerkhofs, M., Vervloessem, T., Bittremieux, M., Bultynck, G., 2019b. Recent advances in uncovering the mechanisms contributing to BIRD-2-induced cell death in B-cell cancer cells. Cell Death Dis. 10, 42.

Khan, A.A., Soloski, M.J., Sharp, A.H., Schilling, G., Sabatini, D.M., Li, S.H., Ross, C.A., Snyder, S.H., 1996. Lymphocyte apoptosis: mediation by increased type 3 inositol 1,4,5-trisphosphate receptor. Science (New York, N.Y.) 273, 503–507.

Khan, M.T., Wagner 2nd, L., Yule, D.I., Bhanumathy, C., Joseph, S.K., 2006. Akt kinase phosphorylation of inositol 1,4,5-trisphosphate receptors. J. Biol. Chem. 281, 3731–3737.

Khan, S.A., Rossi, A.M., Riley, A.M., Potter, B.V., Taylor, C.W., 2013. Subtype-selective regulation of IP(3) receptors by thimerosal via cysteine residues within the IP(3)-binding core and suppressor domain. Biochem. J. 451, 177–184.

Kuchay, S., Giorgi, C., Simoneschi, D., Pagan, J., Missiroli, S., Saraf, A., Florens, L., Washburn, M.P., Collazo-Lorduy, A., Castillo-Martin, M., Cordon-Cardo, C., Sebti, S.M., Pinton, P., Pagano, M., 2017. PTEN counteracts FBXL2 to promote IP3R3- and Ca(2+)-mediated apoptosis limiting tumour growth. Nature 546, 554–558.

Kuchay, S., Saeed, M., Giorgi, C., Li, J., Hoffmann, H.H., Pinton, P., Rice, C.M., Pagano, M., 2018. NS5A promotes constitutive degradation of IP3R3 to counteract apoptosis induced by hepatitis C virus. Cell Rep. 25, 833–840 e3.

Kucukkaya, B., Basoglu, H., Erdag, D., Akbas, F., Susgun, S., Yalcintepe, L., 2019. Calcium homeostasis in cisplatin resistant epithelial ovarian cancer. Gen. Physiol. Biophys. 38, 353–363.

Lee, S.Y., Yoo, H.-S., Choi, H.-S., Chung, K.Y., Seo, M.-D., 2016. Structural and dynamic insights into the subtype-specific IP3-binding mechanism of the IP3 receptor. Biochem. J. 473, 3533–3543.

Lewis, R.S., 2019. Store-operated calcium channels: from function to structure and back again. In: Bultynck, G., Bootman, M.D., Berridge, M.J., Stutzmann, G.E. (Eds.), Calcium Signaling, second ed. Cold Spring Harbor Laboratory Press, New York, pp. 37–54.

Li, C., Wang, X., Vais, H., Thompson, C.B., Foskett, J.K., White, C., 2007. Apoptosis regulation by Bcl-x(L) modulation of mammalian inositol 1,4,5-trisphosphate receptor channel isoform gating. Proc. Natl. Acad. Sci. U. S. A. 104, 12565–12570.

Li, S., Balmain, A., Counter, C.M., 2018. A model for RAS mutation patterns in cancers: finding the sweet spot. Nat. Rev. Cancer 18, 767–777.

Liang, X.H., Jackson, S., Seaman, M., Brown, K., Kempkes, B., Hibshoosh, H., Levine, B., 1999. Induction of autophagy and inhibition of tumorigenesis by beclin 1. Nature 402, 672–676.

Liang, J.Q., Teoh, N., Xu, L., Pok, S., Li, X., Chu, E.S.H., Chiu, J., Dong, L., Arfianti, E., Haigh, W.G., Yeh, M.M., Ioannou, G.N., Sung, J.J.Y., Farrell, G., Yu, J., 2018. Dietary cholesterol promotes steatohepatitis related hepatocellular carcinoma through dysregulated metabolism and calcium signaling. Nat. Commun. 9, 4490.

Liu, Y., Ma, X., Fujioka, H., Liu, J., Chen, S., Zhu, X., 2019. DJ-1 regulates the integrity and function of ER-mitochondria association through interaction with IP3R3-Grp75-VDAC1. Proc. Natl. Acad. Sci. U. S. A. 116 (50), 25322–25328.

Lock, J.T., Alzayady, K.J., Yule, D.I., Parker, I., 2018. All three IP3 receptor isoforms generate Ca2+ puffs that display similar characteristics. Sci. Signal. 11 (561), eaau0344.

Lovy, A., Foskett, J.K., Cardenas, C., 2016. InsP3R, the calcium whisperer: maintaining mitochondrial function in cancer. Mol. Cell. Oncol. 3, e1185563.

Maes, K., Missiaen, L., De Smet, P., Vanlingen, S., Callewaert, G., Parys, J.B., De Smedt, H., 2000. Differential modulation of inositol 1,4,5-trisphosphate receptor type 1 and type 3 by ATP. Cell Calcium 27, 257–267.

Mak, D.O., McBride, S., Foskett, J.K., 2001. Regulation by Ca2+ and inositol 1,4,5-trisphosphate (InsP3) of single recombinant type 3 InsP3 receptor channels. Ca2+ activation uniquely distinguishes types 1 and 3 insp3 receptors. J. Gen. Physiol. 117, 435–446.

Mak, D.O., McBride, S.M., Petrenko, N.B., Foskett, J.K., 2003. Novel regulation of calcium inhibition of the inositol 1,4,5-trisphosphate receptor calcium-release channel. J. Gen. Physiol. 122, 569–581.

Malathi, K., Li, X., Krizanova, O., Ondrias, K., Sperber, K., Ablamunits, V., Jayaraman, T., 2005. Cdc2/cyclin B1 interacts with and modulates inositol 1,4,5-trisphosphate receptor (type 1) functions. J. Immunol. 175, 6205–6210.

Mangla, A., Guerra, M.T., Nathanson, M.H., 2019. Type 3 inositol 1,4,5-trisphosphate receptor: a calcium channel for all seasons. Cell Calcium 85, 102132.

Mannella, C.A., Buttle, K., Rath, B.K., Marko, M., 1998. Electron microscopic tomography of rat-liver mitochondria and their interaction with the endoplasmic reticulum. Biofactors 8, 225–228.

Marchi, S., Marinello, M., Bononi, A., Bonora, M., Giorgi, C., Rimessi, A., Pinton, P., 2012. Selective modulation of subtype III IP(3)R by Akt regulates ER Ca(2)(+) release and apoptosis. Cell Death Dis. 3, e304.

Marchi, S., Lupini, L., Patergnani, S., Rimessi, A., Missiroli, S., Bonora, M., Bononi, A., Corra, F., Giorgi, C., De Marchi, E., Poletti, F., Gafa, R., Lanza, G., Negrini, M., Rizzuto, R., Pinton, P., 2013. Downregulation of the mitochondrial calcium uniporter by cancer-related miR-25. Curr. Biol. 23, 58–63.

Marchi, S., Bittremieux, M., Missiroli, S., Morganti, C., Patergnani, S., Sbano, L., Rimessi, A., Kerkhofs, M., Parys, J.B., Bultynck, G., Giorgi, C., Pinton, P., 2017. Endoplasmic reticulum-mitochondria communication through Ca(2+) signaling: the importance of mitochondria-associated membranes (MAMs). Adv. Exp. Med. Biol. 997, 49–67.

Mataragka, S., Taylor, C.W., 2018. All three IP3 receptor subtypes generate Ca2+ puffs, the universal building blocks of IP3-evoked Ca2+ signals. J. Cell. Sci. 131 (16), jcs220848.

McCormack, J.G., 1985. Characterization of the effects of Ca2+ on the intramitochondrial Ca2+-sensitive enzymes from rat liver and within intact rat liver mitochondria. Biochem. J. 231, 581–595.

Mendes, C.C.P., Gomes, D.A., Thompson, M., Souto, N.C., Goes, T.S., Goes, A.M., Rodrigues, M.A., Gomez, M.V., Nathanson, M.H., Leite, M.F., 2005. The type III inositol 1,4,5-trisphosphate receptor preferentially transmits apoptotic Ca2+ signals into mitochondria. J. Biol. Chem. 280, 40892–40900.

Mignery, G.A., Sudhof, T.C., 1990. The ligand binding site and transduction mechanism in the inositol-1,4,5-triphosphate receptor. EMBO J. 9, 3893–3898.

Missiaen, L., De Smedt, H., Droogmans, G., Casteels, R., 1992. Ca2+ release induced by inositol 1,4,5-trisphosphate is a steady-state phenomenon controlled by luminal Ca2+ in permeabilized cells. Nature 357, 599–602.

Missiaen, L., De Smedt, H., Parys, J.B., Casteels, R., 1994. Co-activation of inositol trisphosphate-induced Ca2+ release by cytosolic Ca2+ is loading-dependent. J. Biol. Chem. 269, 7238–7242.

Missiaen, L., Parys, J.B., Sienaert, I., Maes, K., Kunzelmann, K., Takahashi, M., Tanzawa, K., De Smedt, H., 1998. Functional properties of the type-3 InsP3 receptor in 16HBE14o- bronchial mucosal cells. J. Biol. Chem. 273, 8983–8986.

Missiaen, L., Callewaert, G., De Smedt, H., Parys, J.B., 2001. 2-Aminoethoxydiphenyl borate affects the inositol 1,4,5-trisphosphate receptor, the intracellular Ca2+ pump and the non-specific Ca2+ leak from the non-mitochondrial Ca2+ stores in permeabilized A7r5 cells. Cell Calcium 29, 111–116.

Missiroli, S., Bonora, M., Patergnani, S., Giorgi, C., 2017. Novel function of the tumor suppressor PML at ER-mitochondria sites in the control of autophagy. Oncotarget 8, 81723–81724.

Mitra, K., Wunder, C., Roysam, B., Lin, G., Lippincott-Schwartz, J., 2009. A hyperfused mitochondrial state achieved at G1-S regulates cyclin E buildup and entry into S phase. Proc. Natl. Acad. Sci. U. S. A. 106, 11960–11965.

Miyakawa, T., Maeda, A., Yamazawa, T., Hirose, K., Kurosaki, T., Iino, M., 1999. Encoding of Ca2+ signals by differential expression of IP3 receptor subtypes. EMBO J. 18, 1303–1308.

Monaco, G., Decrock, E., Akl, H., Ponsaerts, R., Vervliet, T., Luyten, T., De Maeyer, M., Missiaen, L., Distelhorst, C.W., De Smedt, H., Parys, J.B., Leybaert, L., Bultynck, G., 2012. Selective regulation of IP3-receptor-mediated Ca2+ signaling and apoptosis by the BH4 domain of Bcl-2 versus Bcl-Xl. Cell Death Differ. 19, 295–309.

Monaco, G., Decrock, E., Nuyts, K., Wagner 2nd, L.E., Luyten, T., Strelkov, S.V., Missiaen, L., De Borggraeve, W.M., Leybaert, L., Yule, D.I., De Smedt, H., Parys, J.B., Bultynck, G., 2013. Alpha-helical destabilization of the Bcl-2-BH4-domain peptide abolishes its ability to inhibit the IP3 receptor. PLoS One 8, e73386.

Monkawa, T., Miyawaki, A., Sugiyama, T., Yoneshima, H., Yamamoto-Hino, M., Furuichi, T., Saruta, T., Hasegawa, M., Mikoshiba, K., 1995. Heterotetrameric complex formation of inositol 1,4,5-trisphosphate receptor subunits. J. Biol. Chem. 270, 14700–14704.

Monteith, G.R., Prevarskaya, N., Roberts-Thomson, S.J., 2017. The calcium-cancer signalling nexus. Nat. Rev. Cancer 17, 367–380.

Montero, J., Letai, A., 2018. Why do BCL-2 inhibitors work and where should we use them in the clinic? Cell Death Differ. 25, 56–64.

Morciano, G., Marchi, S., Morganti, C., Sbano, L., Bittremieux, M., Kerkhofs, M., Corricelli, M., Danese, A., Karkucinska-Wieckowska, A., Wieckowski, M.R., Bultynck, G., Giorgi, C., Pinton, P., 2018. Role of mitochondria-associated ER membranes in calcium regulation in cancer-specific settings. Neoplasia 20, 510–523.

Mound, A., Rodat-Despoix, L., Bougarn, S., Ouadid-Ahidouch, H., Matifat, F., 2013. Molecular interaction and functional coupling between type 3 inositol 1,4,5-trisphosphate receptor and BKCa channel stimulate breast cancer cell proliferation. Eur. J. Cancer 49, 3738–3751.

Mound, A., Vautrin-Glabik, A., Foulon, A., Botia, B., Hague, F., Parys, J.B., Ouadid-Ahidouch, H., Rodat-Despoix, L., 2017. Downregulation of type 3 inositol (1,4,5) -trisphosphate receptor decreases breast cancer cell migration through an oscillatory Ca(2+) signal. Oncotarget 8, 72324–72341.

Nabi, I.R., Watanabe, H., Raz, A., 1992. Autocrine motility factor and its receptor: role in cell locomotion and metastasis. Cancer Metastasis Rev. 11, 5–20.

Nguyen, N., Francoeur, N., Chartrand, V., Klarskov, K., Guillemette, G., Boulay, G., 2009. Insulin promotes the association of heat shock protein 90 with the inositol 1,4, 5-trisphosphate receptor to dampen its Ca2+ release activity. Endocrinology 150, 2190–2196.

Nunn, D.L., Taylor, C.W., 1992. Luminal Ca2+ increases the sensitivity of Ca2+ stores to inositol 1,4,5-trisphosphate. Mol. Pharmacol. 41, 115–119.

Paknejad, N., Hite, R.K., 2018. Structural basis for the regulation of inositol trisphosphate receptors by Ca(2+) and IP3. Nat. Struct. Mol. Biol. 25, 660–668.

Pantazaka, E., Taylor, C.W., 2011. Differential distribution, clustering, and lateral diffusion of subtypes of the inositol 1,4,5-trisphosphate receptor. J. Biol. Chem. 286, 23378–23387.

Parys, J.B., 2018. The multifaceted STAT3: how a transcription factor regulates Ca(2+) signaling via a degradative pathway. Cell Calcium 76, 137–139.

Parys, J.B., Bultynck, G., 2018a. Ca(2+) signaling and cell death: focus on the role of Ca(2+) signals in the regulation of cell death & survival processes in health, disease and therapy. Cell Calcium 70, 1–2.

Parys, J.B., Bultynck, G., 2018b. Calcium signaling in health, disease and therapy. Biochim. Biophys. Acta Mol. Cell Res. 1865, 1657–1659.

Parys, J.B., De Smedt, H., 2012. Inositol 1,4,5-trisphosphate and its receptors. Adv. Exp. Med. Biol. 740, 255–279.

Parys, J.B., Vervliet, T., 2020. New insights in the IP3 receptor and its regulation. Adv. Exp. Med. Biol. 1131, 243–270.

Parys, J.B., Missiaen, L., De Smedt, H., Casteels, R., 1993a. Loading dependence of inositol 1,4,5-trisphosphate-induced Ca2+ release in the clonal cell line A7r5. Implications for the mechanism of quantal Ca2+ release. J. Biol. Chem. 268, 25206–25212.

Parys, J.B., Missiaen, L., De Smedt, H., Droogmans, G., Casteels, R., 1993b. Bell-shaped activation of inositol-1,4,5-trisphosphate-induced Ca2+ release by thimerosal in permeabilized A7r5 smooth-muscle cells. Pflugers Arch. 424, 516–522.

Patel, S., Joseph, S.K., Thomas, A.P., 1999. Molecular properties of inositol 1,4,5-trisphosphate receptors. Cell Calcium 25, 247–264.

Pattingre, S., Tassa, A., Qu, X., Garuti, R., Liang, X.H., Mizushima, N., Packer, M., Schneider, M.D., Levine, B., 2005. Bcl-2 antiapoptotic proteins inhibit Beclin 1-dependent autophagy. Cell 122, 927–939.

Pearce, M.M., Wang, Y., Kelley, G.G., Wojcikiewicz, R.J., 2007. SPFH2 mediates the endoplasmic reticulum-associated degradation of inositol 1,4,5-trisphosphate receptors and other substrates in mammalian cells. J. Biol. Chem. 282, 20104–20115.

Pearce, M.M., Wormer, D.B., Wilkens, S., Wojcikiewicz, R.J., 2009. An endoplasmic reticulum (ER) membrane complex composed of SPFH1 and SPFH2 mediates the ER-associated degradation of inositol 1,4,5-trisphosphate receptors. J. Biol. Chem. 284, 10433–10445.

Pearson, M., Carbone, R., Sebastiani, C., Cioce, M., Fagioli, M., Saito, S., Higashimoto, Y., Appella, E., Minucci, S., Pandolfi, P.P., Pelicci, P.G., 2000. PML regulates p53 acetylation and premature senescence induced by oncogenic Ras. Nature 406, 207–210.

Pierro, C., Cook, S.J., Foets, T.C., Bootman, M.D., Roderick, H.L., 2014. Oncogenic K-Ras suppresses IP(3)-dependent Ca(2)(+) release through remodelling of the isoform composition of IP(3)Rs and ER luminal Ca(2)(+) levels in colorectal cancer cell lines. J. Cell Sci. 127, 1607–1619.

Pierro, C., Sneyers, F., Bultynck, G., Roderick, H.L., 2019. ER Ca(2+) release and store-operated Ca(2+) entry—partners in crime or independent actors in oncogenic transformation? Cell Calcium 82, 102061.

Pinton, P., Ferrari, D., Rapizzi, E., Di Virgilio, F., Pozzan, T., Rizzuto, R., 2001. The Ca2+ concentration of the endoplasmic reticulum is a key determinant of ceramide-induced apoptosis: significance for the molecular mechanism of Bcl-2 action. EMBO J. 20, 2690–2701.

Prakriya, M., Lewis, R.S., 2001. Potentiation and inhibition of Ca(2+) release-activated Ca(2+) channels by 2-aminoethyldiphenyl borate (2-APB) occurs independently of IP(3) receptors. J. Physiol. 536, 3–19.

Prole, D.L., Taylor, C.W., 2016. Inositol 1,4,5-trisphosphate receptors and their protein partners as signalling hubs. J. Physiol. 594, 2849–2866.

Prole, D.L., Taylor, C.W., 2019. Structure and function of IP3 receptors. In: Bultynck, G., Bootman, M.D., Berridge, M.J., Stutzmann, G.E. (Eds.), Calcium Signaling, second ed. Cold Spring Harbor Laboratory Press, New York, pp. 85–102.

Rapizzi, E., Pinton, P., Szabadkai, G., Wieckowski, M.R., Vandecasteele, G., Baird, G., Tuft, R.A., Fogarty, K.E., Rizzuto, R., 2002. Recombinant expression of the voltage-dependent anion channel enhances the transfer of Ca2+ microdomains to mitochondria. J. Cell Biol. 159, 613–624.

Rezuchova, I., Hudecova, S., Soltysova, A., Matuskova, M., Durinikova, E., Chovancova, B., Zuzcak, M., Cihova, M., Burikova, M., Penesova, A., Lencesova, L., Breza, J., Krizanova, O., 2019. Type 3 inositol 1,4,5-trisphosphate receptor has antiapoptotic and proliferative role in cancer cells. Cell Death Dis. 10, 186.

Rimessi, A., Bonora, M., Marchi, S., Patergnani, S., Marobbio, C.M., Lasorsa, F.M., Pinton, P., 2013. Perturbed mitochondrial Ca2+ signals as causes or consequences of mitophagy induction. Autophagy 9, 1677–1686.

Rizzuto, R., Brini, M., Murgia, M., Pozzan, T., 1993. Microdomains with high Ca2+ close to IP3-sensitive channels that are sensed by neighboring mitochondria. Science (New York, N.Y.) 262, 744–747.

Rizzuto, R., Pinton, P., Carrington, W., Fay, F.S., Fogarty, K.E., Lifshitz, L.M., Tuft, R.A., Pozzan, T., 1998. Close contacts with the endoplasmic reticulum as determinants of mitochondrial Ca2+ responses. Science (New York, N.Y.) 280, 1763–1766.

Robb-Gaspers, L.D., Burnett, P., Rutter, G.A., Denton, R.M., Rizzuto, R., Thomas, A.P., 1998. Integrating cytosolic calcium signals into mitochondrial metabolic responses. EMBO J. 17, 4987–5000.

Roberts-Thomson, S.J., Chalmers, S.B., Monteith, G.R., 2019. The calcium-signaling toolkit in cancer: remodeling and targeting. In: Bultynck, G., Bootman, M.D., Berridge, M.J., Stutzmann, G.E. (Eds.), Calcium Signaling, second ed. Cold Spring Harbor Laboratory Press, New York, pp. 479–498.

Rong, Y.P., Aromolaran, A.S., Bultynck, G., Zhong, F., Li, X., McColl, K., Matsuyama, S., Herlitze, S., Roderick, H.L., Bootman, M.D., Mignery, G.A., Parys, J.B., De Smedt, H., Distelhorst, C.W., 2008. Targeting Bcl-2-IP3 receptor interaction to reverse Bcl-2's inhibition of apoptotic calcium signals. Mol. Cell 31, 255–265.

Rong, Y.P., Bultynck, G., Aromolaran, A.S., Zhong, F., Parys, J.B., De Smedt, H., Mignery, G.A., Roderick, H.L., Bootman, M.D., Distelhorst, C.W., 2009. The BH4 domain of Bcl-2 inhibits ER calcium release and apoptosis by binding the regulatory and coupling domain of the IP3 receptor. Proc. Natl. Acad. Sci. U. S. A. 106, 14397–14402.

Rusinek, D., Swierniak, M., Chmielik, E., Kowal, M., Kowalska, M., Cyplinska, R., Czarniecka, A., Piglowski, W., Korfanty, J., Chekan, M., Krajewska, J., Szpak-Ulczok, S., Jarzab, M., Widlak, W., Jarzab, B., 2015. BRAFV600E-associated gene expression profile: early changes in the transcriptome, based on a transgenic mouse model of papillary thyroid carcinoma. PLoS One 10, e0143688.

Rutter, G.A., Denton, R.M., 1988. Regulation of NAD+-linked isocitrate dehydrogenase and 2-oxoglutarate dehydrogenase by Ca2+ ions within toluene-permeabilized rat heart mitochondria. Interactions with regulation by adenine nucleotides and NADH/NAD+ ratios. Biochem. J. 252, 181–189.

Sakakura, C., Hagiwara, A., Fukuda, K., Shimomura, K., Takagi, T., Kin, S., Nakase, Y., Fujiyama, J., Mikoshiba, K., Okazaki, Y., Yamagishi, H., 2003. Possible involvement of inositol 1,4,5-trisphosphate receptor type 3 (IP3R3) in the peritoneal dissemination of gastric cancers. Anticancer Res. 23, 3691–3697.

Sammels, E., Parys, J.B., Missiaen, L., De Smedt, H., Bultynck, G., 2010. Intracellular Ca2+ storage in health and disease: a dynamic equilibrium. Cell Calcium 47, 297–314.

Sano, R., Hou, Y.-C.C., Hedvat, M., Correa, R.G., Shu, C.-W., Krajewska, M., Diaz, P.W., Tamble, C.M., Quarato, G., Gottlieb, R.A., Yamaguchi, M., Nizet, V., Dahl, R., Thomas, D.D., Tait, S.W., Green, D.R., Fisher, P.B., Matsuzawa, S.-I., Reed, J.C., 2012. Endoplasmic reticulum protein BI-1 regulates Ca^{+}-mediated bioenergetics to promote autophagy. Genes Dev. 26, 1041–1054.

Schild, L., Keilhoff, G., Augustin, W., Reiser, G., Striggow, F., 2001. Distinct Ca2+ thresholds determine cytochrome c release or permeability transition pore opening in brain mitochondria. FASEB J. 15, 565–567.

Scorrano, L., De Matteis, M.A., Emr, S., Giordano, F., Hajnoczky, G., Kornmann, B., Lackner, L.L., Levine, T.P., Pellegrini, L., Reinisch, K., Rizzuto, R., Simmen, T., Stenmark, H., Ungermann, C., Schuldiner, M., 2019. Coming together to define membrane contact sites. Nat. Commun. 10, 1287.

Sherry, S.T., Ward, M.H., Kholodov, M., Baker, J., Phan, L., Smigielski, E.M., Sirotkin, K., 2001. dbSNP: the NCBI database of genetic variation. Nucleic Acids Res. 29, 308–311.

Shibao, K., Fiedler, M.J., Nagata, J., Minagawa, N., Hirata, K., Nakayama, Y., Iwakiri, Y., Nathanson, M.H., Yamaguchi, K., 2010. The type III inositol 1,4,5-trisphosphate receptor is associated with aggressiveness of colorectal carcinoma. Cell Calcium 48, 315–323.

Singh, A., Chagtoo, M., Tiwari, S., George, N., Chakravarti, B., Khan, S., Lakshmi, S., Godbole, M.M., 2017a. Inhibition of inositol 1, 4, 5-trisphosphate receptor induce breast cancer cell death through deregulated autophagy and cellular bioenergetics. J. Cell. Biochem. 118, 2333–2346.

Singh, A., Sharma, R.K., Chagtoo, M., Agarwal, G., George, N., Sinha, N., Godbole, M.M., 2017b. 1H NMR metabolomics reveals association of high expression of inositol 1, 4, 5 trisphosphate receptor and metabolites in breast cancer patients. PLoS One 12, e0169330.

Sneyers, F., Rosa, N., Bultynck, G., 2019. Type 3 IP3 receptors driving oncogenesis. Cell Calcium 86, 102141.

Soghoian, D., Jayaraman, V., Silane, M., Berenstein, A., Jayaraman, T., 2005. Inositol 1,4,5-trisphosphate receptor phosphorylation in breast cancer. Tumour Biol. 26, 207–212.

Sotgia, F., Whitaker-Menezes, D., Martinez-Outschoorn, U.E., Flomenberg, N., Birbe, R.C., Witkiewicz, A.K., Howell, A., Philp, N.J., Pestell, R.G., Lisanti, M.P., 2012. Mitochondrial metabolism in cancer metastasis: visualizing tumor cell mitochondria and the "reverse Warburg effect" in positive lymph node tissue. Cell Cycle 11, 1445–1454.

Spruce, B.A., Campbell, L.A., McTavish, N., Cooper, M.A., Appleyard, M.V., O'Neill, M., Howie, J., Samson, J., Watt, S., Murray, K., McLean, D., Leslie, N.R., Safrany, S.T., Ferguson, M.J., Peters, J.A., Prescott, A.R., Box, G., Hayes, A., Nutley, B., Raynaud, F., Downes, C.P., Lambert, J.J., Thompson, A.M., Eccles, S., 2004. Small molecule antagonists of the sigma-1 receptor cause selective release of the death program in tumor and self-reliant cells and inhibit tumor growth in vitro and in vivo. Cancer Res. 64, 4875–4886.

Su, A.I., Wiltshire, T., Batalov, S., Lapp, H., Ching, K.A., Block, D., Zhang, J., Soden, R., Hayakawa, M., Kreiman, G., Cooke, M.P., Walker, J.R., Hogenesch, J.B., 2004. A gene atlas of the mouse and human protein-encoding transcriptomes. Proc. Natl. Acad. Sci. U. S. A. 101, 6062–6067.

Sung, P.J., Tsai, F.D., Vais, H., Court, H., Yang, J., Fehrenbacher, N., Foskett, J.K., Philips, M.R., 2013. Phosphorylated K-Ras limits cell survival by blocking Bcl-xL sensitization of inositol trisphosphate receptors. Proc. Natl. Acad. Sci. U. S. A. 110, 20593–20598.

Szabadkai, G., Bianchi, K., Varnai, P., De Stefani, D., Wieckowski, M.R., Cavagna, D., Nagy, A.I., Balla, T., Rizzuto, R., 2006. Chaperone-mediated coupling of endoplasmic reticulum and mitochondrial Ca2+ channels. J. Cell Biol. 175, 901–911.

Szado, T., Vanderheyden, V., Parys, J.B., De Smedt, H., Rietdorf, K., Kotelevets, L., Chastre, E., Khan, F., Landegren, U., Soderberg, O., Bootman, M.D., Roderick, H.L., 2008. Phosphorylation of inositol 1,4,5-trisphosphate receptors by protein kinase B/Akt inhibits Ca2+ release and apoptosis. Proc. Natl. Acad. Sci. U. S. A. 105, 2427–2432.

Szalai, G., Krishnamurthy, R., Hajnoczky, G., 1999. Apoptosis driven by IP(3)-linked mitochondrial calcium signals. EMBO J. 18, 6349–6361.

Szatkowski, C., Parys, J.B., Ouadid-Ahidouch, H., Matifat, F., 2010. Inositol 1,4, 5-trisphosphate-induced Ca2+ signalling is involved in estradiol-induced breast cancer epithelial cell growth. Mol. Cancer 9, 156.

Taylor, C.W., Konieczny, V., 2016. IP3 receptors: take four IP3 to open. Sci. Signal. 9, pe1.

Taylor, C.W., Genazzani, A.A., Morris, S.A., 1999. Expression of inositol trisphosphate receptors. Cell Calcium 26, 237–251.

Terry, L.E., Alzayady, K.J., Furati, E., Yule, D.I., 2018. Inositol 1,4,5-trisphosphate receptor mutations associated with human disease. Messenger 6, 29–44.

Testa, J.R., Cheung, M., Pei, J., Below, J.E., Tan, Y., Sementino, E., Cox, N.J., Dogan, A.U., Pass, H.I., Trusa, S., Hesdorffer, M., Nasu, M., Powers, A., Rivera, Z., Comertpay, S., Tanji, M., Gaudino, G., Yang, H., Carbone, M., 2011. Germline BAP1 mutations predispose to malignant mesothelioma. Nat. Genet. 43, 1022–1025.

Thillaiappan, N.B., Chavda, A.P., Tovey, S.C., Prole, D.L., Taylor, C.W., 2017. Ca(2+) signals initiate at immobile IP3 receptors adjacent to ER-plasma membrane junctions. Nat. Commun. 8, 1505.

Thomas, D., Lipp, P., Tovey, S.C., Berridge, M.J., Li, W., Tsien, R.Y., Bootman, M.D., 2000. Microscopic properties of elementary Ca2+ release sites in non-excitable cells. Curr. Biol. 10, 8–15.

Tiwari, A., Schneider, M., Fiorino, A., Haider, R., Okoniewski, M.J., Roschitzki, B., Uzozie, A., Menigatti, M., Jiricny, J., Marra, G., 2013. Early insights into the function of KIAA1199, a markedly overexpressed protein in human colorectal tumors. PLoS One 8, e69473.

Tosatto, A., Sommaggio, R., Kummerow, C., Bentham, R.B., Blacker, T.S., Berecz, T., Duchen, M.R., Rosato, A., Bogeski, I., Szabadkai, G., Rizzuto, R., Mammucari, C., 2016. The mitochondrial calcium uniporter regulates breast cancer progression via HIF-1alpha. EMBO Mol. Med. 8, 569–585.

Tu, H., Wang, Z., Nosyreva, E., De Smedt, H., Bezprozvanny, I., 2005. Functional characterization of mammalian inositol 1,4,5-trisphosphate receptor isoforms. Biophys. J. 88, 1046–1055.

Ueasilamongkol, P., Khamphaya, T., Guerra, M.T., Rodrigues, M.A., Gomes, D.A., Kong, Y., Wei, W., Jain, D., Trampert, D.C., Ananthanarayanan, M., Banales, J.M., Roberts, L.R., Farshidfar, F., Nathanson, M.H., Weerachayaphorn, J., 2020. Type 3 inositol 1,4,5-trisphosphate receptor is increased and enhances malignant properties in cholangiocarcinoma. Hepatology 71 (2), 583–599.

Vanlingen, S., Sipma, H., De Smet, P., Callewaert, G., Missiaen, L., De Smedt, H., Parys, J.B., 2000. Ca2+ and calmodulin differentially modulate myo-inositol 1,4, 5-trisphosphate (IP3)-binding to the recombinant ligand-binding domains of the various IP3 receptor isoforms. Biochem. J. 346 (Pt 2), 275–280.

Vautrin-Glabik, A., Botia, B., Kischel, P., Ouadid-Ahidouch, H., Rodat-Despoix, L., 2018. IP3R3 silencing induced actin cytoskeletal reorganization through ARHGAP18/ RhoA/mDia1/FAK pathway in breast cancer cell lines. Biochim. Biophys. Acta Mol. Cell Res. 1865, 945–958.

Vaux, D.L., Cory, S., Adams, J.M., 1988. Bcl-2 gene promotes haemopoietic cell survival and cooperates with c-myc to immortalize pre-B cells. Nature 335, 440–442.

Vermassen, E., Van Acker, K., Annaert, W.G., Himpens, B., Callewaert, G., Missiaen, L., De Smedt, H., Parys, J.B., 2003. Microtubule-dependent redistribution of the type-1 inositol 1,4,5-trisphosphate receptor in A7r5 smooth muscle cells. J. Cell Sci. 116, 1269–1277.

Vermassen, E., Parys, J.B., Mauger, J.P., 2004. Subcellular distribution of the inositol 1,4,5-trisphosphate receptors: functional relevance and molecular determinants. Biol. Cell 96, 3–17.

Vervliet, T., Parys, J.B., Bultynck, G., 2016. Bcl-2 proteins and calcium signaling: complexity beneath the surface. Oncogene 35, 5079–5092.

Vervloessem, T., Yule, D.I., Bultynck, G., Parys, J.B., 2015. The type 2 inositol 1,4,5-trisphosphate receptor, emerging functions for an intriguing Ca(2)(+)-release channel. Biochim. Biophys. Acta 1853, 1992–2005.

Wacquier, B., Combettes, L., Van Nhieu, G.T., Dupont, G., 2016. Interplay between intracellular Ca(2+) oscillations and Ca(2+)-stimulated mitochondrial metabolism. Sci. Rep. 6, 19316.

Wacquier, B., Combettes, L., Dupont, G., 2019. Cytoplasmic and mitochondrial calcium signaling: a two-way relationship. In: Bultynck, G., Bootman, M.D., Berridge, M.J., Stutzmann, G.E. (Eds.), Calcium Signaling, second ed. Cold Spring Harbor Laboratory Press, New York, pp. 183–200.

Wagner 2nd, L.E., Yule, D.I., 2012. Differential regulation of the InsP(3) receptor type-1 and -2 single channel properties by InsP(3), Ca(2)(+) and ATP. J. Physiol. 590, 3245–3259.

Wan, B., LaNoue, K.F., Cheung, J.Y., Scaduto Jr., R.C., 1989. Regulation of citric acid cycle by calcium. J. Biol. Chem. 264, 13430–13439.

Wang, G., Liu, G., Wang, X., Sethi, S., Ali-Fehmi, R., Abrams, J., Zheng, Z., Zhang, K., Ethier, S., Yang, Z.Q., 2012. ERLIN2 promotes breast cancer cell survival by modulating endoplasmic reticulum stress pathways. BMC Cancer 12, 225.

White, C., Li, C., Yang, J., Petrenko, N.B., Madesh, M., Thompson, C.B., Foskett, J.K., 2005. The endoplasmic reticulum gateway to apoptosis by Bcl-X(L) modulation of the InsP3R. Nat. Cell Biol. 7, 1021–1028.

Williams, A., Hayashi, T., Wolozny, D., Yin, B., Su, T.C., Betenbaugh, M.J., Su, T.P., 2016. The non-apoptotic action of Bcl-xL: regulating Ca(2+) signaling and bioenergetics at the ER-mitochondrion interface. J. Bioenerg. Biomembr. 48, 211–225.

Wojcikiewicz, R.J., 1995. Type I, II, and III inositol 1,4,5-trisphosphate receptors are unequally susceptible to down-regulation and are expressed in markedly different proportions in different cell types. J. Biol. Chem. 270, 11678–11683.

Wojcikiewicz, R.J., Furuichi, T., Nakade, S., Mikoshiba, K., Nahorski, S.R., 1994a. Muscarinic receptor activation down-regulates the type I inositol 1,4,5-trisphosphate receptor by accelerating its degradation. J. Biol. Chem. 269, 7963–7969.

Wojcikiewicz, R.J., Tobin, A.B., Nahorski, S.R., 1994b. Muscarinic receptor-mediated inositol 1,4,5-trisphosphate formation in SH-SY5Y neuroblastoma cells is regulated acutely by cytosolic Ca2+ and by rapid desensitization. J. Neurochem. 63, 177–185.

Wright, F.A., Wojcikiewicz, R.J.H., 2016. Chapter 4- inositol 1,4,5-trisphosphate receptor ubiquitination. Prog. Mol. Biol. Transl. Sci. 141, 141–159.

Wu, Z., Bowen, W.D., 2008. Role of sigma-1 receptor C-terminal segment in inositol 1,4,5-trisphosphate receptor activation: constitutive enhancement of calcium signaling in MCF-7 tumor cells. J. Biol. Chem. 283, 28198–28215.

Yager, J.D., Davidson, N.E., 2006. Estrogen carcinogenesis in breast cancer. N. Engl. J. Med. 354, 270–282.

Yang, J., Vais, H., Gu, W., Foskett, J.K., 2016. Biphasic regulation of InsP3 receptor gating by dual Ca2+ release channel BH3-like domains mediates Bcl-xL control of cell viability. Proc. Natl. Acad. Sci. U. S. A. 113, E1953–E1962.

Yang, Y.C., Chang, T.Y., Chen, T.C., Lin, W.S., Chang, S.C., Lee, Y.J., 2017. ITPR3 gene haplotype is associated with cervical squamous cell carcinoma risk in Taiwanese women. Oncotarget 8, 10085–10090.

Yoboue, E.D., Sitia, R., Simmen, T., 2018. Redox crosstalk at endoplasmic reticulum (ER) membrane contact sites (MCS) uses toxic waste to deliver messages. Cell Death Dis. 9, 331.

Yule, D.I., Betzenhauser, M.J., Joseph, S.K., 2010. Linking structure to function: recent lessons from inositol 1,4,5-trisphosphate receptor mutagenesis. Cell Calcium 47, 469–479.

Zannetti, A., Iommelli, F., Fonti, R., Papaccioli, A., Sommella, J., Lettieri, A., Pirozzi, G., Bianco, R., Tortora, G., Salvatore, M., Del Vecchio, S., 2008. Gefitinib induction of in vivo detectable signals by Bcl-2/Bcl-xL modulation of inositol trisphosphate receptor type 3. Clin. Cancer Res. 14, 5209–5219.

Zhang, G., Jiang, G., Wang, C., Zhong, K., Zhang, J., Xue, Q., Li, X., Jin, H., Li, B., 2016. Decreased expression of microRNA-320a promotes proliferation and invasion of non-small cell lung cancer cells by increasing VDAC1 expression. Oncotarget 7, 49470–49480.

Zhong, F., Harr, M.W., Bultynck, G., Monaco, G., Parys, J.B., De Smedt, H., Rong, Y.P., Molitoris, J.K., Lam, M., Ryder, C., Matsuyama, S., Distelhorst, C.W., 2011. Induction of Ca(2)+-driven apoptosis in chronic lymphocytic leukemia cells by peptide-mediated disruption of Bcl-2-IP3 receptor interaction. Blood 117, 2924–2934.

CHAPTER FIVE

On the role of sphingolipids in cell survival and death

Elisabetta Iessi[a,†], Matteo Marconi[a,†], Valeria Manganelli[b], Maurizio Sorice[b], Walter Malorni[a,c,*], Tina Garofalo[b,‡], Paola Matarrese[a,‡]

[a]Center for Gender-Specific Medicine, Oncology Unit, Istituto Superiore di Sanità, Rome, Italy
[b]Department of Experimental Medicine, Sapienza University, Rome, Italy
[c]Department of Biology, University of Rome Tor Vergata, Rome, Italy
*Corresponding author: e-mail address: valter.malorni@uniroma2.it

Contents

† To be considered as first investigators.
‡ To be considered as senior authors.

International Review of Cell and Molecular Biology, Volume 351
ISSN 1937-6448
https://doi.org/10.1016/bs.ircmb.2020.02.004

Abstract

Sphingolipids, universal components of biological membranes of all eukaryotic organisms, from yeasts to mammals, in addition of playing a structural role, also play an important part of signal transduction pathways. They participate or, also, ignite several fundamental subcellular signaling processes but, more in general, they directly contribute to key biological activities such as cell motility, growth, senescence, differentiation as well as cell fate, i.e., survival or death. The sphingolipid metabolic pathway displays an intricate network of reactions that result in the formation of multiple sphingolipids, including ceramide, and sphingosine-1-phosphate. Different sphingolipids, that have key roles in determining cell fate, can induce opposite effects: as a general rule, sphingosine-1-phosphate promotes cell survival and differentiation, whereas ceramide is known to induce apoptosis. Furthermore, together with cholesterol, sphingolipids also represent the basic lipid component of lipid rafts, cholesterol- and sphingolipid-enriched membrane microdomains directly involved in cell death and survival processes. In this review, we briefly describe the characteristics of sphingolipids and lipid membrane microdomains. In particular, we will consider the involvement of various sphingolipids per se and of lipid rafts in apoptotic pathway, both intrinsic and extrinsic, in nonapoptotic cell death, in autophagy, and in cell differentiation. In addition, their roles in the most common physiological and pathological contexts either as pathogenetic elements or as biomarkers of diseases will be considered. We would also hint how the manipulation of sphingolipid metabolism could represent a potential therapeutic target to be investigated and functionally validated especially for those diseases for which therapeutic options are limited or ineffective.

Abbreviations

ALL	acute lymphoblastic leukemia
AD	Alzheimer's disease
APP	amyloid precursor protein
AA	arachidonic acid
Aβ	β-amyloid peptide
BACE1	β-secretase
Cer	ceramide
***Cer*S**	ceramide synthase
CDases	ceramidases
NMIIA	ceramide–RIPK1–nonmuscle myosin IIA
JNK	C-Jun NH_2-terminal kinase
CANX	calnexin
DRs	death receptors
DHA	docosahexaenoic acid
DRMs	detergent-resistant membranes
dh*Cer*	dihydroceramide
dhSph	dihydrosphingolipids
DISC	death-inducing signaling complex
Mdivi-1	3-(2,4-dichloro-5-methoxyphenyl)-2,3-dihydro-2-thioxo-4(1H)-quinazolinone

DLP1	dynamin-like protein 1
ER	endoplasmic reticulum
Gal*Cer*	galacto-glycoconjugate
GD	Gaucher's disease
GBA	glucocerebrosidase gene
Glc*Cer*	glucosylceramides
GSL	glycosphingolipid
GPRC5B	G protein-coupled receptor family C group 5 member B
IRE-1	inositol-requiring element 1
LSD	lipid storage diseases
LMP	lysosomal membrane permeabilization
AIF	mitochondrial-associated apoptosis-inducing factor
MAMs	mitochondria-associated membranes
MA-nSMase	mitochondria-associated neutral sphingomyelinase
MFN2	mitofusin 2
MLKL	mixed lineage protein kinase domain-like protein
NPC	Niemann Pick type C disease
PD	Parkinson disease
PI3P	phosphatidylinositol 3-phosphate
PLAD	pre-ligand assembly domain
PUFA	polyunsaturated fatty acid
PrP	prion protein
PrPC	prion cellular protein
PCD	programmed cell death
ROS	reactive oxygen species
SLs	sphingolipids
SM	sphingomyelin
SMase	sphingomyelinase
SMS	sphingomyelin synthase
S1P	sphingosine-1-phosphate
ST8SIA1	GD3-synthase
TCR	T-cell receptor
IP3R3	3 inositol 1,4,5-trisphosphate receptor
TNF	tumor necrosis factor
TRAIL	TNF-related apoptosis-inducing ligand
VAPs	vesicle-associated proteins
VDAC1	voltage-dependent anion channel 1
WIPI1	WD repeat protein phosphoinositide interacting 1

1. Introduction

Sphingolipids (SLs), first discovered in brain extracts in 1876, are polar lipids in which the molecular skeleton is represented by the sphingosine molecule, an unsaturated long chain amino alcohol. SLs are very similar to phospholipids but are different in the presence of sphingosine instead of glycerol. Ceramide (*Cer*) is the invariant unit of all sphingolipids and constitutes its hydrophobic portion. Depending on the hydrophilic part, there are three different categories of sphingolipids: sphingomyelin (SM), cerebrosides, and gangliosides (Bartke and Hannun, 2009). The SL metabolic pathway displays a complex network of reactions that result in the formation of multiple SLs, including *Cer*, dihydroceramide (dh*Cer*), and sphingosine-1-phosphate (S1P). The reactions of SL catabolism take place in lysosomes and occur thanks to the activity of the hydrolases that have optimal pH values between 3.5 and 5.5.

SLs are important constituents of biological membranes. However, their role is not merely structural. In fact, they are bioactive lipids of relevance in signal transduction pathways playing a central role in several fundamental biological processes such as cell motility, growth, senescence, differentiation, and fate (Hannun and Obeid, 2008). This distinguishes them from other lipids with exclusively structural and/or energetic functions. SLs are found in cellular membranes, lipoproteins, and other lipid-rich structures. They are synthesized in the endoplasmic reticulum (ER) and Golgi apparatus (Hannun and Obeid, 2018) and participate to membrane scrambling among intracellular organelles. Their importance in cell physiology is underscored by the fact that changes in their metabolic flux impact organelle physiology and pathology (Ouasti et al., 2007). In particular, glycosphingolipids, SM and *Cer*, together with phospholipids, are able to regulate vesicle flow and membrane fusion and fission, including plasma membranes and mitochondrial membranes (Bieberich, 2018). For example, it has been reported that deficiency of *Cer* and phosphoethanolamine in *Drosophila* leads to defects at the plasma membrane, resulting in accelerated aging, while the increase of *Cer* affects homeostasis requiring metabolic adaptations that involve glycolytic and lipolytic pathways (Nirala et al., 2013; Rao et al., 2007). Of importance, SLs also represent the basic lipid component, together with cholesterol, of lipid rafts, structural and functional structures that play essential roles in cell life and death.

2. Lipid rafts and mitochondrial raft-like microdomains

Lipid rafts, defined as cholesterol- and SLs-enriched membrane microdomains, are essential components present in plasma membranes in all eukaryotic cells, playing important roles that include cellular stabilization and signaling (Brown and London, 2000; Pike, 2003). Generally, lipid rafts are ranging from 10 to 200 nm, enriched mainly in cholesterol (3–5-fold respect surrounding bilayer) and SLs such as sphingomyelin (SM), and gangliosides, which are typically elevated by 50% compared to the plasma membrane (Pike, 2006; Santos and Preta, 2018). SM and gangliosides are the major SLs present in these small membrane microdomains. SM emerges as an important molecule to generate bioactive SLs by *Cer.* Sphingomyelin synthase (SMS) is the enzyme that generates SM and diacylglycerol from phosphatidylcholine and *Cer.* A large amount of evidence recognized that both SM and SMS are signaling molecules involved in cell migration, apoptosis, autophagy as well as, more in general, in human disorders such as cancer and neurodegenerative disorders (Gabande-Rodriguez et al., 2014; Maceyka and Spiegel, 2014). Rafts are characteristically insoluble in cold, nonionic detergents and can be experimentally isolated as detergent-resistant membranes (DRMs) (Adam et al., 2008). Lipid rafts can act as concentrating platforms for individual receptors that are activated by ligand binding. One subclass of lipid rafts, discovered in the 1950s, was found as cell surface invaginations, named caveolae, which are formed by the polymerization of caveolins (Yamada, 1955). Caveolae are involved in endocytosis of different proteins. Caveolae also play a role in signal transduction, but they can be considered not essential, since several cell types lack caveolin, including lymphocytes and neurons. Lipid rafts are not solely confined to the plasma membrane. Rather, they are present as raft-like microdomain compartments, in internal organelles, including mitochondria (Garofalo et al., 2005, 2015), endoplasmic reticulum (ER) (Boslem et al., 2013; Muniz et al., 2001), and at nuclear level (Cascianelli et al., 2008). For instance, raft-associated GD3 ganglioside, normally localized at the plasma membrane of T lymphocytes, can be redistributed to mitochondrial membranes by microtubules following CD95/Fas triggering, where they contribute to apoptogenic pathways (Sorice et al., 2009). Although rafts from these internal structures have not yet been characterized in a global way, significant progress is evident. Raft-like microdomains are enriched in gangliosides and cholesterol (although with a content lower as compared to plasma membrane), but with a relatively low content of phospholipids (Garofalo et al., 2015). Membrane rafts isolated from mitochondria are envisaged as lateral assemblies of specific lipids and proteins contain the voltage-dependent anion channel 1 (VDAC1) and the fission protein hFis and can recruit other proteins when the cell death program is ignited (Ciarlo et al., 2010; Garofalo et al., 2005). For example, cardiolipin, an essential constituent of functional raft microdomains localized at contact sites between the inner and outer mitochondrial membranes, seems able to regulate apoptosis by integrating signals from a variety of apoptosis-inducing proteins (Scorrano, 2008; Sorice et al., 2009). Raft-like microdomains appear also to be involved in a series of mitochondrial functions, including oxidative phosphorylation, ATP production and membrane "scrambling," participating in cell death pathways and recruitment of proteins to mitochondria. Mitochondrial raft-like microdomains are analogous to lipid rafts on the plasma membrane, since they represent preferential sites on the mitochondrial membrane where key reactions can be catalyzed. The inclusion of proteins into lipid rafts is tightly dependent on lipid composition. Thus, based on this assumption, it was also demonstrated that mitochondrial raft-like microdomains are essential in the extension of mitochondrial interconnected network due to the chemical inhibition of the GTPase activity of the dynamin-like protein 1 (DLP1) by 3-(2,4-dichloro-5-methoxyphenyl)-2,3-dihydro-2-thioxo-4(1H)-quinazolinone (Mdivi-1), a selective inhibitor of mitochondrial fission process (Ciarlo et al., 2018). It is known that mitochondria fusion and fission are strictly related, since the inhibition of one of these processes leads to an increase of the other. Thus, recruitment of mitofusin 2 (MFN2) into lipid rafts through the association with GD3 is mandatory in Mdivi-1-induced mitochondria network organization, i.e., in promoting the fusion of the outer mitochondrial membranes.

The underlying mechanisms coupling SLs and more specifically *Cer* to mitochondrial functions remain poorly understood. Early studies identified several *Cer*-producing enzyme activities in mitochondria and mitochondria-associated membranes (MAMs), which include ceramide synthase (*Cer*S) (Mullen et al., 2012; Senkal et al., 2007), ceramidases (CDases) (Bionda et al., 2004; El Bawab et al., 2000) and, most recently, a mitochondria-associated neutral sphingomyelinase (MA-nSMase) (Wu et al., 2010) (Fig. 1). Finally, the implication of *Cer* in this context should be underscored. In fact, studies using pharmacological inhibitors to block *Cer* generation or performed in genetically modified organisms and/or cells where *Cer* generation is impaired have strongly supported a critical role of *Cer* accumulation in the progression of mitochondrial apoptosis and mitophagy, defining ceramide as a bona fide transducer of mitochondrial function (Hernández-Corbacho et al., 2017).

More recently, the presence and relevance of lipid rafts at ER level has also been suggested. Raft-like microdomains in the ER are characterized by the presence of two proteins that localize at the ER only, the prohibitins erlin-1 and erlin-2 (Browman et al., 2006). In addition, DRMs isolated from

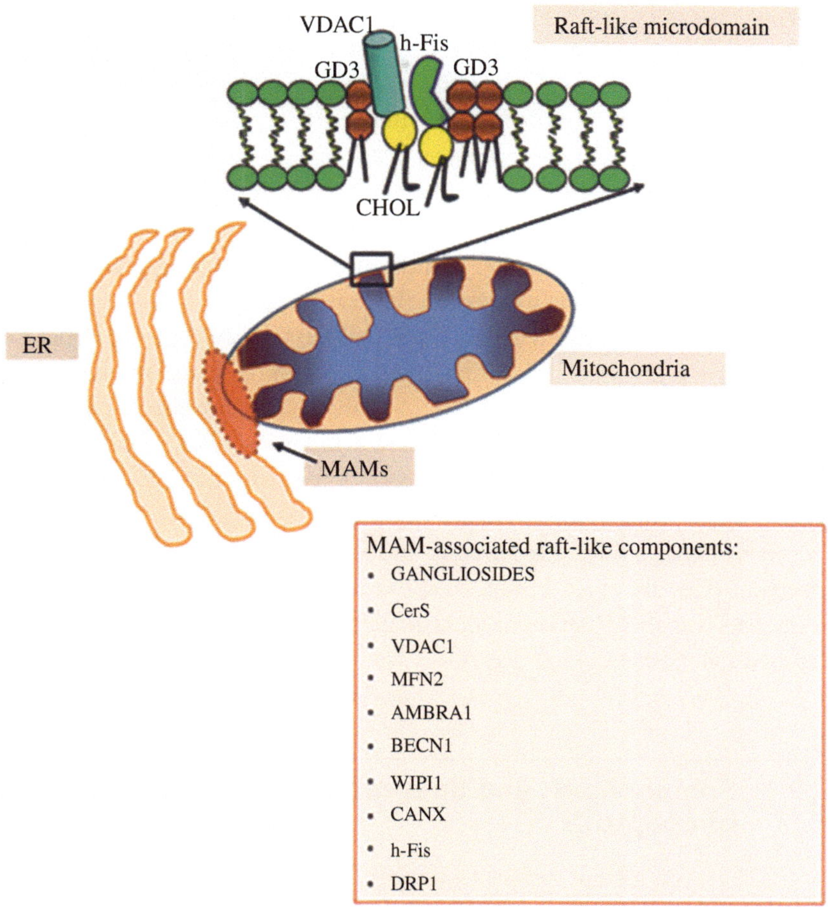

Fig. 1 Schematic drawing depicting raft-like microdomains associated with mitochondria.

the ER sub-structures known as MAM were recently shown to contain the novel ligand-responsive sigma-1 receptor molecular chaperone and the type 3 inositol 1,4,5-trisphosphate receptor (IP3R3) (Hayashi and Fujimoto, 2010). Gangliosides can reside at the MAMs, where their effect on opposite cell fate decisions, survive or die, could depend on their local concentration, structural characteristics and sugar modifications (Annunziata et al., 2018).

Finally, the key role of lipid rafts in nuclear structure and function has also been suggested. In the nuclear membrane and nuclear matrix, lipids regulate fluidity, while they participate in cellular signaling in chromatin-associated activity (Albi and Villani, 2009; Cascianelli et al., 2008). Furthermore, cellular nuclei contain high levels of phosphatidylcholine and SM, which are partially associated with cholesterol and proteins to form lipid–protein complexes. Two pools of cholesterol co-exist in chromatin: a SM-free cholesterol pool that does not change during cellular proliferation and a SM-linked cholesterol pool that can change during the S-phase of the cell cycle in relation to SMase activation (Albi et al., 2003). It is known that nuclear lipids play a role in proliferation, differentiation, and apoptotic process. In the nuclear envelope, GM1, a typical raft glycosphingolipid, was shown to regulate both Ca^{2+} influx into the nucleus and to affect epigenetic regulation of gene expression (Itokazu et al., 2017; Ledeen and Wu, 2008). Moreover, Garofalo et al. (2007) observed that in lymphoblastoid T cells, after CD95/Fas triggering, the mitochondrial raft marker GD3), was detectable at nuclear level. In fact small mitochondria could act as "cargo boats" transporting GD3 into the nucleus, possibly together with other pro-apoptotic molecules, contributing to apoptosis execution (Garofalo et al., 2007). The presence of this molecule in the nucleus has already been reported in neurons, where, during α-amyloid-induced apoptosis, it co-localizes with chromatin (Copani et al., 2002). In addition, a co-localization of GD3 with the nucleus was also observed in neuronal cerebellar granule cells undergoing apoptosis (Melchiorri et al., 2002).

3. Sphingolipids and lipid rafts in the regulation of apoptosis

Cells have several modes to trigger cell death, which can occur in a programmed and highly controlled fashion, known as "programmed cell death" (PCD). These types of cell death (PCD) are involved in the regulation of homeostasis and integrity of tissues in adult unicellular and multicellular organisms as well as in the correct development of organs in mammals (Fuchs and Steller, 2011). In fact, aberrant regulation of these cell death processes can lead to autoimmune diseases (Rossin et al., 2019), neurodegenerative disorders (Radak et al., 2017) or cancer development (Fulda, 2009; Plati et al., 2008).

Apoptosis, the most renowned and morphologically distinct of these programmed death processes, can briefly be described herein as characterized by shrinkage of the cell, condensation and fragmentation of DNA, preservation of organelle structure and plasma membrane integrity, and disintegration of the cell through blebbing of the plasma membrane, which release into the surrounding microenvironment small vesicles, the "apoptotic bodies," engulfed by phagocytes and neighboring cells (Hacker, 2000; Hengartner, 2000; Pistritto et al., 2016). Different stimuli may activate the apoptotic pathway. They include the binding of extracellular ligands to death receptors (DRs) at the cell surface, which stimulates the extrinsic pathway, and a variety of receptor-independent stimuli, such as UV- or gamma-irradiation, free radicals, viral infections, cellular stress, serum or growth factor withdrawal, heat shock, chemotherapy, DNA damage (Pistritto et al., 2016; Ruggieri and Matarrese, 2020). Binding of extracellular death ligands to their cognate receptors (for example, TNF-R1, Fas/CD95, TRAIL-R1, or TRAIL-R2) at the plasma membrane promoted the extrinsic pathway. Their engagement induce receptors aggregation and the subsequent recruitment of the adaptor protein FADD and initiator caspase-8 and -10 molecules, resulting in the formation of the so-called death-inducing signaling complex (DISC). Active caspase-8/10 cleaves and activates the effector caspases (i.e., caspase-3 or -7), and then in turn in the execution of cell death. However, both the extrinsic and intrinsic pathways converge on mitochondria, where permeabilization of the mitochondrial outer membrane occurs, leading to the formation of large pore through which apoptogenic proteins (i.e., cytochrome C) can be released into the cytosol where they trigger the activity of a family of cysteine proteases, i.e., caspases, that lead to proteolysis, cell disassembly, and apoptosis.

Bioactive SLs have emerged as key regulators not only of the intrinsic pathway of apoptosis (Patwardhan et al., 2016) but also of the extrinsic death receptors mediated apoptotic pathways (Bollinger et al., 2005). *Cer* represents a central intermediate in the biosynthetic pathway of SLs (Hannun and Obeid, 2018). SLs, and especially *Cer*, have been linked to the modulation of apoptosis (Ogretmen and Hannun, 2004). Therefore, *Cer* and the enzymes that regulate the generation of ceramides, including *Cer*Ss, SMases, SK1 and SK2, have been studied in regard to cell death, in particular to apoptosis (Galadari et al., 2015). Herein, we will specifically focus on the role of *Cer* and other SLs in regard to the receptor-mediated pathways of apoptosis.

Studies performed in the last two decades, evidenced that a prerequisite for the correct initiation of the apoptotic signaling via death receptors is represented by receptors trimerization at the plasma membrane through the pre-ligand binding assembly extracellular domain (PLAD) (Algeciras-Schimnich et al., 2002; Siegel et al., 2000). This event was extensively described for the signaling pathway induced by the receptor Fas (Siegel et al., 2000), but it was also observed in the signaling induced by TNF (Chan, 2007). It has been reported that once transported at the plasma membrane of T cells Fas molecules trimerize (Algeciras-Schimnich et al., 2002).

Therefore, binding of the trimerized Fas ligand induces receptor aggregation and clustering in lipid rafts (Gajate and Mollinedo, 2015; Simons and Ikonen, 1997; Simons and Toomre, 2000; Sonnino and Prinetti, 2013). Strikingly, surface recruitment of death receptors may also lead to self-activation even in the absence of ligand. Once these clusters are formed, a conformational change of the intracellular part of the receptors occurs (Chan, 2007), enabling the recruitment of FADD and pro-caspase-8 to form the DISC, where caspase-8 is activated and activate effector caspases (i.e., caspase-3 and -7), in turn leading to apoptosis execution (Garofalo et al., 2003; Malorni et al., 2007; Scheel-Toellner et al., 2002). In light of these considerations, the Fas recruitment into the lipid rafts is considered a prerequisite for an efficient apoptosis induction and execution (Gajate and Mollinedo, 2015). According with this, the constitutive localization of TRAIL death receptor 1 (TRAIL-R1), but not of TRAIL death receptor 2 (TRAIL-R2), in lipid rafts was suggested to represent a prerequisite for susceptibility to TRAIL in human cancer B-cell lines (Marconi et al., 2013). In fact, agents able to disaggregate lipid rafts induced TRAIL resistance (Delmas et al., 2004; Marconi et al., 2013), whereas the anticancer agents known to favor the recruitment of DRs into the lipid rafts (e.g., perifosine, a synthetic alkyl-lysophospholipid), significantly increased TRAIL-induced apoptosis (Marconi et al., 2013; Wagner et al., 2007). By contrast, in gastric cancer cells resistant to TRAIL, delfosine upregulated TRAIL-R2 and increased its localization in lipid rafts, making them sensitive to TRAIL (Lim et al., 2016). Several agents, i.e., Oxaliplatin, Epirubicin, β-elemene, Bufalin, Resveratrol, Ursodeoxycholic acid, Edelfosine have been reported to be able to induce death receptor aggregation into lipid rafts, promoting apoptosis (Delmas et al., 2004; Lim et al., 2011, 2016; Xu et al., 2009, 2011, 2018; Yan et al., 2014). Thus, depending on the cell type, the TRAIL-mediated death signal may involve TRAILR1, TRAILR2, or both but, in any case, their localization within the lipid microdomains would seem mandatory. Similarly, the recruitment into membrane lipid rafts of TNFR1 following its engagement by TNF-α has also been observed in T cells (Legler et al., 2003).

Growing evidence hinted that *Cer* could play an essential role in receptor-mediated apoptosis. In fact, *Cer* formed from sphingolipids self-associates through hydrogen bonding to promote raft coalescence and formation of molecular platforms that cluster signal transducer components of DISC. For instance, an increase in the amount of *Cer* molecules was observed at the plasma membrane following Fas- or TRAIL-induced cell death (Dumitru and Gulbins, 2006; Park et al., 2008, 2010). Moreover, low levels of intracellular *Cer* were correlated with resistance to apoptosis induced by death ligands in different tumor cell types (Asakuma et al., 2003; Autelli et al., 2009; Cremesti et al., 2001; Schaefer et al., 2000; White-Gilbertson et al., 2009; Zhang et al., 2009a,b). This is apparently due to the fact that TRAIL, as well as Fas or TNF, are able to promote the activation of SMase that in turn induces the formation of *Cer* and its translocation to the plasma membrane (Colell et al., 2002; Dumitru and Gulbins, 2006; Grassmé et al., 2003). Once at the plasma membrane, ceramides self-associate into lipid rafts to form *Cer*-enriched membrane microdomains that in turn fuse each other's forming *Cer*-enriched membrane platforms (Nurminen et al., 2002). Herein, proteins involved in intracellular signaling cluster together with death receptors (i.e., Fas or TRAILR2), facilitating DISC formation thus amplifying the transduction of death signals into the cells (Dumitru et al., 2007; Grassmé et al., 2001a, b) (Fig. 2A).

All in all, these considerations suggested that *Cer* could enhance Fas-mediated apoptosis, as well as TRAIL-induced cell death, by promoting aggregation of trimerized receptors into lipid rafts (Cremesti et al., 2001; Gajate and Mollinedo, 2015; Park et al., 2010). It seems conceivable that the activation and translocation of SMase at the plasma membrane and consequent formation of *Cer* could depend on caspase-8 activation and could occur at an early stage during the induction of death signaling (Grassmé et al., 2003; Stephan et al., 2017), also involving Syntaxin-4 (Perrotta et al., 2010). However, it has also been suggested that SMase can be activated in late stages of death signaling, once clusters of death receptors are internalized into receptosomes and fuse with intracellular trans-Golgi vesicles carrying inactive SMase molecules (Edelmann et al., 2011; Lee et al., 2006; Schneider-Brachert et al., 2004; Stephan et al., 2017) (Fig. 2B). A further important protein involved in SL metabolism able to modulate cell fate is the *Cer* Synthase-6 (CERS6), an enzyme that catalyzes the synthesis of C16-ceramide molecules, a specific form of *Cer* induced by apoptosis. For example, downregulation of CERS6 seems capable of protecting colon cancer cells from TRAIL-mediated apoptosis (White-Gilbertson et al., 2009). CERS6 downregulation was also correlated with an impaired translocation of active caspase-3 into the nucleus (White-Gilbertson et al., 2009) whereas the increase of C16-ceramide using exogenous C6-ceramide was able to sensitize resistant colon cancer cells to TRAIL activity (White-Gilbertson et al., 2009). Recently, it has also been reported a role for CERS6 in modulating the response to chemotherapy of acute lymphoblastic

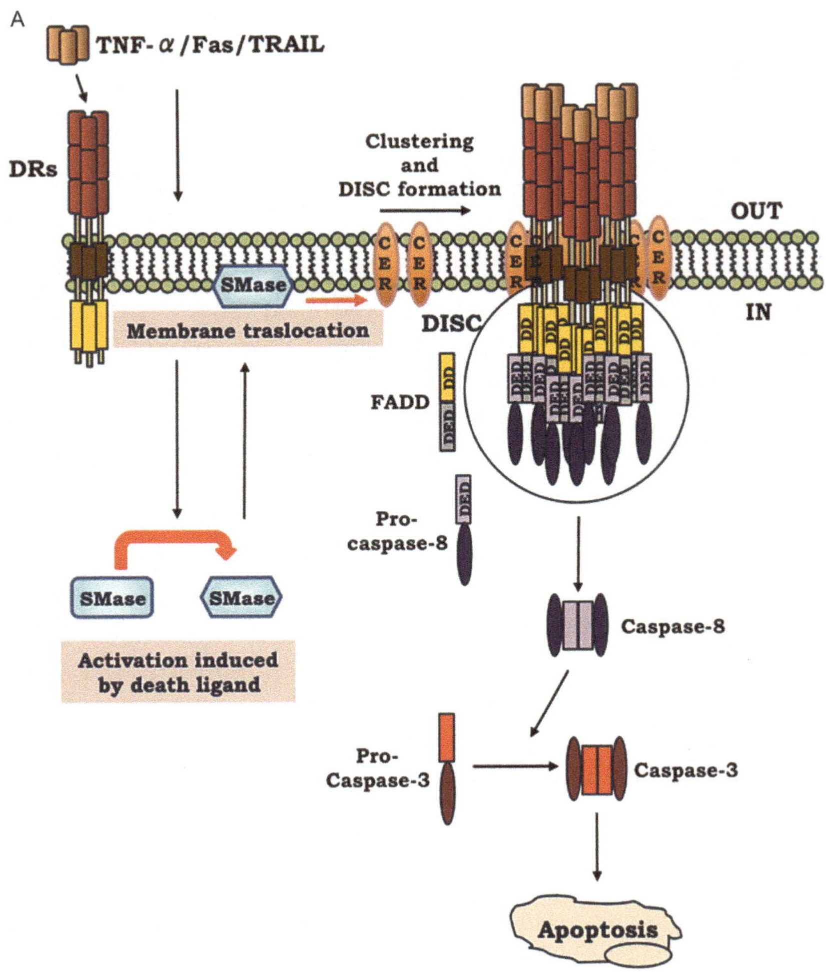

Fig. 2 Molecular events occurring following the activation of receptor-mediated apoptotic signaling pathway. (A) *Early events.* Binding of death ligand to its cognate receptor can induce the activation of SMase molecules that, in turn, translocate at the plasma membrane inducing self-association of ceramide (CER) molecules. These events lead to the formation of CER-enriched membrane microdomains, where DISC formation is highly enhanced. (B) *Late events.* Once the death receptor is internalized into intracellular vesicles, known as "receptosomes," it can fuse with lysosomes containing inactive SMase molecules, stimulating its activation by caspase-7 molecules. CER, ceramides; SMase, sphingomyelinase, DISC, death-inducing-signaling-complex, DD, death domain; DED, death effector domain.

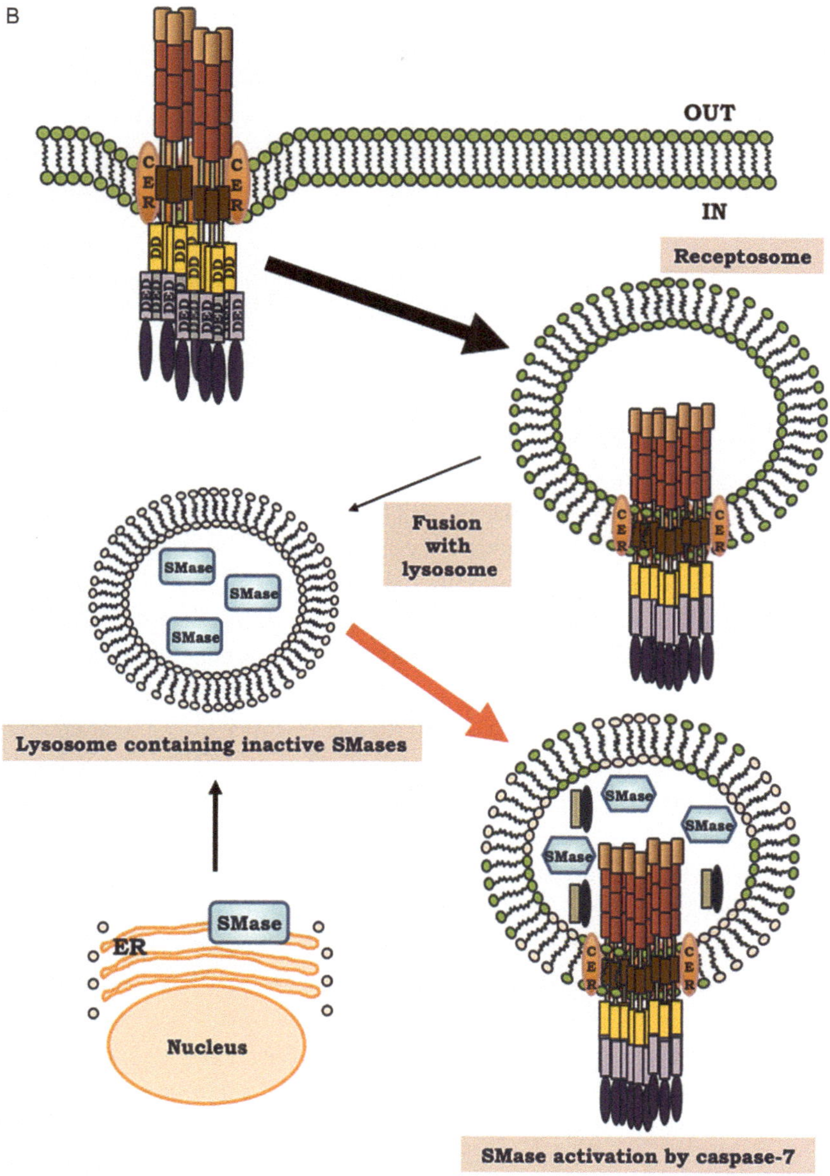

Fig. 2—Cont'd

leukemia (ALL) cells (Verlekar et al., 2018). This function of CERS6 seems to involve the Fas pathway. In fact, the immunoprecipitation of CERS6 in ALL cells demonstrated an association of CERS6 with Fas. These data suggested that the binding of CERS6 to Fas could interfere with FADD association to Fas, then leading to inhibition of DISC formation and, subsequently, of apoptosis (Verlekar et al., 2018). Due to the central role played by *Cer* in the extrinsic pathway of apoptosis, *Cer* analogs have been developed with the aim to mimic endogenous *Cer* (Antoon and Beckman, 2012; Bielawska et al., 2008; Singh et al., 2011). Five of these have shown to be able to increase the sensitivity of human colon cancer cells to Fas-mediated cell death through enhancement of caspase-8 activation (Coe et al., 2016).

4. Sphingolipids in nonapoptotic cell death

4.1 Necroptosis

Even if apoptosis is one of the most investigated forms of programmed cell death, certain SLs have recently been shown to induce a PCD that is independent of caspase activation and is called necroptosis. This particular programmed form of necrosis is dependent on the kinases RIPK1 and RIPK3, which can be activated in response to various stimuli, including Toll-like receptors, virus infection, genotoxic stress, and TNF receptors (Vandenabeele et al., 2010). Necroptosis is involved in a number of diseases, including ischemia/reperfusion injuries and neurodegenerative diseases (Dhuriya and Sharma, 2018). The possible role of SLs in this particular cell death process has emerged in two lipid storage diseases (LSDs): Gaucher disease type 2 and Krabbe disease. Gaucher's disease (GD) is an inherited metabolic disorder caused by mutations in the glucocerebrosidase gene (*GBA*). GD type 2 differs from the type 1 for the presence of neurological signs. These signs are due to a neuronal loss caused by the accumulation of monohexosylceramides, glucosylceramide (Glc*Cer*), and glucosylsphingosine in the brain. In a mouse model of GD type 2, Vitner et al. found that neurodegeneration was not related to an increase in apoptosis cell death or caspase activation. What positively correlated with neuronal loss was instead the presence of elevated levels of RIPK1 and RIPK3 in both neurons and microglia (Vitner et al., 2014). These authors showed that RIP kinases were involved in pathological events of GD type 2 and also in the acute neuropathological changes in Krabbe's disease (severe neurodegenerative SL storage disease caused by the accumulation of a different monohexosyleceramide, the galactoglycoconjugate (Gal*Cer*), and its lyso-derivative psychosine) (Wenger et al., 1997). The necroptosis pathway was not activated in mouse models of other SL storage diseases, including Niemann Pick type C disease (NPC), Sandhoff disease and GM1 gangliosidosis in which the accumulation of cholesterol and Glc*Cer* occurs mainly in lysosomes. At variance, in GD the accumulation of Glu*Cer* mainly occurs in the ER, indicating that the intracellular site of accumulation could be an important determinant in inducing apoptosis or necroptosis (Platt, 2014). *Cer* appears to be involved in the induction of necroptosis also in another pathological model, such as preeclampsia, a serious hypertensive disorder of pregnancy. Bailey et al. (2017) have recently demonstrated a direct role of *Cer* in the activation of the RIP1/RIP3 complex and the consequent activation of necroptosis in human trophoblast cells. These studies could open new avenues for this, currently untreatable, pregnancy disorder.

It may be possible to modulate (enhance or suppress) specific non-apoptotic cell death pathways targeting the lipid-dependent mechanisms of these processes providing novel routes to treat cancer or other pathologies. For example, starting from the evidence that in L929 cell line *Cer*-induced necroptosis in response to tumor necrosis factor α treatment (Strelow et al., 2000), Nganga et al. (2019) have shown that the sphingosine analog drug FTY720 (fingolimod) could induce in both A549 and H1341 human lung cancer cells a form of necroptosis dependent on the formation of *Cer*-enriched membrane pores (ceramidosomes) that were associated with the formation of the ceramide–RIPK1–nonmuscle myosin IIA (NMIIA) complex. Necroptosis was, therefore, ultimately caused by pore formation at the plasma membrane, which results in the release of cytoplasmic material to the extracellular environment. Thus, in this model, FTY720-fingolimod might mimic the function of an endogenous lipid that normally promotes necroptosis of transformed cells to limit tumor formation.

SLs may also play a role in inducing necroptosis due to their structural function within lipid rafts. Chen et al. (2014) demonstrated that the Mixed lineage protein kinase domain-like protein (MLKL), that was identified to function downstream of RIP3K in inducing necroptosis, should form oligomers within lipid rafts of the plasma membrane in order to perform its function. Although there is no definite evidence on the molecular mechanisms by which SLs can act in regulating necroptosis, what seems clear is that their function could be as structural or signaling molecules, with great differences on both the cellular location and the cell type considered.

4.2 Lysosomal cell death

Lysosomal cell death, or lysosome-dependent cell death, is triggered by lysosomal membrane permeabilization (LMP), which results in the translocation of cathepsins to the cytoplasm and subsequent cell death due to degradation events. This peculiar nonapoptotic cell death is relevant for several pathophysiological conditions, including inflammation, tissue remodeling, aging, neurodegeneration, cardiovascular disorders, and intracellular pathogen response (Serrano-Puebla and Boya, 2016). Some lipids and lipid metabolites could trigger LMP. Two works (Taha et al., 2005; Ullio et al., 2012) have suggested in different cellular models (i.e., the rat hepatoma cell line HTC, and the human breast cancer cell line MCF-7) how sphingosine could play a role in this particular process. However, the exact mechanism by which SLs participate in this process is still not entirely clear. Notwithstanding, in the absence of sphingosine a noticeable decrease in LMP is detectable, suggesting that, thanks to its lysosomotropic and detergent properties, sphingosine could play a role in the formation of those specific structures allowing the permeation of lysosomes.

4.3 Parthanatos

Parthanatos is a form of nonapoptotic cell death driven by the hyperactivation of a specific PARP-1's component, involving generation of ROS, increase in mitochondrial permeability, accumulation of poly ADP-ribose polymers and activation of calpain. Parthanatos does not require the mediation of caspases for its execution, but is clearly dependent on the nuclear translocation of the mitochondrial-associated apoptosis-inducing factor (AIF) (Fatokun et al., 2014). Since AIF is a mitochondrial protein, its translocation into the nucleus inevitably relates parthanatos to mitochondrial injury. This particular type of cell death seems to be involved mainly in the context of the central nervous system, with relevance to neurological and neurodegenerative conditions. However, it may also underlie many further disease conditions, including diabetes and inflammatory disorders. Prado Spalm et al. (2019) recently showed that C2-acetylsphingosine (C2-*Cer*), a cell-permeable *Cer*, is able to induce photoreceptor death by activating a caspase-independent mechanism that involves the activation of PARP-1, the loss of mitochondrial membrane potential, calpain activation, and AIF release and translocation. All these represent the typical features of parthanatos. This type of effect induced by C2-*Cer* was not found in other cell types suggesting its possible relevance both from the pathogenetic point of view and in the search for molecular targets for the treatment of these pathologies (Prado Spalm et al., 2019). In fact, there are some checkpoints in the execution of parthanatos that could be taken into account in the perspective of a therapeutic management of some pathologies associated with this type of alterations of nonapoptotic cell death.

4.4 Ferroptosis

Ferroptosis represents an emerging and recently discovered type of regulated cell death characterized by iron-dependent accumulation of ROS occurring through Fe(II)-dependent lipid peroxidation (Dixon et al., 2012). Several proteins responsible for the regulation of iron such as ferritin and transferrin and the cysteine antiporter receptors have been implicated in the regulation of ferroptosis (Mancias et al., 2014). Recent studies suggested that ferroptosis could play an important regulatory role in the onset and progression of many diseases (Buccarelli et al., 2018; Li et al., 2020). Novgorodov et al. (2018) have shown elevations of ceramide species and sphingosine that were preventable by inhibiting acid sphingomyelinase (ASM) activity in a model of primary oligodendrocytes in culture treated with glutamate. The inhibition of the glutamate/cystine antiporter system x_c^- induced by glutamate treatment could trigger ferroptosis in these cells. Oligodendrocytes survival was enhanced by both downregulating ASM expression and blocking ASM activity. These data strongly support the involvement of SLs in the induction of ferroptosis (Novgorodov et al., 2018).

5. Sphingolipids in autophagy

Autophagy is a finely regulated degradative process for damaged or non-essential cellular constituents, such as proteins and organelles, essential to maintain cellular homeostasis (Hamasaki et al., 2013; Mizushima and Komatsu, 2011). During autophagy, cells ensure basal turnover of cellular organelles and constant supply of energy and macromolecular precursors (Ktistakis and Tooze, 2016). Autophagy is induced in response to various stress stimuli, such as nutrient starvation, reactive oxygen species, hypoxia, infection, and drugs. Impairments in autophagy are associated with increased inflammation and the development of a plethora of human diseases (Parzych and Klionsky, 2014). Several phases characterized the autophagic process: initiation, formation, elongation, phagophore formation and fusion with endolysosomal vesicles, during which specific proteins and lipids are actively involved (Mizushima et al., 2011). Thus, it is evident that regulation of autophagy can affect each of these pathways, so researches have focused on the detailed analysis of these complex interactions.

Several species of SL metabolites, mostly GD3, *Cer* and S1P have been demonstrated to act as bioactive signaling molecules as well as key components of microdomains of emerging importance in the different steps of autophagy (Hannun and Obeid, 2008; Matarrese et al., 2014; Taniguchi et al., 2012). In particular, mitochondria-derived glycosphingolipid are critical throughout the autophagic process, from biogenesis to maturation. Because of the opposing roles of certain SL metabolites in the regulation of apoptosis and autophagy, a growing number of experimental data focused on how SLs could be able to mediate the *switch* between cell survival and cell death.

In this field, the role of *Cer* has been considered crucial in autophagy and apoptosis crosstalk regulation. Considering the diversity of individual *Cer* species (Hannun and Obeid, 2011) the high number of reports regarding their ambiguity in cell signaling and biological effects is not surprising. In fact, the de novo biosynthesis pathway leads to a pool of *Cers* with different fatty acyl chains, which has been extensively implicated in inflammation, differentiation as well as in the control of autophagy. However, additional studies have proposed that variations of intracellular *Cer* levels could play a role in the control of cell fate in different models, leading to opposing autophagic pathways: protective (pro-survival) or lethal autophagy (pro-death) depending on metabolic stimuli (Jiang and Ogretmen, 2014). Moreover, it cannot be excluded that the capacity of *Cer* to regulate death or survival could also depend on its different cellular localization including mitochondria, plasma membrane, endosome/lysosome, and endoplasmic reticulum/Golgi apparatus.

Since the regulation of *Cer* levels depends on activation of distinct *CerS* isoform, several papers based on observations derived from human pathological conditions suggested that the increase of these enzymes can strongly contribute to regulate both cell death and/or autophagy processes through the generation of *Cer* with different chain length, varying from C14 to C26 (Ponnusamy et al., 2010). For instance, recently, Brachtendorf et al. (2019) showed that in human colon cancer, chemotherapeutic treatment was able to enhance *CerS* 5 expression in a p53-dependent manner. The enhancement of *CerS* 5 seems to be responsible for the increase of long chain ceramides (C16-*Cer*), which, in turn, can contribute to tumor chemoresistance by influencing early autophagic events and, probably, mitochondrial respiration rate. These data are in agreement with a previous published work (Gosejacob et al., 2016) revealing that *CerS* 5 is essential in glycerophospholipids homeostasis and to maintain C16:0-ceramide pools which contribute to the control of diet-induced obesity by mitophagy (a form of autophagy that selectively acts as quality control of mitochondria through the elimination and recycling of injured mitochondria).

In fact, by using *CerS5* knockout mice, the authors described an impairment of autophagic response after consumption of a high-fat diet as monitored by LC3A lipidation levels (Gosejacob et al., 2016). Thus, they suggested that the loss of *Cer*S5 and, consequently, the impairment of autophagy flux, might apparently lead to the development of obesity and its comorbidities. This is in line with the evidence that autophagy was proposed to play a critical role in normal adipogenesis in vivo because of the requirement of functional specific autophagy-related genes 7 (atg7) (Zhang et al., 2009a, b). Furthermore, *Cer*-induced mitophagy might not merely represent a molecular switch between lethal and pro-survival-related autophagy, but could also have an impact on the regulation of energy metabolism. Thus, mitochondrial dysfunction accompanied by defective mitophagy could lead to toxic, and in some instances lethal, generation of ROS. A good example of the relevance of this matter is represented by the implication of *Cer*S5 in lipid-induced autophagy due to the C14:0-ceramide production in cardiomyocytes, a mechanism of great relevance in cardiac muscle integrity and function (Russo et al., 2012).

Although the mechanism and the function of *Cer*-induced autophagy need to be investigated in more detail, there is evidence (Back et al., 2018) that under nutrient starvation, *Cer* might induce protective autophagy through the deactivation of Akt/mTOR signaling axis, the first negative regulator of autophagy (Yang and Klionsky, 2010), and p38 MAPK activation. By contrast, *Cer* can also be responsible of lethal autophagy. As recently reported by Morad et al. (2019), a decreased phosphorylation of both Akt and mTOR, was evident as an early event in human acute myeloid leukemia (AML) cell line after treatment with C6-ceramide mixed with tamoxifen. This effect was associated to lethal, Atg5-dependent autophagy and participating mitophagy, suggesting a potential role for C6-ceramide when administrated together with tamoxifen in the regulation of chemotherapy resistance in AML.

A further ceramide-induced autophagy pathway is that associated with the increased expression of the autophagy-associated gene beclin1 (Scarlatti et al., 2004); in fact the stimulation with exogenous C2-*Cer* in human colon cancer HT-29 cells was able to induce intracellular long chain *Cer* accumulation, with an increase of Beclin1 expression. Moreover, amino acid deprivation resulted in an increase of *Cer* levels, which suppressed mTOR activity and induced autophagy in a PP1/PP2A-dependent manner. This effect was inhibited by *Cer* synthase inhibitor myriocin. In addition, in neuronal cells, the inhibition of glucosylceramides (Glc*Cer*)-synthase significantly affected this pathway (Shen et al., 2014). Intriguingly, it has also been hypothesized that in certain circumstances the C2-*Cer*-induced autophagy was found to be also protective. In particular, in the human neuroblastoma cell line SH-SY5Y (Fan et al., 2017) C2-*Cer* inhibited the AKT pathway, but simultaneously activated the c-Jun NH2-terminal kinase (JNK) and ERK1/2 pathways. The latter two turned out to play pro-survival roles, at least in SH-SY5Y cells.

Decrease in *Cer*, is also associated to an overexpression of gangliosides. Not surprisingly, neurons are particularly reliant on autophagy, since this process is especially critical for the survival of post-mitotic cells with high-energy demands. Post-mitotic cells are particularly enriched with gangliosides, which are generated mainly via a recycling pathway that is strongly dependent on a correct endosomal/lysosomal processing and transport (Nixon, 2013). Early studies on ganglioside GM1 revealed its protective role against Aβ-induced neurotoxicity in hippocampal tissue, indicating an involvement of GM1 in the initial stage of autophagosome formation through an enhanced expression of Beclin1 and virtually an enhancement of neuronal autophagic activity, although the molecular mechanism remains to be investigated (Dai et al., 2017). Additional studies revealed, in contrast, that accumulation of gangliosides, due to impairment of ganglioside clearance from lysosome in a SPG11 zebrafish model, could lead to accumulation of autophagy markers in lysosomes, contributing to neuronal death. In particular, accumulation of gangliosides was observed when lysosome membrane recycling was inhibited by downregulation of clathrin or dynamin, as responsible of a correct endosomal/lysosomal processing and transport (Boutry et al., 2018). This observation was supported by experimental data demonstrating that inhibition of ganglioside synthesis prevented neurodegeneration and improved motoneuron phenotype. Moreover, Hwang et al. (2010a) reported an interesting contribution of exogenously ganglioside mixture containing GT1b, GD1a and GM1 to autophagic cell death in primary cultures of astrocytes, as revealed by an enhancement of autophagic vacuoles. This effect was inhibited by either a knockdown of beclin-1/Atg-6 or Atg-7 gene expression or by 3-methyladenine, an inhibitor of autophagy. The molecular mechanism(s) underlying this autophagic process involves several signaling components, including generation of reactive oxygen species (ROS), inhibition of the AKT/mTOR pathway, activation of the ERK pathway and integrity of lipid rafts. Interestingly, ganglioside-induced autophagic cell death requires initiation of the NF-κB pathway, which is also causative for inflammatory activation of astrocytes (Hwang et al., 2010b).

6. Lipid rafts in autophagy

Several lines of evidence define autophagy as a dynamic process that involves a dramatic remodeling of intracellular membranes during which core-complex proteins and some core-complex SLs have been suggested to act together as key regulators. In particular, recent progresses in the field (electron microscopy, biochemical and genetic approaches and more recently high-resolution electron tomography studies), indicated the ER membrane as the major source membranes involved in autophagosome biogenesis (Karanasios et al., 2016). However, the properties of the ER domain promoting the autophagosome intermediate are not completely known. It is known that a number of ER localized proteins co-operate in mediating the early steps of autophagosome formation (Hayashi-Nishino et al., 2009). Activation of the ULK complex downstream of mTOR inactivation induces the complex to translocate to the ER and to interact with the tail of anchored ER vesicle-associated proteins (VAPs), an essential condition for nucleation of phagophores. Previous works showed that autophagosome assembles on PI3P-enriched membrane domains that are in dynamic equilibrium with ER, representing a preferential platform for VPS34 complex I recruitment (Karanasios et al., 2013). Although a large family of multiprotein complexes has been identified as essential actors of autophagic pathway, a particular attention was recently given to lipid contribution to the interplay between the ER and mitochondria during autophagy. Recent studies (Annunziata et al., 2018; Vance, 2014) identified particular ER membrane areas tightly associated to outer mitochondria membranes, now called MAMs. The recruitment of autophagy-related core interactors to MAMs was previously reported by a pioneering study of Hamasaki et al. (2013). They suggested that the ATG14 complex as well as DFCP1 re-localize to the MAM fractions during starvation, and that this recruitment could involve the lipid rafts.

An interesting point could be to investigate if the molecular machinery that regulates vesicle trafficking from the ER has a role in autophagosome formation and in particular how proteins and lipids could be transferred to these isolated membranes. It could be assumed that MAMs could act as sites of nonvesicular lipid transfer between the ER and the mitochondria, leading to combined role of proteins and lipids, specifically cholesterol and GSLs, in allowing the formation of MAMs and their physiological function(s). On these regard, recent studies (Hayashi-Nishino et al., 2009;

He and Levine, 2010; Saita et al., 2013) provide some insight on this issue suggesting that the autophagic process induced upon starvation could be the result of a complex framework of events under the control of both core-complex proteins and some core-complex SLs. It means that lipid rafts, which have been detected within MAM, could represent a physical and functional platform operating during the early steps of the autophagic process (Garofalo et al., 2016). Accordingly, changes of organelle curvature leading to the formation of vesicles during autophagy could be facilitated by the SL constituents (Carlsson and Simonsen, 2015). A direct implication of ganglioside GD3 to the biogenesis and maturation of autophagic vacuoles has already been reported by Matarrese et al. (2014) under nutrient starvation. In particular, in a model of human primary fibroblasts, it has been revealed that GD3, a core component of lipid raft-like microdomains, could be detected in immature autophagosomes associated with phosphatidylinositol 3-phosphate (PI3P) and LC3-II as well as in autolysosomes associated with LAMP1. Accordingly, disruption of mitochondrial dynamics by *Cer*S-inhibitor fumonisin B1 significantly impaired these associations and consequently prevented autophagosome biogenesis and maturation. Moreover, knocking down of GD3 synthase was found to significantly inhibit autophagy, suggesting that the block of SL biogenesis may affect PI3P synthesis and autophagosome formation. This is not surprising, since it has been reported that during autophagosome formation, the clearance of PI3P is critical in regulating both the size and the rate of production of autophagosomes (Cebollero et al., 2012). In fact, PI3P production appears to be involved in the generation of a platform that is essential for the recruitment of specific effectors required for membrane trafficking events occurring during autophagy (Karanasios et al., 2013; Koyama-Honda et al., 2013).

Interestingly, Garofalo et al. (2016) also observed that some key autophagic proteins, such as AMBRA1 and WD repeat protein phosphoinositide interacting 1 (WIPI1) were recruited to the BECN1 complex and re-localized to lipid rafts at MAM level by interacting with ganglioside GD3 and Calnexin (CANX) and these associations were mainly evident in the early phases of autophagic process. Importantly, the disassembly of signaling molecules from lipid rafts by knocking down ST8SIA1 (GD3-synthase) or MFN2 expression impaired both AMBRA1 and WIPI1 interaction with lipid raft components at MAM level, hindering autophagic flux. These data suggest that GD3 could strongly be implicated in the early steps of membrane scrambling at MAM level and lead to autophagosome biogenesis and maturation. Lipid rafts could thus play an important role in regulation of all the autophagic machinery by controlling the association and dissociation of the interactor proteins of relevance in various steps of autophagosome formation.

7. Sphingolipids and lipid microdomains in differentiation

As mentioned above, SLs, universal components of biological membranes of all eukaryotic organisms, from yeasts to mammals, in addition to playing a structural role, also play an important part in signal transduction, thus participating in all cellular functions, including differentiation (Hannun and Obeid, 2018).

SLs and lipid microdomains are indispensable in the development and differentiation of the nervous system, both central and peripheral, as well as in neuronal differentiation of mesenchymal stem cells (Mattei et al., 2015). Interestingly, nervous system is second only to the adipose tissue, as far as lipid content is concerned (Adibhatla and Hatcher, 2008; Aureli et al., 2015), and the amount of lipids detectable in the brain represents more than 50% of its dry weight. Lipids found in the brain are SLs, glycerophospholipids, and cholesterol, which are present in almost equal ratios (Zhang and Liu, 2015). These lipids are involved in developmental, maintenance, and in almost all cellular processes of the brain. In fact, they act as signaling molecules, source of energy, contribute to synaptogenesis, neurogenesis, impulse conduction, and many other functions (Hussain et al., 2013). All the events associated with the development, maintenance, and activities of the nervous system depend on the various combination of lipids present in the different lipid rafts of the neuronal cells, and each alteration of the lipid metabolism that alters the composition of these specialized membrane areas often represents a marker of a neuronal pathology (Hussain et al., 2019). Lipid membrane microdomains present in neuronal cells essentially contain cholesterol and GSLs (Saher et al., 2011). Most of the cholesterol is present in the brain as myelin and represents a determining factor in the processes of myelination and synaptogenesis, as well as a limiting factor in the maturation of the brain as it also contributes to the neurotransmission (Liu et al., 2010). In particular, it has been observed that the glia-derived cholesterol promotes the development and differentiation of synapses making it indispensable for neuronal differentiation and plasticity (Goritz et al., 2005), and experimental lines of evidence show that pre- and postsynaptic areas are rich in cholesterol that maintains and organizes the synaptic proteins. The level of cholesterol in the brain appears particularly critical since neuronal functions are impaired either by its deficiency or by its excessive level, and must therefore be finely monitored (Suzuki et al., 2007).

Gangliosides also have a recognized role in neuronal differentiation and the pattern of synthesized gangliosides changes according to the stages of brain development so that some of them are considered useful markers of differentiation (Yanagisawa et al., 2005; Yanagisawa and Yu, 2007).

In particular, GD1a has been shown to induce early neural differentiation, while GD3 is also involved in the maturation process. By contrast, GD1b seems to be responsible for the more advanced stages of differentiation by increasing the formation of actin-rich dendrites (Ryu et al., 2017). The mandatory role of gangliosides in the development and differentiation of the nervous system is further supported by the fact that mutations affecting genes involved in their metabolism lead to both cognitive and motor disorders of clinical relevance (Ryu et al., 2017).

A further cell types in which lipid rafts are considered of great relevance are immune system cells. In particular, T cells require distinct species of gangliosides during differentiation and activation. In fact, the initial events of T-cell activation involve the movement of the T-cell receptor (TCR) into lipid rafts. It has been proposed that the repertoire selection from immature thymocytes to mature T-cell subsets is accompanied by selective expression of gangliosides in individual T-cell subsets. The selection of gangliosides seems to represent an indispensable process in the formation of distinct and functional lipid rafts in mature T cells. For example, ganglioside composition of rafts necessary for signaling events mediated by TCR and costimulatory molecules differ between CD4+ T cells and CD8+ T cells (Nagafuku et al., 2012). In particular, by liquid chromatography (LC)-MS analysis, Nagafuku et al. demonstrated that thymocytes and peripheral CD4+ T and CD8+ T cells express comparable levels of GM1 and GD1b, whereas GD1a was not detectable in any of the three cell types, and that GalNAcGM1b and GM1b levels were significantly higher in the CD8+ T-cell subset. By using GM3S and GM2/GD2S null mice these authors revealed a distinct expression of gangliosides between CD4+ T cells and CD8+ T cells. This knowledge lays the foundations for the development of a therapeutic strategy to treat immune diseases by targeting specific T-cell subpopulations by modulating ganglioside composition of lipid rafts.

A cross talk between the CD3 signaling and GM synthesis pathway was also observed in splenic T lymphocytes during functional maturation following mild activation. In lymphocytes derived from mice, Brumeanu et al. (2007) observed that CD3 signaling was able to upregulate de novo synthesis of GM without affecting the cholesterol biosynthesis. This activity can lead to a change of the proportion of GM and cholesterol in lipid rafts accounting for the stage of functional maturation. In the same vein Zhu et al. (2011) reported that lowering the levels of major components of lipid rafts, i.e., SLs, can attenuate TCR signaling, T-cell activation, and differentiation to the Th17 lineage.

It has also been reported an important role of S1P in orchestrating the fate of photoreceptors in retina (Crousillac et al., 2009; Simón et al., 2019). In particular, S1P can act as a key molecule in determining the final number of photoreceptors, first regulating the proliferation of photoreceptor neuroblasts, and subsequently promoting their survival and making their differentiation progress. The mechanisms responsible for S1P actions in retinal cells have not yet been clarified. Some studies suggested that S1P could mediate an increase in calcium levels in dendrites and cytosol (Borges et al., 2008; Crousillac et al., 2009) promoting either external Ca^{2+} influx or its release from intracellular stores. Thus, S1P might participate in neurotransmitter release and signaling in the inner retina (Crousillac et al., 2009). While on the one hand scientific evidence identifies lipid dysmetabolism as a possible cause of retinal degeneration, on the other hand it suggests the manipulation of lipid metabolism as a possible therapeutic strategy to prevent retinal degeneration.

SLs, in particular *Cer* and S1P, represent important signaling molecules involved in the differentiation of skin cells, i.e., keratinocytes. It has been observed that S1P can induce an increase of intracellular Ca^{2+} and cell cycle arrest, which constitute the first stages of keratinocyte differentiation. Moreover, through the modulation of calcium, S1P could regulate the expression of genes involved in the formation of adherent junctions between neighboring cells indispensable for the formation of the epidermal barrier (Borodzicz et al., 2016). According with this, a dysmetabolism of SLs has also been observed in psoriatic patients, in which, in comparison with control subjects, a lower serum *Cer* concentration and a higher S1P concentration were found (Myśliwiec et al., 2017). If this dysmetabolism simply represents a pathological marker or if it could play a pathogenic role still remains to be elucidated.

8. Sphingolipids and diseases

The involvement of SLs and their dysregulated metabolism in the pathogenetic mechanisms of human diseases and their progression is well established. Sphingolipid biochemistry and signaling in human diseases has very recently been considered, since implication of dihydrosphingolipids (dhSph) in the pathophysiology of different disease has been suggested, as elegantly reviewed by Megaye et al. (2019). For example dihydroceramide (dhCer), produced as a result of the addition of fatty acyl-CoAs of differing chain lengths to dihydrosphingosine-1-phosphate (dhSph) by the ceramide synthase (CerS), has been implicated in predicting type 2 diabetes in obese individuals (Blachnio-Zabielska et al., 2012). *Cer* metabolism is affected by obesity and diabetes in human adipose tissue (Blachnio-Zabielska et al., 2012). In addition, dhS1P plasma levels were found altered after major cardiac event such as a myocardial infarction (Knapp et al., 2012) and in hepato-renal toxicity (Riley and Voss, 2006).

Taking into account the key role of alterations occurring in programmed cell death and autophagic pathways in human diseases and the implication of SLs in both these cell processes described above, the "interplay" between SLs and cell metabolism has been proposed to play a role in the onset and progression of several human diseases. In addition, since cell fate pathways and its disturbances have been identified not only as pathogenetic determinants but, also, as important therapeutic targets, the impact and the role of SLs have been investigated in some detail in several pathological conditions. Here below we will mention some paradigmatic examples of human diseases in which SLs and their metabolic intermediates modifying cell fate can play key roles.

8.1 Metabolic diseases

Several works indicate that SLs can play a key role in the pathogenesis of several diseases including lifestyle-associated diseases, such as obesity, hypertension, stroke, type 2 diabetes mellitus, and myocardial infarction (Borodzicz et al., 2015). Evidence has been provided that sphingomyelin synthase 2-driven diacylglycerol generation is the crucial lipid for insulin resistance (Hirabayashi et al., 2013). Moreover, the interplay of the orphan receptor G protein-coupled receptor family C group 5 member B (GPRC5B) with plasma membrane SM synthase may contribute to high-fat-induced insulin resistance (Kim et al., 2018). It has also been hypothesized that an imbalance in GM3 species levels could contribute to chronic inflammation playing a pathogenetic role in metabolic disorders (Inokuchi et al., 2017) whereas *Cer* has been suggested to be involved in the cell senescence in adipocytes, hepatocytes, and myoblasts, also leading to insulin resistance (Trayssac et al., 2018). A role for autophagy has been suggested since many years for metabolic diseases, e.g., in terms of protection against glucose intolerance after a high-fat diet (He et al., 2012). In addition, obesity and insulin resistance seem to be associated with a downregulation of autophagy, whereas deletion of autophagy proteins in adipose tissue can promote insulin sensitivity (Singh et al., 2009; Yang et al., 2010).

8.2 Cancer

The field of investigation in which the impact of SLs in cell fate has more intensively carried out can be referred as to cancer. This is due to the relevance of apoptotic and autophagic cell programs either as concerns the onset or the treatment of tumors. A plethora of works deals in fact with the implication of disturbances of different types of cell death in the pathogenesis of tumors as well as with the search for drugs capable of inducing cell death, respectively. Since sphingolipids and their metabolism have been associated with processes determining cell fate, they have been extensively studied in this matter. More recent works and hypotheses stem thus on the implication of tumor suppressors and oncogenes affecting SL metabolism, modifying the balance between pro-senescence and anti-senescence SLs. For example, some in vitro studies suggest that the p53's tumor-suppressive effects could be mediated, at least in part, through the downregulation of sphingosine kinase 1, leading to decreased sphingosine-1-phosphate levels and increased *Cer* and SM levels, which then mediate apoptosis. Furthermore, the conversion of *Cer*, a pro-apoptotic lipid, to sphingosine-1-phosphate, a pro-survival lipid, can in turn modulate cancer cell fate. In the field of cancer, several studies have been devoted to acidic ceramidase that has been observed as overexpressed in many tumor types such as prostate (Saad et al., 2007), head and neck squamous cell carcinoma (Elojeimy et al., 2007), and acute myeloid leukemia (Tan et al., 2017). Importantly, it has also been shown that the overexpression of this enzyme is associated with more aggressive cancers and a higher resistance to chemotherapy (Beckham et al., 2013; Saad et al., 2007). Something special also deals with the most promising anticancer treatment, i.e., immunotherapy. In fact, targeting of the SL network could promote the anti-tumor immune response and overcome resistance to immune checkpoint blockade therapy. Conversely, altered SL metabolism seems also able to promote immune escape in certain cancers, e.g., in melanoma cells (Garandeau et al., 2019). Hence, more in general, SL interaction with the adaptive immune system has been suggested to be of great relevance. In few words, the study of the implication of SLs and their metabolites in cancer appears as mandatory in order to improve our knowledge of the mechanisms underlying cancer growth as well as cancer control.

8.3 Neurodegenerative diseases

Neural development and brain function are strictly dependent on sphingolipids and their subcellular activity (Olsen and Færgeman, 2017; van Echten-Deckert and Alam, 2018). Their accumulation, as glucosylceramide, has been observed in several diseases, for example in the brain of mouse models of Gaucher disease (Vardi et al., 2016), whereas a protective effect of sphingolipids against Sindbis and West Nile virus infections has been hypothesized. This is apparently due to the induction of a type I interferon response counteracting viral infection. Mutations of GBA2, the nonlysosomal glucosylceramidase, have been associated in humans with hereditary spastic paraplegia, autosomal recessive cerebellar ataxia and Marinesco–Sjögren like syndrome (Woeste et al., 2019).

Raft-like microdomains may play a relevant role in the early events of age-dependent neurodegenerative diseases. In particular in neurons, early lipid impairment of these microdomains may affect the conformation and intracellular signaling of raft integrated proteins involved in pathological events. In fact, the analyses of lipid rafts isolated from human frontal cortex in nondemented subjects (24–85 year old), demonstrated that lipid rafts undergo significant alterations of specific lipid classes during aging. In particular, the major changes involve polyunsaturated fatty acids, polar lipids (mainly phosphatidylinositol, sphingomyelin, sulfatides, and cerebrosides), and neutral lipids (particularly cholesterol and sterol esters). Very interestingly, this "senescence" phenomenon of lipid raft show a clear gender difference, being much more pronounced in women, especially after menopause, than in men. Thus, it can hypothesize that the gender specificity of lipid metabolism may be responsible, at least in part, of the higher prevalence of cognitive decline in postmenopausal women (Díaz et al., 2018; Gabelli and Codem, 2015).

Therefore, the characterization of lipid raft patterns in different brain areas may be useful to identify early neurodegeneration events, even in asymptomatic stages, thus contributing to the development of new tools of early diagnosis of neurodegeneration. Although the relevance of sphingolipids in brain function has been known for decades, the molecular details of their roles and the manipulation of sphingolipid metabolism is emerging as a new challenge for pharmacology studies in the field of neurological disorders. Determination of brain lipid alterations and their effects on lipid/protein and protein/protein interactions may contribute to identify new potential pharmacological targets.

8.3.1 Alzheimer's disease

AD is a progressive neurodegenerative disease that represents the majority of all cases of dementia. This pathology induces an irreversible neurological disorder that causes cognitive and behavioral impairment. At the molecular level, β-amyloid peptide (Aβ), and τ protein have been identified as the main agents causing Alzheimer's disease (Rauk, 2009). Although amyloid precursor protein (APP) itself is not a raft protein, a significant proportion of APP is localized within lipid rafts (Cheng et al., 2007), following protein palmitoylation. Therefore, it is generally considered that APP cleavage is modulated within lipid raft microenvironment (Lemkul and Bevan, 2011).

The particular protein/lipid composition is known to influence Aβ release (Ikeda et al., 2011) and aggregation (Tong et al., 2018). In addition, cholesterol, a major component in lipid rafts, has an important role in APP processing and Aβ production (Beel et al., 2010; Wahrle et al., 2002). Indeed, this molecule plays a key role in the integration of APP within lipid rafts, thus promoting its trafficking into these microdomains (Mayeux, 2003). Wahrle et al. (2002) observed that Γ-secretase activity could be modulated following cholesterol content modifications at the cell plasma membrane. The correlation between raft cholesterol and Γ-secretase levels also enhances the cleavage of APP, depending on the availability of β-secretase (BACE1), thereby increasing Aβ production in lipid rafts (Beel et al., 2010). BACE1 is located within lipid rafts, where it is stabilized by three palmitoylated residues (Kalvodova et al., 2005). Conversely, depletion of membrane cholesterol leads to an increase of α-secretase cleavage of APP (Wahrle et al., 2002). Overall, these data demonstrate the importance of cholesterol in APP processing and Aα production, a fact that may have important consequences in AD. In addition, some studies have reported alterations in cholesterol levels in specific brain areas with extensive Aβ deposits and neurofibrillary tangles (NFTs) related to AD. Furthermore, it has been demonstrated that membrane cholesterol deregulation prompts the integration of Aβ peptide into nerve membranes (Wahrle et al., 2002). Indeed, modulation of cholesterol levels is considered a relevant factor of lipid raft physico-chemical properties and function (Cecchi et al., 2009). Conversely, Aβ oligomers can also perturb cholesterol levels in these microdomains (Han et al., 2011). Also, MAM has been observed as a regulatory compartment for lipid regulation of cholesterol trafficking (Tambini et al., 2016). In this concern, Tambini et al. showed that lipoproteins containing ApoE4 can upregulate MAM function to a significantly greater degree than those containing ApoE3 (Vaya and Schipper, 2007).

Moreover, cholesterol derivatives, such as oxysterols (the oxidized derivative of cholesterol) may also be involved in AD pathogenesis (Marwarha et al., 2013). Indeed, 2 of these oxysterol molecules, 24-hydroxycholesterol (24-OH Chol) and 27-hydrocholesterol (27-OH Chol), have been found to increase BACE1 activity at least in neuroblastoma cells.

A further relevant factor is represented by sphingomyelin. Indeed, low levels of sphingomyelin can be detected in AD brains, accompanied by an increased concentration of ceramides (Soreghan et al., 2003). The higher production of *Cer* seems to be consequent to an increase of sphingomyelinase (SMase) activity that hydrolyzes the sphingomyelin, generating sphingomyelin depletion and promoting an abnormal APP processing and cellular trafficking (Farooqui, 2012). Enhanced levels of ceramides directly increase Aβ through stabilization of β-secretase. Evidence also supports important roles of ceramides in neuronal apoptosis. Ceramides may initiate a cascade of biochemical pathways, which lead to neuronal death by different mechanisms, including depolarization and permeabilization of mitochondria, increased production of reactive oxygen species (ROS), cytochrome *c* release, Bcl-2 depletion, and caspase-3 activation (Jazvinšćak et al., 2015).

8.3.2 Parkinson's disease

Parkinson's disease (PD) is characterized by a progressive loss of dopaminergic neurons. The main alteration is the presence of intraneuronal inclusions, termed Lewy bodies, formed by random and abnormal aggregation of α-synuclein protein.

Alterations in sphingolipid metabolism may also be involved in PD pathological progression, since membrane rearrangements of different lipid classes, such as gangliosides, PUFA, cholesterol, saturated fatty acids and phospholipids may influence α-synuclein distribution and oligomerization (Martinez et al., 2007). Specifically, it has been observed that GM1 ganglioside inhibits fibrillation of α-synuclein (Ariga, 2014). This observation is in agreement with previous studies claiming an implication of gangliosides in synucleinopathies (Badawy et al., 2018; Colasanti et al., 2014). In experimental models some modifications were observed in the levels of other ganglioside species (GD1a, GD1b, and GT1b), particularly in aged mice (Martinez et al., 2007). These data suggested that gangliosides may participate in the partitioning and conformational structure of α-synuclein. Indeed, this protein has been shown to be integrated in lipid rafts, upon binding with phospholipids (Kubo et al., 2005). This event also requires the presence of monounsaturated or polyunsaturated fatty acids, such as oleic acid or DHA, respectively, that are particularly abundant in lipid rafts (Gedalya et al., 2009). Moreover, distinct *n*-3 and *n*-6 PUFA proportions appear to be crucial for α-synuclein conformation, a phenomenon that may stabilize the α-helix secondary structure of this protein (Gedalya et al., 2009). Thus, PUFA proportion in lipid raft structure stability may play a significant role in α-synuclein configuration and homeostasis at the plasma membrane. Indeed, α-synuclein association with lipid raft moieties is a requirement for the synaptic localization, vesicular transporter and recycling of α-synuclein through PUFA association (Gedalya et al., 2009). In particular, α-synuclein/PUFA binding in lipid rafts seems to be mandatory, since it contributes to synapse vesicle recycling following neuronal stimulation (Madeira et al., 2011). It has also been shown that α-synuclein interacts with several proteins, including caveolin-1 (Fortin, 2004) and CD55 (Ferrucci et al., 2008), suggesting a pleiotropic mechanism for this protein within membrane microdomains. In the last few years, it was shown that α-synuclein could be localized in MAMs (Guardia-Laguarta et al., 2014, 2015). In particular, the association of α-synuclein with MAMs is affected by pathogenic mutations of this protein in familiar PD. As a consequence, a reduced ER-mitochondria connectivity, a decrease in MAM functions and an increase of mitochondrial fragmentation were observed (Guardia-Laguarta et al., 2014).

8.3.3 Prion-associated neuropathology

Since Prion protein (PrP) is a GPI-anchored protein, most of PrPC (prion cellular protein), as well as PrPSc (prion scrapie protein), were found in lipid rafts, as reported in neuronal (Mattei et al., 2002), and in lymphocytic cells (Mattei et al., 2004). This localization is required for conversion of PrPC to the transmissible spongiform encephalopathy-associated protease-resistant isoform (Vey et al., 1996). The refolding process of PrPC into PrPSc is generally accompanied by increase in β-sheet structure and capability to aggregate into oligomers (Prusiner, 2013). Several studies suggested that PrPC is also involved in the regulation of presynaptic copper concentration, intracellular calcium homeostasis, lymphocyte activation, astrocyte proliferation, and cellular resistance to oxidative stress (Watt et al., 2005). Recent studies showed the involvement of PrPC in apoptotic signaling pathways (Zhang et al., 2006). In this context, Mattei et al. (2011) showed that CD95/Fas triggering induced a redistribution of PrPC to mitochondrial raft-like microdomains, as well as at ER-mitochondria associated membranes, as revealed by immunoelectron microscopy observations. In particular, it was suggested that lipid rafts could contribute to define the metabolic fate of PrPC. Indeed, if raft-embedded PrPC is part of the complex framework normally contributing to the death of the cell, a defective trafficking of PrPC from and toward lipid rafts could also affect normal PrPC catabolism, also leading to the formation of the 17-kDa polypeptide hydrolysis to form the PrP 27–30 scrapie isoform.

8.4 Aging

The functional decline associated with brain aging is characterized by small changes including loss of dendrites, reduction and morphological modifications of spines density, as well as changes in the molecular profile of synapses (Mattson and Magnus, 2006). Moreover, the concentrations of several lipid species in the human brain have been shown to decrease during senescence. In particular, aging induces a decrease of neural gangliosides GM1 and GD1a levels and, to a lower extent, GD1b and GT1b in the human frontal cortex (Kracun et al., 1992). In addition, several studies have reported anomalies in lipid rafts composition and function. In particular, it has been reported that lipid raft composition undergoes significant alterations of specific lipid classes (cholesterol, sterol esters, PUFA, sphingomyelin, sulfatides, and cerebrosides) and phospholipid-bound fatty acids (especially DHA and AA) in correlation with brain aging (Díaz et al., 2018). Interestingly, the impairment in the proportion of distinct lipid classes follows a gradual temporal behavior. For instance, saturated fatty acids appear to increase their levels in lipid rafts after 70 years, whereas there is a decay in the main n-6 and n-3 PUFA (AA and DHA, respectively) by the age of 80 years (Cabré et al., 2018). As a consequence of these changes of lipid classes, several protein-protein interactions and protein multicomplex rearrangements within lipid rafts were affected, thereby modifying different transduction pathways involved in neuronal physiology (Colin et al., 2016; Marin et al., 2017; Martín et al., 2010; Yamamoto et al., 2004). Nevertheless, the specific contribution of lipid raft patterns to the aging process remains to be completely elucidated.

9. Conclusions

The extreme complexity of SLs metabolism makes their study particularly difficult both in the individual tissue districts, where some lipids can perform specific functions, and in the whole organism, e.g., in the blood flow. Biochemical and flow cytometric methods contributed to partially clarify the importance of some sphingolipid/protein associations in the physiological and pathological processes responsible for determining cell fate, such as apoptosis and autophagy. In particular in neurons, early lipid impairment of these microdomains may affect the conformation and intracellular signaling of raft integrated proteins involved in pathological events. Therefore, the characterization of lipid raft patterns may be useful to identify early neurodegeneration events, even in asymptomatic stages, thus contributing to the development of new tools of early diagnosis of neurodegeneration. A further paradigmatic example is that of cancer cell apoptosis. For example, if the recruitment into the lipid rafts of death receptors is mandatory for lethal signaling, it can be hypothesized that the use of drugs capable of favoring or inducing the embedding of death receptors into lipid rafts could be of great interest in the treatment of certain forms of cancer. Only the comprehension of these mechanisms in different physiological and pathological contexts will allow to get fruitful information for the use of SLs as potential biomarkers, as pathogenetic determinants of diseases or, why not, as therapeutic innovative targets.

Acknowledgments

This work was supported in part by: the Italian Association for Cancer Research (18526 to P.M.), and Peretti Foundation (PM).

Conflict of interest

The authors declare that they have no conflict of interest.

Human and animal rights

This article does not contain any studies with human participants or animals performed by any of the authors.

References

Adam, R.M., Yang, W., Di Vizio, D., Mukhopadhyay, N.K., Steen, H., 2008. Rapid preparation of nuclei-depleted detergent-resistant membrane fractions suitable for proteomics analysis. BMC Cell Biol. 9, 30.

Adibhatla, R.M., Hatcher, J.F., 2008. Altered lipid metabolism in brain injury and disorders. Subcell. Biochem. 49, 241–268.

Albi, E., Villani, M., 2009. Nuclear lipid microdomains regulate cell function. Commun. Integr. Biol. 2, 23–24.

Albi, E., Cataldi, S., Rossi, G., Magni, M.V., 2003. A possible role of cholesterol sphingomyelin/phosphatidylcholine in nuclear matrix during rat liver regeneration. J. Hepatol. 38, 623–628.

Algeciras-Schimnich, A., Shen, L., Barnhart, B.C., Murmann, A.E., Burkhardt, J.K., Peter, M.E., 2002. Molecular ordering of the initial signaling events of CD95. Mol. Cell. Biol. 22, 207–220.

Annunziata, I., Sano, R., d'Azzo, A., 2018. Mitochondria-associated ER membranes (MAMs) and lysosomal storage diseases. Cell Death Dis. 9, 328.

Antoon, J.W., Beckman, B.S., 2012. Anti-proliferative effects of the novel ceramide analog (S)-2-(benzylideneamino)-3-hydroxy-Ntetrade-cylpropanamide in chemoresistant cancer. Bioorg. Med. Chem. Lett. 22, 2624–2628.

Ariga, T., 2014. Pathogenic role of ganglioside metabolism in neurodegenerative diseases. J. Neurosci. Res. 1242, 1227–1242.

Asakuma, J., Sumitomo, M., Asano, T., Hayakawa, M., 2003. Selective Akt inactivation and tumor necrosis actor-related apoptosis-inducing ligand sensitization of renal cancer cells by low concentrations of paclitaxel. Cancer Res. 63, 1365–1370.

Aureli, M., Grassi, S., Prioni, S., Sonnino, S., Prinetti, A., 2015. Lipid membrane domains in the brain. Biochim. Biophys. Acta 1851, 1006–1016.

Autelli, R., Ullio, C., Prigione, E., Crepaldi, S., Schiavone, N., Brunk, U.T., Capaccioli, S., Baccino, F.M., Bonelli, G., 2009. Divergent pathways for TNF and C2-ceramide toxicity in HTC hepatoma cells. Biochim. Biophys. Acta 1793, 1182–1190.

Back, M.J., Ha, H.C., Fu, Z., Choi, J.M., Piao, Y., Won, J.H., Jang, J.M., Shin, I.C., Kim, D.K., 2018. Activation of neutral sphingomyelinase 2 by starvation induces cell-protective autophagy via an increase in Golgi-localized ceramide. Cell Death Dis. 9, 1–18.

Badawy, M.M.S., Okada, T., Kajimoto, T., Hirase, M., Matovelo, S., Nakamura, S., Yoshida, D., Ijuin, T., Nakamura, S., 2018. Extracellular α-synuclein drives sphingosine 1-phosphate receptor subtype 1 out of lipid rafts, leading to impaired inhibitory G-protein signaling. J. Biol. Chem. 293, 8208–8216.

Bailey, L.J., Alahari, S., Tagliaferro, A., Post, M., Caniggia, I., 2017. Augmented trophoblast cell death in preeclampsia can proceed via ceramide-mediated necroptosis. Cell Death Dis. 8, e2590.

Bartke, N., Hannun, Y.A., 2009. Bioactive sphingolipids: metabolism and function. J. Lipid Res. 50, S91–S96.

Beckham, T.H., Cheng, J.C., Lu, P., Marrison, S.T., Norris, J.S., Liu, X., 2013. Acid ceramidase promotes nuclear export of PTEN through sphingosine 1-phosphate mediated Akt signaling. PLoS One 8, e76593.

Beel, A.J., Sakakura, M., Barrett, P.J., Sanders, C.R., 2010. Direct binding of cholesterol to the amyloid precursor protein: an important interaction in lipid-Alzheimer's disease relationships? Biochim. Biophys. Acta Mol. Cell Biol. Lipids 1801, 975–982.

Bieberich, E., 2018. Sphingolipids and lipid rafts: novel concepts and methods of analysis. Chem. Phys. Lipids 216, 114–131.

Bielawska, A., Bielawski, J., Szulc, Z.M., Mayroo, N., Liu, X., Bai, A., Elojeimy, S., Rembiesa, B., Pierce, J., Norris, J.S., Hannun, Y.A., 2008. Novel analogs of D-e-MAPP and B13. Part 2: signature effects on bioactive sphingolipids. Bioorg. Med. Chem. 16, 1032–1045.

Bionda, C., Portoukalian, J., Schmitt, D., Rodriguez-Lafrasse, C., Ardail, D., 2004. Subcellular compartmentalization of ceramide metabolism: MAM (mitochondria-associated membrane) and/or mitochondria? Biochem. J. 382, 527–533.

Blachnio-Zabielska, A.U., Pulka, M., Baranowski, M., Nikolajuk, A., Zabielski, P., Gorska, M., Gorski, J., 2012. Ceramide metabolism is affected by obesity and diabetes in human adipose tissue. J. Cell. Physiol. 227, 550–557.

Bollinger, C.R., Teichgraber, V., Gulbins, E., 2005. Ceramide-enriched membrane domains. Biochim. Biophys. Acta 1746, 284–294.

Borges, S., Lindstrom, S., Walters, C., Warrier, A., Wilson, M., 2008. Discrete influx events refill depleted Ca2+ stores in a chick retinal neuron. J. Physiol. 586, 605–626.

Borodzicz, S., Czarzasta, K., Kuch, M., Cudnoch-Jedrzejewska, A., 2015. Sphingolipids in cardiovascular diseases and metabolic disorders. Lipids Health Dis. 14, 55.

Borodzicz, S., Rudnicka, L., Mirowska-Guzel, D., Cudnoch-Jedrzejewska, A., 2016. The role of epidermal sphingolipids in dermatologic diseases. Lipids Health Dis. 15, 13. https://doi.org/10.1186/s12944-016-0178-7. Review. PubMed PMID: 26786937. PubMed Central PMCID: PMC4717587.

Boslem, E., Weir, J.M., MacIntosh, G., Sue, N., Cantley, J., Meikle, P.J., Biden, T.J., 2013. Alteration of endoplasmic reticulum lipid rafts contributes to lipotoxicity in pancreatic beta-cells. J. Biol. Chem. 288, 26569–26582.

Boutry, M., Branchu, J., Lustremant, C., Pujol, C., Pernelle, J., Matusiak, R., Seyer, A., Poirel, M., Chu-Van, E., Pierga, A., Dobrenis, K., Puech, J.P., Caillaud, C., Durr, A., Brice, A., Colsch, B., Mochel, F., El Hachimi, K.H., Stevanin, G., Darios, F., 2018. Inhibition of lysosome membrane recycling causes accumulation of gangliosides that contribute to neurodegeneration. Cell Rep. 23, 3813–3826.

Brachtendorf, S., Wanger, R.A., Birod, K., Thomas, D., Trautmann, S., Wegner, M.S., Fuhrmann, D.C., Brüne, B., Geisslinger, G., Grösch, S., 2019. Chemosensitivity of human colon cancer cells is influenced by a p53-dependent enhancement of ceramide synthase 5 and induction of autophagy. Int. J. Biochem. Cell Biol. 112, 95–106.

Browman, D.T., Resek, M.E., Zajchowski, L.D., Robbins, S.M., 2006. Erlin-1 and erlin-2 are novel members of the prohibitin family of proteins that define lipid-raft-like domains of the ER. J. Cell Sci. 119, 3149–3160.

Brown, D.A., London, E., 2000. Structure and function of sphingolipid- and cholesterol-rich membrane rafts. J. Biol. Chem. 275, 17221–17224.

Brumeanu, T.D., Preda-Pais, A., Stoica, C., Bona, C., Casares, S., 2007. Differential partitioning and trafficking of GM gangliosides and cholesterol-rich lipid rafts in thymic and splenic CD4 T cells. Mol. Immunol. 44, 530–540.

Buccarelli, M., Marconi, M., Pacioni, S., De Pascalis, I., D'Alessandris, Q.G., Martini, M., Ascione, B., Malorni, W., Larocca, L.M., Pallini, R., Ricci-Vitiani, L., Matarrese, P., 2018. Inhibition of autophagy increases susceptibility of glioblastoma stem cells to temozolomide by igniting ferroptosis. Cell Death Dis. 9, 841.

Cabré, R., Naudí, A., Dominguez-Gonzalez, M., Jové, M., Ayala, V., Mota-Martorell, N., Pradas, I., Nogueras, L., Rué, M., Portero-Otín, M., Ferrer, I., Pamplona, R., 2018. Lipid profile in human frontal cortex is sustained throughout healthy adult life span to decay at advanced ages. J. Gerontol. A Biol. Sci. Med. Sci. 73, 703–710.

Carlsson, S.R., Simonsen, A., 2015. Membrane dynamics in autophagosome biogenesis. J. Cell Sci. 128, 193–205.

Cascianelli, G., Villani, M., Tosti, M., Marini, F., Bartoccini, E., Magni, M.V., Albi, E., 2008. Lipid microdomains in cell nucleus. Mol. Biol. Cell 19, 5289–5295.

Cebollero, E., van der Vaart, A., Zhao, M., Rieter, E., Klionsky, D.J., Helms, J.B., Reggiori, F., 2012. Phosphatidylinositol-3-phosphate clearance plays a key role in autophagosome completion. Curr. Biol. 22, 1545–1553.

Cecchi, C., Nichino, D., Zampagni, M., Bernacchioni, C., Evangelisti, E., Pensal, A., Liguri, G., Gliozzi, A., Stefani, M., Relini, A., 2009. A protective role for lipid raft cholesterol against amyloid-induced membrane damage in human neuroblastoma cells. Biochim. Biophys. Acta Mol. Cell Biol. Lipids 1788, 2204–2216.

Chan, F.K., 2007. Three is better than one: pre-ligand receptor assembly in the regulation of TNF receptor signaling. Cytokine 37, 101–107. Epub 2007 Apr 20. Review. PubMed PMID: 17449269. PubMed Central PMCID: PMC1965282.

Chen, X., Li, W., Ren, J., 2014. Translocation of mixed lineage kinase domain-like protein to plasma membrane leads to necrotic cell death. Cell Res. 24, 105–121.

Cheng, H., Vetrivel, K.S., Gong, P., Meckler, X., Parent, A., Thinakaran, G., 2007. Mechanisms of disease: new therapeutic strategies for Alzheimer's disease—targeting APP processing in lipid rafts. Nat. Clin. Pract. Neurol. 3, 374–382.

Ciarlo, L., Manganelli, V., Garofalo, T., Matarrese, P., Tinari, A., Misasi, R., Malorni, W., Sorice, M., 2010. Association of fission proteins with mitochondrial raft-like domains. Cell Death Differ. 17, 1047–1058.

Ciarlo, L., Vona, R., Manganelli, V., Gambardella, L., Raggi, C., Marconi, M., Malorni, W., Sorice, M., Garofalo, T., Matarrese, P., 2018. Recruitment of mitofusin 2 into "lipid rafts" drives mitochondria fusion induced by Mdivi-1. Oncotarget 9, 18869–18884.

Coe, G.L., Redd, P.S., Paschall, A.V., Lu, C., Gu, L., Cai, H., Albers, T., Lebedyeva, I., Liu, K., 2016. Ceramide mediates FasL-induced caspase 8 activation in colon carcinoma cells to enhance FasL-induced cytotoxicity by tumor-specific cytotoxic T lymphocytes. Sci. Rep. 6, 30816.

Colasanti, T., Vomero, M., Alessandri, C., Barbati, C., Maselli, A., Camperio, C., Conti, F., Tinari, A., Carlo-Stella, C., Tuosto, L., Benincasa, D., Valesini, G., Malorni, W., Pierdominici, M., Ortona, E., 2014. Role of alpha-synuclein in autophagy modulation of primary human T lymphocytes. Cell Death Dis. 5, e1265.

Colell, A., Morales, A., Fernandez-Checa, J.C., Garcia-Ruiz, C., 2002. Ceramide generated by acidic sphingomyelinase contributes to tumor necrosis factor-α-mediated apoptosis in human colon HT-29 cells through glycosphingolipids formation. Possible role of ganglioside GD3. FEBS Lett. 526, 135–141.

Colin, J., Gregory-Pauron, L., Lanhers, M.C., Claudepierre, T., Corbier, C., Yen, F.T., Malaplate-Armand, C., Oster, T., 2016. Membrane raft domains and remodeling in aging brain. Biochimie 130, 178–187.

Copani, A., Melchiorri, D., Caricatole, A., Martini, F., Sale, P., Carnevale, R., Gradini, R., Sortino, M.A., Lenti, L., De Maria, R., Nicoletti, F., 2002. Beta-amyloid induced synthesis of the gangliosides GD3 is a requisite for cell cycle reactivation and apoptosis in neurons. J. Neurosci. 22, 3963–3968.

Cremesti, A., Paris, F., Grassmé, H., Holler, N., Tschopp, J., Fuks, Z., Gulbins, E., Kolesnick, R., 2001. Ceramide enables fas to cap and kill. J. Biol. Chem. 276, 23954–23961.

Crousillac, S., Colonna, J., McMains, E., Dewey, J.S., Gleason, E., 2009. Sphingosine-1-phosphate elicits receptor-dependent calcium signaling in retinal amacrine cells. J. Neurophysiol. 102, 3295–3309.

Dai, R., Zhang, S., Duan, W., Wei, R., Chen, H., Cai, W., Yang, L., Wang, Q., 2017. Enhanced autophagy contributes to protective effects of GM1 ganglioside against Abeta1-42-induced neuro-toxicity and cognitive deficits. Neurochem. Res. 42, 2417–2426.

Delmas, D., Rebe, C., Micheau, O., Athias, A., Gambert, P., Grazide, S., Laurent, G., Latruffe, N., Solary, E., 2004. Redistribution of CD95, DR4 and DR5 in rafts accounts for the synergistic toxicity of resveratrol and death receptor ligands in colon carcinoma cells. Oncogene 23, 8979–8986.

Dhuriya, Y.K., Sharma, D., 2018. Necroptosis: a regulated inflammatory mode of cell death. J. Neuroinflammation 15, 199–208.
Díaz, M., Fabelo, N., Ferrer, I., Marín, R., 2018. "Lipid raft aging" in the human frontal cortex during nonpathological aging: gender influences and potential implications in Alzheimer's disease. Neurobiol. Aging 67, 42–52.
Dixon, S.J., Lemberg, K.M., Lamprecht, M.R., Skouta, R., Zaitsev, E.M., Gleason, C.E., Patel, D.N., Bauer, A.J., Cantley, A.M., Yang, W.S., Morrison 3rd., B., Stockwell, B.R., 2012. Ferroptosis: an iron-dependent form of nonapoptotic cell death. Cell 149, 1060–1072.
Dumitru, C.A., Gulbins, E., 2006. TRAIL activates acid sphingomyelinase via a redox mechanism and releases ceramide to trigger apoptosis. Oncogene 25, 5612–5625.
Dumitru, C.A., Carpinteiro, A., Trarbach, T., Hengge, U.R., Gulbins, E., 2007. Doxorubicin enhances TRAIL-induced cell death via ceramide-enriched membrane platforms. Apoptosis 12, 1533–1541.
Edelmann, B., Bertsch, U., Tchikov, V., Winoto-Morbach, S., Perrotta, C., Jakob, M., Adam-Klages, S., Kabelitz, D., Schütze, S., 2011. Caspase-8 and caspase-7 sequentially mediate proteolytic activation of acid sphingomyelinase in TNF-R1 receptosomes. EMBO J. 30, 379–394.
El Bawab, S., Roddy, P., Qian, T., Bielawska, A., Lemasters, J.J., Hannun, Y.A., 2000. Molecular cloning and characterization of a human mitochondrial ceramidase. J. Biol. Chem. 275, 21508–21513.
Elojeimy, S., Liu, X., McKillop, J.C., El-Zawahry, A.M., Holman, D.H., Cheng, J.Y., Meacham, W.D., Mahdy, A.E., Saad, A.F., Turner, L.S., Cheng, J., Day, T., Dong, J.Y., Bielawska, A., Hannun, Y.A., Norris, J.S., 2007. Role of acid ceramidase in resistance to FasL: therapeutic approaches based on acid ceramidase inhibitors and FasL gene therapy. Mol. Ther. 15, 1259–1263.
Fan, C., Liu, Y., Zhao, M., Zhan, R., Cui, W., Jin, H., Teng, Y., Lv, P., Zheng, L., Huang, Y., 2017. Autophagy inhibits C2-ceramide-mediated cell death by decreasing the reactive oxygen species levels in SH-SY5Y cells. Neurosci. Lett. 651, 198–206.
Farooqui, A.A., 2012. Lipid mediators and their metabolism in the nucleus: implications for Alzheimer's disease. J. Alzheimers Dis. 30, 163–178.
Fatokun, A.A., Dawson, V.L., Dawson, T.M., 2014. Parthanatos: mitochondrial-linked mechanisms and therapeutic opportunities. Br. J. Pharmacol. 171, 2000–2016.
Ferrucci, L., Giallauria, F., Guralnik, J.M., 2008. Epidemiology of aging. Radiol. Clin. North Am. 46, 643–652.
Fortin, D.L., 2004. Lipid rafts mediate the synaptic localization of α-synuclein. J. Neurosci. 24, 6715–6723.
Fuchs, Y., Steller, H., 2011. Programmed cell death in animal development and disease. Cell 147, 742–758.
Fulda, S., 2009. Tumor resistance to apoptosis. Int. J. Cancer 124, 511–515. https://doi.org/10.1002/ijc.24064. Review. PubMed PMID: 19003982.
Gabande-Rodriguez, E., Boya, P., Labrador, V., Dotti, C.G., Ledesma, M.D., 2014. High sphingomyelin levels induce lysosomal damage and autophagy dysfunction in Niemann Pick disease type A. Cell Death Differ. 21, 864–875.
Gabelli, C., Codem, A., 2015. Gender differences in cognitive decline and Alzheimer's disease Ital. J. Gend. Specif. Med. 1, 21–28.
Gajate, C., Mollinedo, F., 2015. Lipid rafts and raft-mediated supra-molecular entities in the regulation of CD95 death receptor apoptotic signaling. Apoptosis 20, 584–606.
Galadari, S., Rahman, A., Pallichankandy, S., Thayyullathil, F., 2015. Tumor suppressive functions of ceramide: evidence and mechanisms. Apoptosis 20, 689–711.
Garandeau, D., Noujarède, J., Leclerc, J., Imbert, C., Garcia, V., Bats, M.L., Rambow, F., Gilhodes, J., Filleron, T., Meyer, N., Brayer, S., Arcucci, S., Tartare-Deckert, S.,

Ségui, B., Marine, J.C., Levade, T., Bertolotto, C., Andrieu-Abadie, N., 2019. Targeting the sphingosine 1-phosphate axis exerts potent antitumor activity in BRAFi-resistant melanomas. Mol. Cancer Ther. 18, 289–300.

Garofalo, T., Misasi, R., Mattei, V., Giammarioli, A.M., Malorni, W., Pontieri, G.M., Pavan, A., Sorice, M., 2003. Association of the death-inducing signaling complex with microdomains after triggering through CD95/Fas. Evidence for caspase-8-ganglioside interaction in T cells. J. Biol. Chem. 278, 8309–8315.

Garofalo, T., Giammarioli, A.M., Misasi, R., Tinari, A., Manganelli, V., Gambardella, L., Pavan, A., Malorni, W., Sorice, M., 2005. Lipid microdomains contribute to apoptosis-associated modifications of mitochondria in T cells. Cell Death Differ. 12, 1378–1389.

Garofalo, T., Tinari, A., Matarrese, P., Giammarioli, A.M., Manganelli, V., Ciarlo, L., Misasi, R., Sorice, M., Malorni, W., 2007. Do mitochondria act as "cargo boats" in the journey of GD3 to the nucleus during apoptosis? FEBS Lett. 581, 3899–3903.

Garofalo, T., Manganelli, V., Grasso, M., Mattei, V., Ferri, A., Misasi, R., Sorice, M., 2015. Role of mitochondrial raft-like microdomains in the regulation of cell apoptosis. Apoptosis 20, 621–634.

Garofalo, T., Matarrese, P., Manganelli, V., Marconi, M., Tinari, A., Gambardella, L., Faggioni, A., Misasi, R., Sorice, M., Malorni, W., 2016. Evidence for the involvement of lipid rafts localized at the ER-mitochondria associated membranes in autophagosome formation. Autophagy 12, 917–935.

Gedalya, T.B., Loeb, V., Israeli, E., Altschuler, Y., Selkoe, D.J., Sharon, R., 2009. α-Synuclein and polyunsaturated fatty acids promote clathrin-mediated endocytosis and synaptic vesicle recycling. Traffic 10, 218–234.

Goritz, C., Mauch, D.H., Pfrieger, F.W., 2005. Multiple mechanisms mediate cholesterol-induced synaptogenesis in a CNS neuron. Mol. Cell. Neurosci. 29, 190–201.

Gosejacob, D., Jäger, P.S., Vom Dorp, K., Frejno, M., Carstensen, A.C., Köhnke, M., Degen, J., Dörmann, P., Hoch, M., 2016. Ceramide synthase 5 is essential to maintain C16:0-ceramide pools and contributes to the development of diet-induced obesity. J. Biol. Chem. 291, 6989–7003.

Grassmé, H., Schwarz, H., Gulbins, E., 2001a. Molecular mechanisms of ceramide-mediated CD95 clustering. Biochem. Biophys. Res. Commun. 284, 1016–1030.

Grassmé, H., Jekle, A., Riehle, A., Schwarz, H., Berger, J., Sandhoff, K., Kolesnick, R., Gulbins, E., 2001b. CD95 signaling via ceramide-rich membrane rafts. J. Biol. Chem. 276, 20589–20596.

Grassmé, H., Cremesti, A., Kolesnick, R., Gulbins, E., 2003. Ceramide-mediated clustering is required for CD95-DISC formation. Oncogene 22, 5457–5470.

Guardia-Laguarta, C., Area-Gomez, E., Rub, C., Liu, Y., Magranè, J., Becker, D., Voos, W., Schon, E.A., Przedborski, S., 2014. α-Synuclein is localized to mitochondria-associated ER membranes. J. Neurosci. 34, 249–259.

Guardia-Laguarta, C., Area-Gomez, E., Schon, E.A., Przedborski, S., 2015. Novel subcellular localization for α-synuclein: possible functional consequences. Front. Neuroanat. 9, 17.

Hacker, G., 2000. The morphology of apoptosis. Cell Tissue Res. 301, 5–17.

Hamasaki, M., Furuta, N., Matsuda, A., Nezu, A., Yamamoto, A., Fujita, N., Oomori, H., Noda, T., Haraguchi, T., Hiraoka, Y., Amano, A., Yoshimori, T., 2013. Autophagosomes form at ER-mitochondria contact sites. Nature 495, 389–393. https://doi.org/10.1038/nature11910. Epub 2013 Mar 3. PubMed PMID: 23455425.

Han, X., Rozen, S., Boyle, S.H., Hellegers, C., Cheng, H., Burke, J.R., Welsh-Bohmer, K.A., Doraiswamy, P.M., Kaddurah-Daouk, R., 2011. Metabolomics in early Alzheimer's disease: identification of altered plasma sphingolipidome using shotgun lipidomics. PLoS One 6, e21643.

Hannun, Y.A., Obeid, L.M., 2008. Principles of bioactive lipid signalling: lessons from sphingolipids. Nat. Rev. Mol. Cell Biol. 9, 139–150.
Hannun, Y.A., Obeid, L.M., 2011. Many ceramides. J. Biol. Chem. 286, 27855–27862.
Hannun, Y.A., Obeid, L.M., 2018. Sphingolipids and their metabolism in physiology and disease. Nat. Rev. Mol. Cell Biol. 19, 175–191.
Hayashi, T., Fujimoto, M., 2010. Detergent-resistant microdomains determine the localization of sigma-1 receptors to the endoplasmic reticulum-mitochondria junction. Mol. Pharmacol. 77, 517–528.
Hayashi-Nishino, M., Fujita, N., Noda, T., Yamaguchi, A., Yoshimori, T., Yamamoto, A., 2009. A subdomain of the endoplasmic reticulum forms a cradle for autophagosome formation. Nat. Cell Biol. 11, 1433–1437.
He, C., Levine, B., 2010. The beclin 1 interactome. Curr. Opin. Cell Biol. 22, 140–149.
He, C., Bassik, M.C., Moresi, V., Sun, K., Wei, Y., Zou, Z., An, Z., Loh, J., Fisher, J., Sun, Q., Korsmeyer, S., Packer, M., May, H.I., Hill, J.A., Virgin, H.W., Gilpin, C., Xiao, G., Bassel-Duby, R., Scherer, P.E., Levine, B., 2012. Exercise-induced BCL2-regulated autophagy is required for muscle glucose homeostasis. Nature 481, 511–515.
Hengartner, M.O., 2000. Apoptosis: corralling the corpses. Cell 104, 325–328.
Hernández-Corbacho, M.J., Salama, M.F., Canals, D., Senkal, C.E., Obeid, L.M., 2017. Sphingolipids in mitochondria. Biochim. Biophys. Acta. 1862, 56–68.
Hirabayashi, S., Baranski, T.J., Cagan, R.L., 2013. Transformed Drosophila cells evade diet-mediated insulin resistance through wingless signaling. Cell 154, 664–675.
Hussain, G., Schmitt, F., Loeffler, J.P., de Aguilar, J.L.G., 2013. Fatting the brain: a brief of recent research. Front. Cell. Neurosci. 7, 1–14.
Hussain, G., Wang, J., Rasul, A., Anwar, H., Imran, A., Qasim, M., Zafar, S., Kamran, S.K.S., Razzaq, A., Aziz, N., Ahmad, W., Shabbir, A., Iqbal, J., Baig, S.M., Sun, T., 2019. Role of cholesterol and sphingolipids in brain development and neurological diseases. Lipids Health Dis. 18, 26.
Hwang, J., Lee, S., Lee, J.T., Kwon, T.K., Kim, D.R., Kim, H., Park, H.C., Suk, K., 2010a. Gangliosides induce autophagic cell death in astrocytes. Br. J. Pharmacol. 159, 586–603.
Hwang, J., Lee, H.J., Lee, W.H., Suk, K., 2010b. NF-κB as a common signaling pathway in ganglioside-induced autophagic cell death and activation of astrocytes. J. Neuroimmunol. 226, 66–72.
Ikeda, K., Yamaguchi, T., Fukunaga, S., Hoshino, M., Matsuzaki, K., 2011. Mechanism of amyloid β-protein aggregation mediated by GM1 ganglioside clusters. Biochemistry 50, 6433–6440.
Inokuchi, J.I., Go, S., Yoshikawa, M., Strauss, K., 2017. Gangliosides and hearing. Biochim. Biophys. Acta Gen. Subj. 1861, 2485–2493.
Itokazu, Y., Tsai, Y.T., Yu, R.K., 2017. Epigenetic regulation of ganglioside expression in neural stem cells and neuronal cells. Glycoconj. J. 34, 749–756.
Jazvinšćak, J.M., Hof, P.R., Šimić, G., 2015. Ceramides in Alzheimer's disease: key mediators of neuronal apoptosis induced by oxidative stress and Aβ accumulation. Oxid. Med. Cell Longev. 2015, 346783.
Jiang, W., Ogretmen, B., 2014. Autophagy paradox and ceramide. Biochim. Biophys. Acta 1841, 783–792.
Kalvodova, L., Kahya, N., Schwille, P., Ehehalt, R., Verkade, P., Drechsel, D., Simons, K., 2005. Lipids as modulators of proteolytic activity of BACE: involvement of cholesterol, glycosphingolipids, and anionic phospholipids in vitro. J. Biol. Chem. 280, 36815–36823.
Karanasios, E., Stapleton, E., Manifava, M., Kaizuka, T., Mizushima, N., Walker, S.A., Ktistakis, N.T., 2013. Dynamic association of the ULK1 complex with omegasomes during autophagy induction. J. Cell Sci. 126, 5224–5238.

Karanasios, E., Walker, S.A., Okkenhaug, H., Manifava, M., Hummel, E., Zimmermann, H., Ahmed, Q., Domart, M.C., Collinson, L., Ktistakis, N.T., 2016. Autophagy initiation by ULK complex assembly on ER tubulovesicular regions marked by ATG9 vesicles. Nat. Commun. 7, 12420.

Kim, Y.J., Greimel, P., Hirabayashi, Y., 2018. GPRC5B-mediated sphingomyelin synthase 2 phosphorylation plays a critical role in insulin resistance. Science 8, 250–266.

Knapp, M., Zendzian-Piotrowska, M., Blachnio-Zabielska, A., Zabielski, P., Kurek, K., Gorski, J., 2012. Myocardial infarction differentially alters sphingolipid levels in plasma, erythrocytes and platelets of the rat. Basic Res. Cardiol. 107, 294.

Koyama-Honda, I., Itakura, E., Fujiwara, T.K., Mizushima, N., 2013. Temporal analysis of recruitment of mammalian ATG proteins to the autophagosome formation site. Autophagy 9, 1491–1499.

Kracun, I., Rosner, H., Drnovsek, V., Vukelic, Z., Cosovic, C., Trbojevic-Cepe, M., Kubat, M., 1992. Gangliosides in the human brain development and aging. Neurochem. Int. 20, 421–431.

Ktistakis, N.T., Tooze, S.A., 2016. Digesting the expanding mechanisms of autophagy. Trends Cell Biol. 26, 624–635.

Kubo, S., Nemani, V.M., Chalkley, R.J., Anthony, M.D., Hattori, N., Mizuno, Y., Edwards, R.H., Fortin, D.L.A., 2005. Combinatorial code for the interaction of alpha-synuclein with membranes. J. Biol. Chem. 280, 31664–31672.

Ledeen, R.W., Wu, G., 2008. Nuclear sphingolipids: metabolism and signaling. J. Lipid Res. 49, 1176–1186.

Lee, K.H., Feig, C., Tchikov, V., Schickel, R., Hallas, C., Schütze, S., Peter, M.E., Chan, A.C., 2006. The role of receptor internalization in CD95 signaling. EMBO J. 25, 1009–1023.

Legler, D.F., Micheau, O., Doucey, M.A., Tschopp, J., Bron, C., 2003. Recruitment of TNF receptor 1 to lipid rafts is essential for TNF alpha-mediated NF-kappaB activation. Immunity 18, 655–664.

Lemkul, J.A., Bevan, D.R., 2011. Lipid composition influences the release of Alzheimer's amyloid β-peptide from membranes. Protein Sci. 20, 1530–1545.

Li, J., Cao, F., Yin, H.L., Huang, Z.J., Lin, Z.T., Mao, N., Sun, B., Wang, G., 2020. Ferroptosis: past, present and future. Cell Death Dis. 11, 88.

Lim, S.C., Duong, H.Q., Choi, J.E., Lee, T.B., Kang, J.H., Oh, S.H., Han, S.I., 2011. Lipid raft-dependent death receptor 5 (DR5) expression and activation are critical for ursodeoxycholic acid-induced apoptosis in gastric cancer cells. Carcinogenesis 32, 723–731.

Lim, S.C., Parajuli, K.R., Han, S.I., 2016. The alkyllysophospholipid edelfosine enhances TRAIL-mediated apoptosis in gastric cancer cells through death receptor 5 and the mitochondrial pathway. Tumour Biol. 37, 6205–6216.

Liu, J.P., Tang, Y., Zhou, S., Toh, B.H., McLean, C., Li, H., 2010. Cholesterol involvement in the pathogenesis of neurodegenerative diseases. Mol. Cell. Neurosci. 43, 33–42.

Maceyka, M., Spiegel, S., 2014. Sphingolipid metabolites in inflammatory disease. Nature 510, 58–67.

Madeira, A., Yang, J., Zhang, X., Vikeved, E., Nilsson, A., Andrén, P.E., Svenningsson, P., 2011. Caveolin-1 interacts with alpha-synuclein and mediates toxic actions of cellular alpha-synuclein overexpression. Neurochem. Int. 59, 280–289.

Malorni, W., Giammarioli, A., Garofalo, T., Sorice, M., 2007. Dynamics of lipid raft components during lymphocyte apoptosis: the paradigmatic role of GD3. Apoptosis 12, 941–949.

Mancias, J.D., Wang, X., Gygi, S.P., Harper, J.W., Kimmelman, A.C., 2014. Quantitative proteomics identifies NCOA4 as the cargo receptor mediating ferritinophagy. Nature 509, 105–109.

Marconi, M., Ascione, B., Ciarlo, L., Vona, R., Garofalo, T., Sorice, M., Gianni, A.M., Locatelli, S.L., Carlo-Stella, C., Malorni, W., Matarrese, P., 2013. Constitutive localization of DR4 in lipid rafts is mandatory for TRAIL-induced apoptosis in B-cell hematologic malignancies. Cell Death Dis. 17, e863.

Marin, R., Fabelo, N., Martín, V., Garcia-Esparcia, P., Ferrer, I., Quinto-Alemany, D., Díaz, M., 2017. Anomalies occurring in lipid profiles and protein distribution in frontal cortex lipid rafts in dementia with lewy bodies disclose neurochemical traits partially shared by Alzheimer's and Parkinson's diseases. Neurobiol. Aging 49, 52–59.

Martín, V., Fabelo, N., Santpere, G., Puig, B., Marín, R., Ferrer, I., Díaz, M., 2010. Lipid alterations in lipid rafts from Alzheimer's disease human brain cortex. J. Alzheimers Dis. 19, 489–502.

Martinez, Z., Zhu, M., Han, S., Fink, A.L., 2007. GM1 specifically interacts with alpha-synuclein and inhibits fibrillation. Biochem. Pharmacol. 46, 1868–1877.

Marwarha, G., Raza, S., Prasanthi, J.R.P., Ghribi, O., 2013. Gadd153 and NF-KB crosstalk regulates 27-hydroxycholesterol-induced increase in BACE1 and β-amyloid production in human neuroblastoma SH-SY5Y cells. PLoS One 8, e70773.

Matarrese, P., Garofalo, T., Manganelli, V., Gambardella, L., Marconi, M., Grasso, M., Tinari, A., Misasi, R., Malorni, W., Sorice, M., 2014. Evidence for the involvement of GD3 ganglioside in autophagosome formation and maturation. Autophagy 10, 750–765.

Mattei, V., Garofalo, T., Misasi, R., Gizzi, C., Mascellino, M.T., Dolo, V., Pontieri, G.M., Sorice, M., Pavan, A., 2002. Association of cellular prion protein with gangliosides in plasma membrane microdomains of neural and lymphocytic cells. Neurochem. Res. 27, 743–749.

Mattei, V., Garofalo, T., Misasi, R., Circella, A., Manganelli, V., Lucania, G., Pavan, A., Sorice, M., 2004. Prion protein is a component of the multimolecular signaling complex involved in T cell activation. FEBS Lett. 560, 14–18.

Mattei, V., Matarrese, P., Garofalo, T., Tinari, A., Gambardella, L., Ciarlo, L., Manganelli, V., Tasciotti, V., Misasi, R., Malorni, W., 2011. Recruitment of cellular prion protein to mitochondrial raft-like microdomains contributes to apoptosis execution. Mol. Biol. Cell 22, 4842–4853.

Mattei, V., Santacroce, C., Tasciotti, V., Martellucci, S., Santilli, F., Manganelli, V., Piccoli, L., Misasi, R., Sorice, M., Garofalo, T., 2015. Role of lipid rafts in neuronal differentiation of dental pulp-derived stem cells. Exp. Cell Res. 339, 231–240.

Mattson, M.P., Magnus, T., 2006. Ageing and neuronal vulnerability. Nat. Rev. Neurosci. 7, 278–294.

Mayeux, R., 2003. Epidemiology of neurodegeneration. Annu. Rev. Neurosci. 26, 81–104.

Megaye, R.R., Savira, F., Hua, Y., Kelly, D.J., Reid, C., Flynn, B., Liew, D., Wang, B.H., 2019. The role of dihydrosphingolipids in disease. Cell. Mol. Life Sci. 76, 1107–1134.

Melchiorri, D., Martini, F., Lococo, E., Gradini, R., Barletta, E., De Maria, R., Caricasole, A., Nicoletti, F., Lenti, L., 2002. An early increase in the disialoganglioside GD3 contributes to the development of neuronal apoptosis in culture. Cell Death Differ. 9, 609–615.

Mizushima, N., Komatsu, M., 2011. Autophagy: renovation of cells and tissues. Cell 147, 728–741.

Mizushima, N., Yoshimori, T., Ohsumi, Y., 2011. The role of Atg proteins in autophagosome formation. Annu. Rev. Cell Dev. Biol. 27, 107–132.

Morad, S.A.F., MacDougall, M.R., Abdelmageed, N., Kao, L.P., Feith, D.J., Tan, S.F., Kester, M., Loughran Jr., T.P., Wang, H.G., Cabot, M.C., 2019. Pivotal role of mitophagy in response of acute myelogenous leukemia to a ceramide-tamoxifen-containing drug regimen. Exp. Cell Res. 381, 256–264.

Mullen, T.D., Hannun, Y.A., Obeid, L.M., 2012. Ceramide synthases at the centre of sphingolipid metabolism and biology. Biochem. J. 441, 789–802.

Muniz, M., Morsomme, P., Riezman, H., 2001. Protein sorting upon exit from the endoplasmic reticulum. Cell 104, 313–320.

Myśliwiec, H., Baran, A., Harasim-Symbor, E., Choromańska, B., Myśliwiec, P., Milewska, A.J., Chabowski, A., Flisiak, I., 2017. Increase in circulating sphingosine-1-phosphate and decrease in ceramide levels in psoriatic patients. Arch. Dermatol. Res. 309, 79–86.

Nagafuku, M., Okuyama, K., Onimaru, Y., Suzuki, A., Odagiri, Y., Yamashita, T., Iwasaki, K., Fujiwara, M., Takayanagi, M., Ohno, I., Inokuchi, J., 2012. CD4 and CD8 T cells require different membrane gangliosides for activation. Proc. Natl. Acad. Sci. U. S. A. 109, E336–E342.

Nganga, R., Oleinik, N., Kim, J., 2019. Receptor-interacting Ser/Thr kinase 1 (RIPK1) and myosin IIA-dependent ceramidosomes form membrane pores that mediate blebbing and necroptosis. J. Biol. Chem. 294, 502–519.

Nirala, N.K., Rahman, M., Walls, S.M., Singh, A., Zhu, L.J., Bamba, T., Fukusaki, E., Srideshikan, S.M., Harris, G.L., Ip, Y.T., Bodmer, R., Acharya, U.R., 2013. Survival response to increased ceramide involves metabolic adaptation through novel regulators of glycolysis and lipolysis. PLoS Genet. 9, e1003556.

Nixon, R.A., 2013. The role of autophagy in neurodegenerative disease. Nat. Med. 19, 983–997.

Novgorodov, S.A., Voltin, J.R., Gooz, M.A., Li, L., Lemasters, J.J., Gudz, T.I., 2018. Acid sphingomyelinase promotes mitochondrial dysfunction due to glutamate-induced regulated necrosis. J. Lipid Res. 59, 312–329.

Nurminen, T.A., Holopainen, J.M., Zhao, H., Kinnunen, P.K., 2002. Observation of topical catalysis by sphingomyelinase coupled to microspheres. J. Am. Chem. Soc. 124, 12129–12134. PubMed PMID: 12371852.

Ogretmen, B., Hannun, Y.A., 2004. Biologically active sphingolipids in cancer pathogenesis and treatment. Nat. Rev. Cancer 4, 604–616.

Olsen, A.S.B., Færgeman, N.J., 2017. Sphingolipids: membrane microdomains in brain development, function and neurological diseases. Open Biol. 7, 170069.

Ouasti, S., Matarrese, P., Paddon, R., Khosravi-Far, R., Sorice, M., Tinari, A., Malorni, W., Degli Esposti, M., 2007. Death receptor ligation triggers membrane scrambling between golgi and mitochondria. Cell Death Differ. 14, 453–461.

Park, M.A., Zhang, G., Martin, A.P., Hamed, H., Mitchell, C., Hylemon, P.B., Graf, M., Rahmani, M., Ryan, K., Liu, X., Spiegel, S., Norris, J., Fisher, P.B., Grant, S., Dent, P., 2008. Vorinostat and sorafenib increase ER stress, autophagy and apoptosis via ceramide-dependent CD95 and PERK activation. Cancer Biol. Ther. 7, 1648–1662.

Park, M.A., Mitchell, C., Zhang, G., Yacoub, A., Allegood, J., Häussinger, D., Reinehr, R., Larner, A., Spiegel, S., Fisher, P.B., Voelkel-Johnson, C., Ogretmen, B., Grant, S., Dent, P., 2010. Vorinostat and sorafenib increase CD95 activation in gastrointestinal tumor cells through a Ca(2+)-de novo ceramide-PP2A-reactive oxygen species-dependent signaling pathway. Cancer Res. 70, 6313–6324.

Parzych, K.R., Klionsky, D.J., 2014. An overview of autophagy: morphology, mechanism, and regulation. Antioxid. Redox Signal. 20, 460–473.

Patwardhan, G.A., Beverly, L.J., Siskind, L.J., 2016. Sphingolipids and mitochondrial apoptosis. J. Bioenerg. Biomembr. 48, 153–168.

Perrotta, C., Bizzozero, L., Cazzato, D., Morlacchi, S., Assi, E., Simbari, F., Zhang, Y., Gulbins, E., Bassi, M.T., Rosa, P., Clementi, E., 2010. Syntaxin 4 is required for acid sphingomyelinase activityand apoptotic function. J. Biol. Chem. 285, 40240–40251.

Pike, L.J., 2003. Lipid rafts: bringing order to chaos. J. Lipid Res. 44, 655–667.

Pike, L.J., 2006. Rafts defined: a report on the Keystone symposium on lipid rafts and cell function. J. Lipid Res. 47, 1597–1598.

Pistritto, G., Trisciuoglio, D., Ceci, C., Garufi, A., D'Orazi, G., 2016. Apoptosis as anticancer mechanism: function and dysfunction of its modulators and targeted therapeutic strategies. Aging 8, 603–619.

Plati, J., Bucur, O., Khosravi-Far, R., 2008. Dysregulation of apoptotic signaling in cancer. Molecular mechanisms and therapeutic opportunities. J. Cell. Biochem. 104, 1124–1149.

Platt, F.M., 2014. Sphingolipid lysosomal storage disorders. Nature 510, 68–75.

Ponnusamy, S., Meyers-Needham, M., Senkal, C.E., Saddoughi, S.A., Sentelle, D., Selvam, S.P., Salas, A., Ogretmen, B., 2010. Sphingolipids and cancer: ceramide and sphingosine-1-phosphate in the regulation of cell death and drug resistance. Future Oncol. 6, 1603–1624.

Prado Spalm, F.H., Vera, M.S., Dibo, M.J., Simón, M.V., Politi, L.E., Rotstein, N.P., 2019. Ceramide induces the death of retina photoreceptors through activation of parthanatos. Mol. Neurobiol. 56, 4760–4777.

Prusiner, S.B., 2013. Biology and genetics of prions causing neurodegeneration. Annu. Rev. Genet. 47, 601–623.

Radak, D., Katsiki, N., Resanovic, I., Jovanovic, A., Sudar-Milovanovic, E., Zafirovic, S., Mousad, S.A., Isenovic, E.R., 2017. Apoptosis and acute brain ischemia in ischemic stroke. Curr. Vasc. Pharmacol. 15, 115–122.

Rao, R.P., Yuan, C., Allegood, J.C., Rawat, S.S., Edwards, M.B., Wang, X., Merrill Jr., A.H., Acharya, U., Acharya, J.K., 2007. Ceramide transfer protein function is essential for normal oxidative stress response and lifespan. Proc. Natl. Acad. Sci. U. S. A. 104, 11364–11369.

Rauk, A., 2009. The chemistry of Alzheimer's disease. Chem. Soc. Rev. 38, 2698–2715.

Riley, R., Voss, K.A., 2006. Differential sensitivity of rat kidney and liver to fumonisin toxicity: organ-specific differences in toxin accumulation and sphingoid base metabolism. Toxicol. Sci. 92, 335–345.

Rossin, A., Miloro, G., Hueber, A.O., 2019. TRAIL and FasL functions in cancer and autoimmune diseases: towards an increasing complexity. Cancers 11, pii: E639.

Ruggieri, A., Matarrese, P., 2020. Male and female cells: same stress, different response. Ital. J. Gend. Specif. Med. 6, 1–2.

Russo, S.B., Baicu, C.F., Van Laer, A., Geng, T., Kasiganesan, H., Zile, M.R., Cowart, L.A., 2012. Ceramide synthase 5 mediates lipid-induced autophagy and hypertrophy in cardiomyocytes. J. Clin. Invest. 122, 3919–3930.

Ryu, J.S., Ko, K., Ko, K., Kim, J.S., Kim, S.U., Chang, K.T., Choo, Y.K., 2017. Roles of gangliosides in the differentiation of mouse pluripotent stem cells to neural stem cells and neural cells. Mol. Med. Rep. 16, 987–993.

Saad, A.F., Meacham, W.D., Bai, A., Anelli, V., Elojeimy, S., Mahdy, A.E., Turner, L.S., Cheng, J., Bielawska, A., Bielawski, J., Keane, T.E., Obeid, L.M., Hannun, Y.A., Norris, J.S., Liu, X., 2007. The functional effects of acid ceramidase overexpression in prostate cancer progression and resistance to chemotherapy. Cancer Biol. Ther. 9, 1455–1460.

Saher, G., Quintes, S., Nave, K.A., 2011. Cholesterol: a novel regulatory role in myelin formation. Neuroscientist 11, 79–93.

Saita, S., Shirane, M., Nakayama, K.I., 2013. Selective escape of proteins from the mitochondria during mitophagy. Nat. Commun. 4, 1410.

Santos, A.L., Preta, G., 2018. Lipids in the cell: organization regulates function. Cell. Mol. Life Sci. 75, 1909–1927.

Scarlatti, F., Bauvy, C., Ventruti, A., Sala, G., Cluzeaud, F., Vandewalle, A., Ghidoni, R., Codogno, P., 2004. Ceramide-mediated macroautophagy involves inhibition of protein kinase B and up-regulation of beclin 1. J. Biol. Chem. 279, 18384–18391.

Schaefer, J.T., Barthlen, W., Schweizer, P., 2000. Ceramide induces apoptosis in neuroblastoma cell cultures resistant to CD95 (Fas/APO-1)-mediated apoptosis. J. Pediatr. Surg. 35, 473–479.

Scheel-Toellner, D., Wang, K., Singh, R., Majeed, S., Raza, K., Curnow, S.J., Salmon, M., Lord, J.M., 2002. The death-inducing signaling complex is recruited to lipid rafts in Fas-induced apoptosis. Biochem. Biophys. Res. Commun. 297, 876–879.

Schneider-Brachert, W., Tchikov, V., Neumeyer, J., Jakob, M., Winoto-Morbach, S., Held-Feindt, J., Heinrich, M., Merkel, O., Ehrenschwender, M., Adam, D., Mentlein, R., Kabelitz, D., Schütze, S., 2004. Compartmentalization of TNF receptor 1 signaling: internalized TNF receptosomes as death signaling vesicles. Immunity 21, 415–428.

Scorrano, L., 2008. Caspase-8 goes cardiolipin: a new platform to provide mitochondria with microdomains of apoptotic signals? J. Cell Biol. 183, 579–581.

Senkal, C.E., Ponnusamy, S., Rossi, M.J., Bialewski, J., Sinha, D., Jiang, J.C., Jazwinski, S.M., Hannun, Y.A., Ogretmen, B., 2007. Role of human longevity assurance gene 1 and C18-ceramide in chemotherapy-induced cell death in human head and neck squamous cell carcinomas. Mol. Cancer Ther. 6, 712–722.

Serrano-Puebla, A., Boya, P., 2016. Lysosomal membrane permeabilization in cell death: new evidence and implications for health and disease. Ann. N. Y. Acad. Sci. 1371, 30–44.

Shen, W., Henry, A.G., Paumier, K.L., Li, L., Mou, K., Dunlop, J., Berger, Z., Hirst, W.D., 2014. Inhibition of glucosylceramide synthase stimulates autophagy flux in neurons. J. Neurochem. 129, 884–894.

Siegel, R.M., Frederiksen, J.K., Zacharias, D.A., Chan, F.K., Johnson, M., Lynch, D., Tsien, R.Y., Lenardo, M.J., 2000. Fas preassociation required for apoptosis signaling and dominant inhibition by pathogenic mutations. Science 288, 2354–2357.

Simón, M.V., Prado Spalm, F.H., Vera, M.S., Rotstein, N.P., 2019. Sphingolipids as emerging mediators in retina degeneration. Front. Cell. Neurosci. 13, 246.

Simons, K., Ikonen, E., 1997. Functional rafts in cell membranes. Nature 387, 569–572.

Simons, K., Toomre, D., 2000. Lipid rafts and signal transduction. Nat. Rev. Mol. Cell Biol. 1, 31–39.

Singh, R., Xiang, Y., Wang, Y., Baikati, K., Cuervo, A.M., Luu, Y.K., Tang, Y., Pessin, J.E., Schwartz, G.J., Czaja, M.J., 2009. Autophagy regulates adipose mass and differentiation in mice. J. Clin. Invest. 119, 3329–3339.

Singh, A., Ha, H.J., Park, J., Kim, J.H., Lee, W.K., 2011. 3,4-Disubstituted oxazolidin-2-ones as constrained ceramide analogs with anticancer activities. Bioorg. Med. Chem. 19, 6174–6181.

Sonnino, S., Prinetti, A., 2013. Membrane domains and the 'lipid raft' concept. Curr. Med. Chem. 20, 4–21.

Soreghan, B., Thomas, S.N., Yang, A.J., 2003. Aberrant sphingomyelin/ceramide metabolic-induced neuronal endosomal/lysosomal dysfunction: potential pathological consequences in age-related neurodegeneration. Adv. Drug Deliv. Rev. 55, 1515–1524.

Sorice, M., Manganelli, V., Matarrese, P., Tinari, A., Misasi, R., Malorni, W., Garofalo, T., 2009. Cardiolipin enriched raft-like microdomains are essential activating platforms for apoptotic signals on mitochondria. FEBS Lett. 583, 2447–2450.

Stephan, M., Edelmann, B., Winoto-Morbach, S., Janssen, O., Bertsch, U., Perrotta, C., Schütze, S., Fritsch, J., 2017. Role of caspases in CD95-induced biphasic activation of acid sphingomyelinase. Oncotarget 8, 20067–20085.

Strelow, A., Bernardo, K., Adam-Klages, S., Linke, T., Sandhoff, K., Krönke, M., Adam, D., 2000. Overexpression of acid ceramidase protects from tumor necrosis factor-induced cell death. J. Exp. Med. 192, 601–612.

Suzuki, S., Kiyosue, K., Hazama, S., Ogura, A., Kashihara, M., Hara, T., 2007. Brain-derived neurotrophic factor regulates cholesterol metabolism for synapse development. J. Neurosci. 27, 6417–6427.

Taha, T.A., Kitatani, K., Bielawski, J., Cho, W., Hannun, Y.A., Obeid, L.M., 2005. Tumor necrosis factor induces the loss of sphingosine kinase-1 by a cathepsin B-dependent mechanism. J. Biol. Chem. 280, 17196–17202.

Tambini, M.D., Pera, M., Kanter, E., Yang, H., Guardia-Laguarta, C., Holtzman, D., Sulzer, D., Area-Gomez, E., Schon, E.A., 2016. ApoE4 upregulates the activity of mitochondria-associated ER membranes. EMBO Rep. 17, 27–36.

Tan, S.F., Pearson, J.M., Feith, D.J., Loughran Jr., T.P., 2017. The emergence of acid ceramidase as a therapeutic target for acute myeloid leukemia. Expert Opin. Ther. Targets 21, 583–590.

Taniguchi, M., Kitatani, K., Kondo, T., Hashimoto-Nishimura, M., Asano, S., Hayashi, A., Mitsutake, S., Igarashi, Y., Umehara, H., Takeya, H., Kigawa, J., Okazaki, T., 2012. Regulation of autophagy and its associated cell death by "sphingolipid rheostat": reciprocal role of ceramide and sphingosine 1-phosphate in the mammalian target of rapamycin pathway. J. Biol. Chem. 287, 39898–39910.

Tong, B.C.K., Wu, A.J., Li, M., Cheung, K.H., 2018. Calcium signaling in Alzheimer's disease & therapies. Biochim. Biophys. Acta Mol. Cell Res. 1865, 1745–1760.

Trayssac, M., Hannun, Y.A., Obeid, L.M., 2018. Role of sphingolipids in senescence: implication in aging and age-related diseases. J. Clin. Invest. 128, 2702–2712.

Ullio, C., Casas, J., Brunk, U.T., 2012. Sphingosine mediates TNFα-induced lysosomal membrane permeabilization and ensuing programmed cell death in hepatoma cells. J. Lipid Res. 53, 1134–1143.

van Echten-Deckert, G., Alam, S., 2018. Sphingolipid metabolism—an ambiguous regulator of autophagy in the brain. Biol. Chem. 399, 837–850.

Vance, J.E., 2014. MAM (mitochondria-associated membranes) in mammalian cells: lipids and beyond. Biochim. Biophys. Acta 1841, 595–609.

Vandenabeele, P., Galluzzi, L., Vanden Berghe, T., Kroemer, G., 2010. Molecular mechanisms of necroptosis: an ordered cellular explosion. Nat. Rev. Mol. Cell Biol. 11, 700–714.

Vardi, A., Zigdon, H., Meshcheriakova, A., Klein, A.D., Yaacobi, C., Eilam, R., Kenwood, B.M., Rahim, A.A., Massaro, G., Merrill, A.H.J., Vitner, E.B., Futerman, A.H., 2016. Delineating pathological pathways in a chemically induced mouse model of Gaucher disease. J. Pathol. 239, 496–509.

Vaya, J., Schipper, H.M., 2007. Oxysterols, cholesterol homeostasis, and Alzheimer disease. J. Neurochem. 102, 1727–1737.

Verlekar, D., Wei, S.J., Cho, H., Yang, S., Kang, M.H., 2018. Ceramide synthase-6 confers resistance to chemotherapy by binding to CD95/Fas in T-cell acute lymphoblastic leukemia. Cell Death Dis. 9, 925.

Vey, M., Pilkuhn, S., Wille, H., Nixon, R., DeArmond, S.J., Smart, E.J., Anderson, R.G.W., Taraboulos, A., Prusiner, S.B., 1996. Subcellular colocalization of the cellar and scrapie prion proteins in caveole- like membranous domains. Proc. Natl. Acad. Sci. U. S. A. 93, 14945–14949.

Vitner, E.B., Salomon, R., Farfel-Becker, T., Meshcheriakova, A., Ali, M., Klein, A.D., Platt, F.M., Cox, T.M., Futerman, A.H., 2014. RIPK3 as a potential therapeutic target for Gaucher's disease. Nat. Med. 20, 204–208.

Wagner, K.W., Punnoose, E.A., Januario, T., Lawrence, D.A., Pitti, R.M., Lancaster, K., 2007. Death-receptor O-glycosylation controls tumor-cell sensitivity to the proapoptotic ligand Apo2L/TRAIL. Nat. Med. 13, 1070–1077.

Wahrle, S., Das, P., Nyborg, A.C., McLendon, C., Shoji, M., Kawarabayashi, T., Younkin, L.H., Younkin, S.G., Golde, T.E., 2002. Cholesterol-dependent γ-secretase activity in buoyant cholesterol-rich membrane microdomains. Neurobiol. Dis. 9, 11–23.

Watt, N.T., Taylor, D.R., Gillot, A., Thomas, D.A., Perera, W.S., Hooper, N.M., 2005. Reactive oxygen species-mediated beta-cleavage of the prion protein in the cellular response to oxidative stress. J. Biol. Chem. 280, 35914–35921.
Wenger, D.A., Rafi, M.A., Luzi, P., 1997. Molecular genetics of Krabbe disease (globoid cell leukodystrophy): diagnostic and clinical implications. Hum. Mutat. 10, 268–279.
White-Gilbertson, S., Mullen, T., Senkal, C., Lu, P., Ogretmen, B., Obeid, L., Voelkel-Johnson, C., 2009. Ceramide synthase 6 modulates TRAIL sensitivity and nuclear translocation of active caspase-3 in colon cancer cells. Oncogene 28, 1132–1141.
Woeste, M.A., Stern, S., Raju, D.N., Grahn, E., Dittmann, D., Gutbrod, K., Dörmann, P., Hansen, J.N., Schonauer, S., Marx, C.E., Hamzeh, H., Körschen, H.G., Aerts, J.M.F.G., Bönigk, W., Endepols, H., Sandhoff, R., Geyer, M., Berger, T.K., Bradke, F., Wachten, D., 2019. Species-specific differences in nonlysosomal glucosylceramidase GBA2 function underlie locomotor dysfunction arising from loss-of-function mutations. J. Biol. Chem. 294, 3853–3871.
Wu, B.X., Rajagopalan, V., Roddy, P.L., Clarke, C.J., Hannun, Y.A., 2010. Identification and characterization of murine mitochondria-associated neutral sphingomyelinase (MA-nSMase), the mammalian sphingomyelin phosphodiesterase 5. J. Biol. Chem. 285, 17993–18002.
Xu, L., Qu, X., Zhang, Y., Hu, X., Yang, X., Hou, K., Teng, Y., Zhang, J., Sada, K., Liu, Y., 2009. Oxaliplatin enhances TRAIL-induced apoptosis in gastric cancer cells by CBL-regulated death receptor redistribution in lipid rafts. FEBS Lett. 583, 943–948.
Xu, L., Qu, X., Luo, Y., Zhang, Y., Liu, J., Qu, J., Zhang, L., Liu, Y., 2011. Epirubicin enhances TRAIL-induced apoptosis in gastric cancer cells by promoting death receptor clustering in lipid rafts. Mol. Med. Rep. 4, 407–411.
Xu, L., Guo, T., Qu, X., Hu, X., Zhang, Y., Che, X., Song, H., Gong, J., Ma, R., Li, C., Fan, Y., Ma, Y., Hou, K., Wu, P., Dong, H., Liu, Y., 2018. β-elemene increases the sensitivity of gastric cancer cells to TRAIL by promoting the formation of DISC in lipid rafts. Cell Biol. Int. 42, 1377–1385.
Yamada, E., 1955. The fine structure of the gall bladder epithelium of the mouse. J. Biophys. Biochem. Cytol. 1, 445–458.
Yamamoto, N., Igbabvoa, U., Shimada, Y., Ohno-iwashita, Y., Kobayashi, M., Wood, W.G., Fujita, S.C., Yanagisawa, K., 2004. Accelerated Aβ aggregation in the presence of GM1-ganglioside-accumulated synaptosomes of aged ApoE4-knock-in mouse brain. FEBS Lett. 569, 135–139.
Yan, S., Qu, X., Xu, L., Che, X., Ma, Y., Zhang, L., Teng, Y., Zou, H., Liu, Y., 2014. Bufalin enhances TRAIL-induced apoptosis by redistributing death receptors in lipid rafts in breast cancer cells. Anticancer Drugs 25, 683–689.
Yanagisawa, M., Yu, R.K., 2007. The expression and functions of glycoconjugates in neural stem cells. Glycobiology 17, 57R–74R.
Yanagisawa, M., Taga, T., Nakamura, K., Ariga, T., Yu, R.K., 2005. Characterization of glycoconjugate antigens in mouse embryonic neural precursor cells. J. Neurochem. 95, 1311–1320.
Yang, Z., Klionsky, D.J., 2010. Mammalian autophagy: core molecular machinery and signaling regulation. Curr. Opin. Cell Biol. 22, 124–131.
Yang, L., Li, P., Fu, S., Calay, E.S., Hotamisligil, G.S., 2010. Defective hepatic autophagy in obesity promotes ER stress and causes insulin resistance. Cell Metab. 11, 467–478.
Zhang, J., Liu, Q., 2015. Cholesterol metabolism and homeostasis in the brain. Protein Cell 6, 254–264. Springer.
Zhang, Y., Qin, K., Wang, J., Hung, T., Zhao, R.Y., 2006. Dividing roles of prion protein in staurosporine-mediated apoptosis. Biochem. Biophys. Res. Commun. 349, 759–768.
Zhang, Y., Goldman, S., Baerga, R., Zhao, Y., Komatsu, M., Jin, S., 2009a. Adipose-specific deletion of autophagy-related gene 7 (atg7) in mice reveals a role in adipogenesis. Proc. Natl. Acad. Sci. U. S. A. 106, 19860–19865.

Zhang, Y., Yoshida, T., Zhang, B., 2009b. TRAIL induces endocytosis of its death receptors in MDA-MB-231 breast cancer cells. Cancer Biol. Ther. 8, 917–922.

Zhu, Y., Gumlaw, N., Karman, J., Zhao, H., Zhang, J., Jiang, J.L., Maniatis, P., Edling, A., Chuang, W.L., Siegel, C., Shayman, J.A., Kaplan, J., Jiang, C., Cheng, S.H., 2011. Lowering glycosphingolipid levels in CD4+ T cells attenuates T cell receptor signaling, cytokine production, and differentiation to the Th17 lineage. J. Biol. Chem. 286, 14787–14794.

Further reading

Algeciras-Schimnich, A., Peter, M.E., 2003. Actin dependent CD95 internalization is specific for Type I cells. FEBS Lett. 546, 185–188.

Algeciras-Schimnich, A., Pietras, E.M., Barnhart, B.C., Legembre, P., Vijayan, S., Holbeck, S.L., Peter, M.E., 2003. Two CD95 tumor classes with different sensitivities to antitumor drugs. Proc. Natl. Acad. Sci. U. S. A. 100, 11445–11450.

Feig, C., Tchikov, V., Schütze, S., Peter, M.E., 2007. Palmitoylation of CD95 facilitates formation of SDS-stable receptor aggregates that initiate apoptosis signaling. EMBO J. 26, 221–231.

Rotstein, N.P., Miranda, G.E., Abrahan, C.E., German, O.L., 2010. Regulating survival and development in the retina: key roles for simple sphingolipids. J. Lipid Res. 51, 1247–1262.

Schutze, S., Tchikov, V., Schneider-Brachert, W., 2008. Regulation of TNFR1 and CD95 signalling by receptor compartmentalization. Nat. Rev. Mol. Cell Biol. 9, 655–662.

CHAPTER SIX

A lipid perspective on regulated cell death

Hector Flores-Romero, Uris Ros, Ana J. García-Sáez*
Interfaculty Institute of Biochemistry, Eberhard-Karls-Universität Tübingen, Tübingen, Germany
*Corresponding author: e-mail address: ana-jesus.garcia-saez@uni-tuebingen.de

Contents

Abstract

Lipids are fundamental to life as structural components of cellular membranes and for signaling. They are also key regulators of different cellular processes such as cell division, proliferation, and death. Regulated cell death (RCD) requires the engagement of lipids and lipid metabolism for the initiation and execution of its killing machinery. The permeabilization of lipid membranes is a hallmark of RCD that involves, for each kind of cell death, a unique lipid profile. While the permeabilization of the mitochondrial outer membrane allows the release of apoptotic factors to the cytosol during apoptosis, permeabilization of the plasma membrane facilitates the release of intracellular content in other nonapoptotic types of RCD like necroptosis and ferroptosis. Lipids and lipid membranes are important accessory molecules required for the activation of protein executors of cell death such as BAX in apoptosis and MLKL in necroptosis. Peroxidation of membrane phospholipids and the subsequent membrane destabilization is a prerequisite to ferroptosis. Here, we discuss how lipids are essential players in apoptosis, the most common form of RCD, and also their role in necroptosis and ferroptosis. Altogether, we aim to highlight the contribution of lipids and membrane dynamics in cell death regulation.

International Review of Cell and Molecular Biology, Volume 351
ISSN 1937-6448
https://doi.org/10.1016/bs.ircmb.2019.11.004

1. Introduction

Regulated cell death (RCD) is fundamental for the development and tissue homeostasis of multicellular organisms. It can be triggered under several physiological conditions and the imbalance between cell death and proliferation has been linked to many diseases, including autoimmunity, viral infections, neurodegenerative diseases and cancer (Nagata and Tanaka, 2017). Given the relevance of RCD to different human pathologies, its mechanisms of action and modulation by small molecules, is of profound therapeutic interest. Apoptosis constitutes the most frequent form of RCD in vertebrates. Well-defined morphological features are linked with apoptotic cell death, such as membrane blebbing, pyknotic nuclei formation, phosphatidylserine (PS) exposure on the outer leaflet of the plasma membrane, etc. A variety of additional nonapoptotic forms of RCD have recently been described, including necroptosis and ferroptosis, amongst others. These forms of RCD involve disruption of the plasma membrane (PM), cellular swelling and release of damage-associated molecular patterns (DAMPs), and have an inflammatory phenotype (Linkermann et al., 2013; Vince and Silke, 2016; Wallach et al., 2016; Zhang et al., 2018). While membrane permeabilization appears as common theme in RCD, each form of cell death involves unique membrane-related changes that trigger and regulate their distinct biochemical machinery (Zhang et al., 2018). This is not surprising considering that lipids are essential structural biomolecules that are important for signaling in cell death (Friedmann Angeli et al., 2019; Phan et al., 2019; Tonnus et al., 2019).

Lipids have a key role as structural components in the membrane, where cell death-inducing proteins elicit pore formation. Each lipid species has a concrete form and shape, built by a defined combination of core structures, head groups and acyl chains. The unique physiochemical properties of specific lipids do not only modulate their location and the properties of membranes where they are integrated, but also their interactions with different mediators of RCD (Agmon and Stockwell, 2017; Bleicken et al., 2017; Flores-Romero et al., 2018; Magtanong et al., 2016). Lipids can enhance cell death by, for example, modulating the affinities between molecular executors and regulators (Kuwana et al., 2002). Indeed, in apoptosis, lipids constitute the building blocks of the proteolipidic apoptotic pore (Gonzalvez et al., 2005, 2008; Jalmar et al., 2013). Moreover, membrane lipids play an important role in membrane dynamics during RCD

(Cosentino and Garcia-Saez, 2014; Ugarte-Uribe and Garcia-Saez, 2017). However, despite intense research efforts to study lipids at the molecular level, the functions of individual lipids in many cellular processes, including cell death, remain poorly defined. Here, we discuss current knowledge about the role of different lipids in apoptosis and two of the best-characterized forms of RCD, necroptosis and ferroptosis.

2. A lipid view for mitochondrial apoptosis

2.1 Mitochondrial changes in apoptosis

Mitochondria are key organelles in eukaryotic cells and play a fundamental role in stress sensing for cellular adaptation and in cell fate decisions. Beyond its importance in ATP production, mitochondrial dysfunction is involved in several human pathologies including neurodegenerative diseases (e.g., Parkinson's, Alzheimer's), and cancer (Flannery and Trushina, 2019; Grunewald et al., 2019; Vyas et al., 2016). The mitochondrion is thought to have a bacterial-evolutionary origin, resulting in the ultimate symbiont, which still retains its own DNA, RNA, protein and lipid synthesizing machinery (Horvath and Daum, 2013). In addition, mitochondria have a unique membrane system, which is the basis for their intricate functions (Harner et al., 2011). This organelle is surrounded by an envelope composed of the outer (MOM) and inner (MIM) mitochondrial membranes, which differ in size, shape and composition (Fig. 1A). The MOM is smooth, fluid and enriched in porins such as the voltage-dependent anion channel (VDAC), which renders the membrane relatively permeant to small molecules and ionic species (up to 3–5 kDa). In contrast, the MIM is largely impermeable and constitutes the main barrier between the cytoplasm and the mitochondrial matrix. The MIM is divided into the inner boundary membrane (IBM), which is in continuous apposition to the MOM, and the cristae membrane (CM), which invaginates and protrudes into the matrix (Reichert and Neupert, 2002). The MIM contains proteins and protein complexes involved in key mitochondrial functions, such as the components of the electron transport chain (Ow et al., 2008), and is enriched in cardiolipin (CL), a mitochondria-specific phospholipid (Ardail et al., 1990).

Under healthy conditions, mitochondria participate in multiple metabolic pathways, of which energy (ATP) production is probably the most significant. Mitochondria can assume a wide variety of morphologies presumably related to the adaptation of this organelle to the cellular energetic

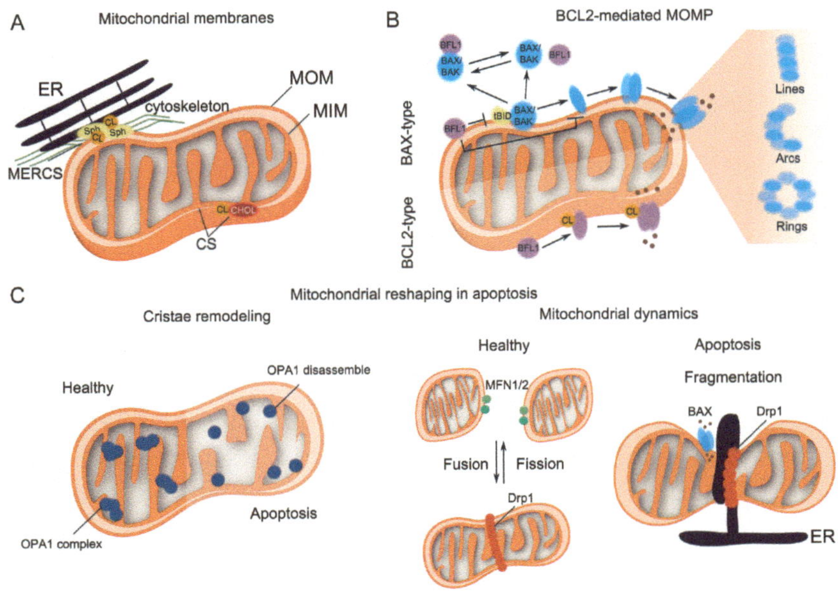

Fig. 1 Mitochondrial features in apoptotic cell death. (A) Mitochondrial membranes. Lipid transfer occurs either at contact sites with the ER (through the MERCS) or at MIM/MOM contact sites (CS). (B) BCL2-mediated MOMP. Top. Activation of BAX/BAK at the MOM by tBID induces BAX/BAK oligomerization (lines, arcs and rings) and pore formation with the consequent release of apoptogenic factors. The apoptotic repressor BFL1, blocks this process by either interacting with tBID or with BAX/BAK. Bottom. Under cellular stress, BFL1 can switch its antiapoptotic phenotype, directly eliciting rather than inhibiting membrane permeabilization. (C) Mitochondrial reshaping in apoptosis. Left. Cristae remodeling. Under healthy conditions, OPA1 oligomers are the main responsible for the maintenance of the cristae junction integrity. Upon apoptosis, the disassemble of OPA1 complexes leads to cristae remodeling and cyt *c* release. Right. Mitochondrial dynamics in apoptosis. Under healthy conditions mitochondria undergo fusion and fission events in a dynamic equilibrium. During apoptosis mitochondrial dynamics are altered favoring mitochondria fission over fusion. Drp1-mediated mitochondrial fragmentation is thought to occur at the MERCS, where the ER constriction wrapping around mitochondria facilitates the formation of Drp1 and Mfn2 oligomers and mitochondrial division. These points represent areas with a lot of membrane tension that are likely to be the place where BAX/BAK elicit MOMP.

status and requirements (Mannella, 2006). These reticular organelles exhibit high plasticity and constantly undergo fission and fusion, thereby segregating damaged mitochondrial fragments as well as sharing and redistributing mitochondrial DNA (mtDNA) and other components. Perturbations to mitochondrial dynamics can have tremendous consequences, as manifested in a large number of diseases linked to imbalance between fusion and fission (Ong and Hausenloy, 2017; Shah et al., 2017; Suarez-Rivero et al., 2016). Moreover, mitochondria establish intimate contacts with the cytoskeleton and the endoplasmic reticulum (ER) (Fig. 1A). Interactions between the mitochondria and cytoskeleton are essential for normal mitochondrial morphology, motility and cellular distribution (Boldogh and Pon, 2006). The close contacts with the ER are referred to as mitochondria-ER contact sites (MERCS) and constitute signaling platforms that participate in several cellular processes (Giorgi et al., 2009; Vance, 1990).

In vertebrates, apoptosis arises through two main pathways: the extrinsic or death receptor pathway and the "intrinsic" or mitochondrial pathway, being the former the most frequent alternative (Green and Kroemer, 2004). This pathway can be triggered by a variety of stress stimuli including radiation (UV and gamma rays), heat, viral virulence factors, growth-factor deprivation, DNA damage, oxidative stress and the activation of oncogenic factors (Flores-Romero and Garcia-Saez, 2019b; Hwang et al., 2019; Plati et al., 2011). Mitochondria undergo dramatic changes in their structure and function during intrinsic apoptosis. However, the exact link between mitochondrial morphology and apoptotic cell death remains ill-defined (Cosentino and Garcia-Saez, 2014; Martinou and Youle, 2011; Suen et al., 2008; Ugarte-Uribe and Garcia-Saez, 2017).

The crucial event in the intrinsic mitochondrial pathway is MOM permeabilization (MOMP) that allows the release of so-called apoptotic factors occluded in the intermembrane space to the cytosol (Fig. 1B). MOMP usually constitutes "the point of no return" in apoptotic signaling (Green and Kroemer, 2004; Kroemer et al., 2007). Indeed, MOMP typically leads to cell death irrespective of caspase activity, which is thought to be the result of progressive decline in mitochondrial function. However, cells with an incomplete MOMP are still able to survive, which can have pathological consequences (Tait et al., 2010).

Dissipation of membrane potential (Zamzami et al., 1995), alterations in lipid transfer between mitochondria and the ER (Hoppins and Nunnari, 2012) as well as between the MOM and the MIM (Kagan et al., 2005), and cristae remodeling (Scorrano et al., 2002), are other important apoptosis-related mitochondrial processes have been described to occur prior to or simultaneously with MOMP (Fig. 1). In this section, we revise the most important changes in mitochondrial morphology and function during apoptosis, with a focus on the relevance of lipids in regulating and triggering these processes.

2.2 Mitochondrial lipid profile as a cell fate indicator

Mitochondria play a fundamental role in lipid metabolism, including degradation and synthesis of fatty acids and the production of complex lipids. The ER, through intimate contact with the mitochondria, is responsible for the transfer of lipids and lipid precursors to mitochondria (Fig. 1A) (Csordas et al., 2006, 2018; Hajnoczky and Csordas, 2010; Hirabayashi et al., 2017). MERCS are CL-enriched contact areas that mediate apoptosis, autophagy and other cellular processes (Ardail et al., 1990; Hajnoczky et al., 2006). Upon apoptosis induction, MERCS increase in number and area, increasing communication between the ER and mitochondria and enhancing signaling pathways (Csordas et al., 2010).

Under normal conditions, the main components of mitochondrial membranes are glycerophospholipids and to a lesser extent sphingolipids and sterols. Despite being highly dynamic organelles, mitochondrial lipid composition does not vary significantly between different conditions and cell types, suggesting that major changes are not tolerated (Fleischer et al., 1967; Osman et al., 2011; Van Meer et al., 2008). Regarding the relative abundance of glycerophospholipids in mitochondria, a consensus view exists for phosphatidylcholine (PC), phosphatidylethanolamine (PE), and phosphatidylinositol (PI) levels, constituting the 55%, 25% and 10%, respectively, while the other lipidic components are less defined (Colbeau et al., 1971; De Kroon et al., 1997). Importantly, the only asymmetrically-distributed lipid is CL, which represents approximately 5% of the total lipid content of the MOM and appears to be enriched in the MIM to around 20% (Ardail et al., 1990). Some studies reported patches in the MOM and in membrane contact sites (CS) that are enriched with as high as 25% of CL (Ardail et al., 1990; Gebert et al., 2009) (Fig. 1A).

Far from being inert structural elements, mitochondrial lipids can affect the function of specific mitochondrial proteins that are implicated in the apoptotic pathway, either through specific binding interactions or through changes in the physical properties of the mitochondrial membrane (Unsay et al., 2013). MOM lipid composition is critical for the function of the BCL2 family of proteins, and therefore perturbation of the lipid composition in the mitochondria can lead to phenotypes associated with apoptotic cell death (Canals and Hannun, 2013; Claypool and Koehler, 2012; Ueda, 2015; Van Brocklyn and Williams, 2012).

2.2.1 *CL-fingerprint in healthy and apoptotic conditions*

CL is an important lipid constituent of the mitochondria and possesses unique structural properties (e.g., two negative charges, a relatively small head group and four acyl chains). It is, therefore, not surprising that CL is implicated in many mitochondrial functions such as; normal organelle ultrastructure, mitochondrial dynamics and energy metabolism (Schlame and Ren, 2009; Tamura et al., 2009). Different lines of evidence indicate that the net content of CL at the MOM increases during apoptosis (Kagan et al., 2004, 2005). Phospholipid transport between the MOM and MIM was proposed to occur at CS between both membranes (Ardail et al., 1990; Simbeni et al., 1991). Although the CL translocation process is still unclear, some possible candidates have emerged during the last decades to fulfill this action: (i) mitochondrial kinases (Epand et al., 2007; Schlattner et al., 2013), (ii) mitochondrial phospholipid scramblase-3 (PLS3) (Liu et al., 2003a) and (iii) tBID (a proapoptotic member of BCL2 family proteins) (Esposti et al., 2001; Gonzalvez et al., 2005; Scorrano et al., 2002). Considering that in the presence of calcium, CL forms highly curved inverted hexagonal structures (Gonzalvez and Gottlieb, 2007; Grijalba et al., 1999; Ortiz et al., 1999; Unsay et al., 2013), calcium accumulation could cause significant membrane stress in CL-enriched regions of mitochondrial membranes, thereby promoting the redistribution of CL to the surface of the mitochondria through nonbilayer CS (Schug and Gottlieb, 2009).

On the other hand, accumulation of anionic lipids in the membrane can promote the binding of proteins with polycationic motifs (Yeung et al., 2006). It is therefore conceivable that an increase of negative charge in the MOM, mainly due to CL accumulation, during early stages of apoptosis (Heit et al., 2011) could lead to the recruitment of certain proteins, including BCL2 family members that modulate the final outcome of the cell. On this subject, a CL-binding motif has been proposed for tBID, which consists of two consecutive lysines that might establish electrostatic interactions with CL and serve as a recruiter to mitochondria (Gonzalvez et al., 2010; Petit et al., 2009).

CL has also been reported to laterally segregate into defined (nano) domains in bacteria, which suggests the formation of similar platforms in mitochondria (Kawai et al., 2004; Sorice et al., 2009). Due to the particular properties of CL, these CL-enriched areas potentially form a unique environment with increased membrane fluidity and decreased mechanical stability, which would be able to handle high membrane curvature and form nonlamellar structures. Furthermore, different types of CL modifications have been associated with apoptotic cell death. For instance, the peroxidized version of CL (CLox) weakens the interaction of cytochrome *c* with the MIM, a process that may contribute to ease MOMP (Kagan et al., 2005; Li et al., 2019b; Ott et al., 2002). Altogether, these evidences support the concept that CL potentially induces mitochondrial membrane alterations that facilitate bilayer structure remodeling, deformation, and ultimately permeabilization.

2.2.2 Sphingolipids and BCL2 proteins

Beyond their structural function, sphingolipids participate in diverse cellular processes, such as in the regulation of cell proliferation, migration and differentiation and in the pathobiology of some diseases like autoimmune and neurodegenerative disorders and cancer (van Brocklyn and Williams, 2012). Sphingosine (Sph) and sphingosine-1-P (S1P), ceramide (CER) and ceramide-1-P (CER-1P), are recognized as key players in many relevant physiological functions (Arana et al., 2010; Gomez-Munoz, 2006).

Although these lipids are minor components of MOM under healthy conditions, they are frequently associated with apoptosis (Chipuk et al., 2012; Czubowicz et al., 2019; Dadsena et al., 2019; Moro et al., 2019; Patwardhan et al., 2016). CER levels increase in response to a variety of apoptotic stimuli such as DNA damage, TNFα or FAS ligand (Cremesti et al., 2001; Obeid et al., 1993). It seems that CER generation occurs before the activation of the apoptotic cascade (Selzner et al., 2001) and that it affects bioenergetics, ROS production and MOMP (Birbes et al., 2001; Colombini, 2010, 2017; Siskind, 2005). Another group of sphingolipids that is related to apoptosis are gangliosides. GD3, a CER-based glycolipid, has been reported to accumulate in mitochondria in response to CER or TNFα treatment enhancing caspase activation and cytochrome *c* release (Rippo et al., 2000). Regarding the BCL2 family proteins, CER increases BAX translocation to mitochondria and MOMP (Birbes et al., 2005). Finally, sphingolipid metabolites like S1P and hexadecenal (HEX), cooperate with BAX-type proteins in promoting MOMP (Chipuk et al., 2012).

2.2.3 Mitochondrial cholesterol in apoptosis

Cholesterol (CHOL) plays an important role defining the physical properties, structure and function of biological membranes. Connected to apoptotic RCD, cancer cells exhibit CHOL accumulation in the mitochondrial membrane, which contributes to diminishing cellular capacity for inducing MOMP and resistance to chemotherapy (Baggetto et al., 1992; Crain et al., 1983; Daum, 1985; Montero et al., 2008; Zhang and Wang, 2015). Several mechanisms were proposed to explain the apoptosis-inhibiting capacity that CHOL exhibits in mammal cells. Five decades ago, it was reported that high CHOL levels prevent mitochondria from swelling (Graham and Green,

1970). Moreover, the ability of CHOL to form ordered domains has been linked to apoptosis modulation by inhibiting BAX membrane insertion and activation (Garofalo et al., 2005; Lucken-Ardjomande et al., 2008). On the other hand, raft-like lipid nanodomains enriched in gangliosides and CL, but with low CHOL content, have been detected on mitochondria upon apoptotic stimulation (Garofalo et al., 2003, 2005; Malorni et al., 2007; Sorice et al., 2009). These raft-like domains seem to be also enriched in specific proteins, including BCL2 family proteins such as tBID and BAX, the fission protein hFis1 and VDAC-1 that may participate in the generation of a death signaling platform (Dadsena et al., 2019; Malorni et al., 2007; Sorice et al., 2009).

2.3 Lipid role in BCL2-regulated MOMP

The proteins of the BCL2 family are key regulators of the mitochondrial pathway of apoptosis. Under cell stress, apoptotic effectors BAX and BAK accumulate at distinct foci on the mitochondrial surface where they undergo a conformational change, oligomerize, and form pores on the MOM (Fig. 1B) (Czabotar et al., 2013; Flores-Romero et al., 2017; Salvador-Gallego et al., 2016; Salvador-Gil et al., 2017; Subburaj et al., 2015). It was recently reported that BAX can promote MIM herniation through the MOM, inducing mtDNA release, which is linked with cGAS-STING signaling-mediated immunogenic effects (McArthur et al., 2018; Riley et al., 2018). The molecular mechanisms of BAX and BAK assembly and mitochondrial permeabilization, as well as the regulation of their activation and activity by other BCL2 proteins, remain poorly understood. The role of mitochondrial lipids as potential modulators of the proteins of the BCL2 family, either directly by modulating their interaction network or indirectly by altering membrane properties, has gained attention in recent years. In the following, we discuss how mitochondrial lipids modulate the BCL2 interactome and cell fate.

2.3.1 BCL2 interaction network

The members of the BCL2 family of proteins are typically classified into three groups based on their impact on cell viability and the presence of up to four conserved BCL2 homology (BH) motifs: (i) BCL2-type repressors, which contain all four BH motifs and primarily function by inhibiting MOMP (BCL2, BCLXL, MCL1 and others); (ii) BAX-type effectors, which contain BH1–BH3 motifs and directly elicit MOMP (BAX, BAK and perhaps BOK); and (iii) BH3-only activators (BID, BIM, BAD, and others), which instigate the function of the effectors (Aouacheria et al., 2013; Birkinshaw and Czabotar, 2017; Czabotar et al., 2014). Despite the importance of this classification, many unanswered questions exist concerning the molecular mechanisms by which these proteins elicit their biological function (Flores-Romero and Garcia-Saez, 2019a). Chief among them is their structural organization and the hierarchy of interactions between pro- and antiapoptotic BCL2 proteins at the MOM, where these proteins carry out their function (Kale et al., 2018). Understanding the function of BCL2 proteins in their membrane environment is key relevant, given that during apoptosis most, if not all, BCL2 family proteins are targeted to the MOM (Lindsay et al., 2011). Moreover, the membrane and its constituting lipids modulate the affinities between the different family members (Bleicken et al., 2017; Garcia-Saez et al., 2009; Pecot et al., 2016) or by altering their canonical phenotype or function, for example, switching their antiapoptotic nature to proapoptotic activity (Flores-Romero et al., 2018).

A single view about how BCL2-type proteins inhibit apoptosis is still lacking. Early models proposed that antiapoptotic proteins repress apoptosis by forming stable neutralizing heterodimers with either BH3-only activators (indirect model, MODE 1) or BAX-type proteins (direct model, MODE 2) (Chen et al., 2005). Recent models shifted toward a bimodal interaction mode of action where BCL2-type proteins stably block both classes of proapoptotic counterparts, rather than only one of them. However, these bimodal models differ regarding binding preferences within the BCL2 interaction network. According to the embedded-together model, insertion into the MOM triggers a conformational change that affects interaction surfaces in BCL2-type repressors that are required for neutralization of proapoptotic proteins, thereby altering the affinity between family members (Leber et al., 2007). The unified model also recognizes the role of the lipid membrane in antiapoptotic inhibition and postulates that BCL2-type proteins bind preferably BAX-type effectors over BH3-only proteins (MODE 2 > MODE 1) (Llambi et al., 2011). In stark contrast, the interconnected hierarchical model gives no particular importance to the MOM and to noncanonical interactions and proposes that BCL2-type proteins bind preferably to the activators than to apoptotic effectors (MODE 1 > MODE 2) (Bleicken et al., 2017; Chen et al., 2015; Landeta et al., 2014).

An additional antiapoptotic mode of action for the BCL2 network has been recently reported. Termed retrotranslocation or MODE 0, the model is based on the notion that antiapoptotic proteins keep BAX-type proteins inactive by continuously retrotranslocating them from the mitochondria into the cytosol (Fig. 1B) (Edlich et al., 2011; Schellenberg et al., 2013; Todt et al., 2015). These models, however, do not consider an enigmatic property shared by all BCL2-type proteins, which is their ability to promote—rather than inhibit—apoptosis under specific conditions (Basanez et al., 2001; Flores-Romero et al., 2018; Landeta et al., 2014; Lin et al., 2004).

2.3.2 MOMP by BAX-type effectors

Irrespective of the mode of action, all models agree that BAX/BAK activation leads to the formation of the apoptotic pore (Fig. 1B). The structural changes driving BAX-type proteins from an inactive structure to a fully activated conformation in the mitochondria were considered the "holy grail" of apoptosis research (Youle and Strasser, 2008). These events are usually divided into (i) early activation steps involving TM dislodgement and N terminal exposure, (ii) BH3 exposure, which occurs due to BAX reorganization in two different parts (core and latch domains), (iii) oligomerization and (iv) pore formation. Currently, it is mostly accepted that BAX/BAK mediate MOMP through the formation of toroidal pores of tunable size composed of both protein and lipid molecules (Basanez et al., 1999, 2012; Bleicken et al., 2013a; Cosentino and Garcia-Saez, 2017; Terrones et al., 2004). Importantly, all these events appear to be regulated, at least partially, by the mitochondrial membrane (Fig. 2).

(i) *Early activation steps.* During apoptosis the N-terminus of BAX and BAK undergoes conformational changes (e.g., partial rotation of the $\alpha 1$ helix). In the specific case of BAX, N-terminus rearrangement is accompanied by the membrane insertion of the $\alpha 9$ helix (TM) and the process has been reported to be dependent on the mitochondrial membrane composition (Kim et al., 2009; Lai et al., 2019; Nechushtan et al., 1999; Yethon et al., 2003).

(ii) *BH3 exposure.* Solution structures of multidomain BCL2 proteins revealed that the hydrophobic face of helix $\alpha 2$ (BH3 domain) is oriented toward the protein interior (Moldoveanu et al., 2006; Muchmore et al., 1996; Suzuki et al., 2000). Therefore, for the BH3 into groove interaction, which drives BCL2 interaction network, it is reasonable to assume that the BH3 domain needs to change its conformation in the membrane to engage in homo- or hetero-dimerization reactions (Cartron et al., 2005; Czabotar et al., 2013; Dewson et al., 2008; Moldoveanu et al., 2013). As previously mentioned, membrane lipid composition can regulate BH3 into groove

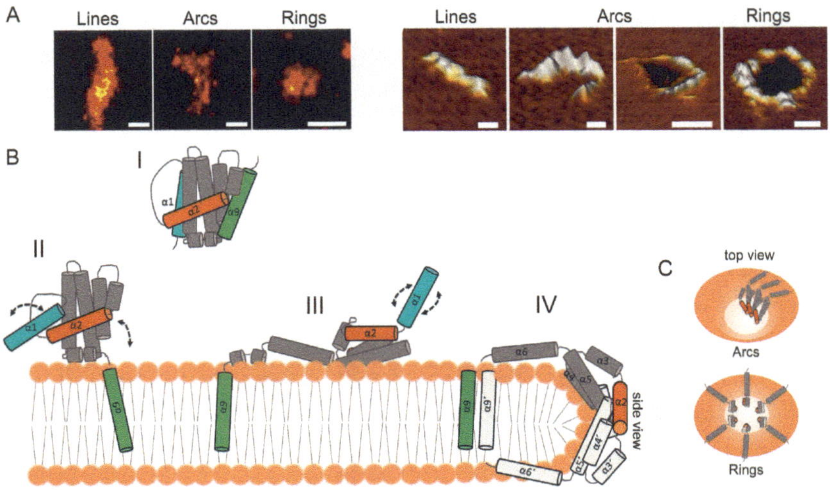

Fig. 2 Pore formation by BAX. (A) BAX oligomers assembly into lines, arcs and rings, visualized in cells by super-resolved microscopy (left) or in membrane mimetic systems by atomic force microscopy (right). (B and C) Toroidal pore model for BAX. (B) The asymmetric insertion of protein exclusively into one of the membrane leaflets leads to an increase in membrane tension (I–III), making it unstable and inducing the bending of the membrane (IV). (I) Protein disposition in solution. BAX is represented with nine cylinders corresponding to its nine α-helixes. (II) BAX/BAK early activation steps: including TM dislodgement and N terminal exposure (depicted in green and cyan respectively). (III) BAX/BAK reorganization in two different parts (dimerization and piercing domains) and BH3 domain exposure (depicted in orange). (IV) Oligomerization and pore formation, structural representation of membrane embedded BAX/BAK in the context of toroidal pore one monomer is showed in gray (α1–9) and the other is depicted in dark gray (α1′–9′). The relative orientation of the helices 9 remains unresolved. In the clamp model (IV), the dimerization domain is located at the edge of a membrane pore induced while α6 helices are located on opposite sides of the membrane. (C) In these structures, a portion of the pore could remain protein-free filled with highly curved lipids. BAX molecules partially line the pore ring forming arcs or rings.

homo- or hetero-dimerization reactions (Bleicken et al., 2017; Flores-Romero et al., 2018; Garcia-Saez et al., 2009).

(iii) *Oligomerization.* Deep structural changes induce a conformation of BAX that inserts extensively into the lipid bilayer and perforates it; a step that appears to be concomitant with BAX assembly into oligomers (Bleicken et al., 2014; Westphal et al., 2014). Recently, it has been reported that the assembly of BAX oligomers into lines, arcs and rings is associated with its apoptotic activity (Figs. 1B and 2) (Grosse et al., 2016; Salvador-Gallego et al., 2016). The role of the membrane in the regulation of BAX/BAK oligomerization remains still unclear. However, considering that the MOM is critical for the active conformation of BAX, it is conceivable that BAX assembly is regulated by mitochondrial membranes and lipids.

(iv) *Pore formation.* The nature of the membrane pores formed by activated BAX/BAK has been highly controversial. Currently, it is widely accepted that BAX/BAK oligomers and membrane lipids are key components of the apoptotic pore, which has a proteolipidic nature with tunable size (Bleicken et al., 2013a) (Fig. 2). In the toroidal pore model for BAX, the asymmetric insertion of protein exclusively into one of the membrane leaflets leads to an increase in membrane tension, making the structure unstable and thereby inducing the opening of a highly curved pore (Fuertes, 2011; Fuertes et al., 2010; Garcia-Saez et al., 2004; Unsay et al., 2013). In this scenario, BAX not only destabilizes the membrane, but also reduces the line tension at the pore edge, lowering the energy barrier for pore stabilization (Basanez et al., 1999). Therefore, the toroidal pore induced by BAX is affected by modifying the physical properties of the membrane such as intrinsic curvature, fluidity and order (Basanez et al., 1999, 2002; Montero et al., 2008; Terrones et al., 2004). A peptide derived from the helix $\alpha5$ of BAX, which is reported to recapitulate the full-length pore-forming activity, forms pores with lipid molecules in the lumen as demonstrated by X-ray diffraction (Qian et al., 2008) and conductance experiments (Garcia-Saez et al., 2005). These data are also in concordance with the transbilayer lipid movement detected when BAX induces membrane permeabilization (Epand et al., 2003; Garcia-Saez et al., 2006; Terrones et al., 2004).

In addition, specific interactions between BAX-type proteins and lipids were also reported in the literature. First, CL seems to play a role in BAX proapoptotic function. In vitro experiments demonstrated a less potent BAX-induced pore formation in the absence of this phospholipid (Gonzalvez et al., 2005; Kuwana et al., 2002; Lovell et al., 2008). However, the impact of the contribution of the CL for BAX activation was recently challenged, as cells lacking CL (but with an exacerbated presence of PG, a CL precursor) were apparently able to compensate the absence of this lipid and undergo apoptosis similar to wild type cells (Raemy et al., 2016). In addition, CER production was shown to increase BAX translocation to mitochondria, thereby stimulating MOMP (Birbes et al., 2005; Ganesan et al., 2010). BAK, but not BAX, was suggested to increase CER synthesis, which would result in a synergy between BAK and BAX (Beverly et al., 2013; Siskind et al., 2010). Moreover, results obtained using pharmacological inhibitors of sphingolipid enzymes in isolated mitochondria suggest that the mitochondrion-specific sphingosine kinase 2 (generating S1P) and S1P lyase (producing HEX) enhance apoptosis triggered by BAK and BAX, respectively (Chipuk et al., 2012; Siskind et al., 2010).

2.3.3 MOMP by BCL2-type "repressors"

According to the current view, antiapoptotic proteins inhibit their proapoptotic counterparts by a canonical BH3 into groove interaction, in which the BH3 domain of the proapoptotic protein is inserted into the hydrophobic pocket of the antiapoptotic protein, in solution and at the membrane (Bleicken et al., 2017; Flores-Romero et al., 2018; Liu et al., 2003b; Petros et al., 2000). Other groups previously reported that this inhibition occurs in the membrane due to the insertion of the helices α5 and α6 of BCL2 and BCLXL, which prevents BAX activation by hairpin–hairpin interactions (Billen et al., 2008; Dlugosz et al., 2006; Shamas-Din et al., 2013; Zhang et al., 2004). However, other evidences link extensive membrane insertion of antiapoptotic proteins to a phenotypic switch, where rather than acting as inhibitors, these proteins gain a pore-forming activity (Fig. 1B) (Basanez et al., 2001; Cheng et al., 1997; Clem et al., 1998; Flores-Romero et al., 2018; Landeta et al., 2014; Lin et al., 2004; Menoret et al., 2010; Michels et al., 2004). This dual role of BCL2-type proteins remains poorly understood but could have implications for therapy, as increased expression of the antiapoptotic BCL2 members is often associated to chemotherapeutic resistance (Campbell and Tait, 2018). Membrane pores formed in vitro by "pro-survival" BCL2 type proteins appear generally smaller than pores formed by the canonical effectors BAX and BAK, which may limit their proapoptotic activity to the release of small-size content (Bleicken et al., 2013b; Flores-Romero et al., 2018). Interestingly, mitochondrial lipids such as CL were reported to modulate the antiapoptotic capacity of BFL1 and MCL1 in both model membranes and in cells (Flores-Romero et al., 2018; Landeta et al., 2014). When membrane CL content is increased, BFL1 and MCL1 lose their inhibitory capacity and gain an ability to permeabilize lipid membranes. As in the case of BAX, the α5 helix of MCL1 and BFL1 seems to be the smallest domain for their membrane destabilizing function (Flores-Romero et al., 2018; Landeta et al., 2014).

2.4 Connecting mitochondrial dynamics and morphology with cell fate via the BCL2 family of proteins

The mitochondrial network is a highly dynamic structure that undergoes constant remodeling via continuous cycles of fission and fusion, in both healthy and apoptotic conditions. Mitochondria can undergo remodeling in response to cell stress, changes in energy demand or fluctuations in intracellular calcium levels (Abate et al., 2019; Singh et al., 2019; Valera-Alberni and Canto, 2018). Mitochondrial morphology (e.g., fragmentation, elongation, branching, and reticulation) refers to the static picture of mitochondrial shape and size, while mitochondrial dynamics refers to the sum of phenomena occurring simultaneously (Aouacheria et al., 2017; Bordt et al., 2017). These aspects are discussed below and schematized in Fig. 1C.

2.4.1 Mitochondrial morphology and dynamics in apoptosis. Fragmentation: Cause or consequence?

During apoptosis, and close in time to MOMP, mitochondria typically undergo massive fragmentation (Fig. 1C) (Frank et al., 2001; Martinou and Youle, 2006). In vertebrates, the essential protein that mediates mitochondrial fission is a large GTPase termed dynamin-related protein 1 (Drp1) (Fonseca et al., 2019). This protein shuttles between the cytosol and MOM and when fission is required, it accumulates in patches on the mitochondria to promote membrane constriction and division. Drp1 interacts with specific adaptors at the mitochondrial surface (e.g., Mff, MiD49 and MiD51) and oligomerizes into a spiral-like scaffold that leads to mitochondrial division (Gandre-Babbe and van der Bliek, 2008; Otera et al., 2010; Palmer et al., 2011; Strack and Cribbs, 2012).

The localization of the adaptor proteins and Drp1 at constriction points is not random, but it seems to be mainly associated with MERCS (Friedman et al., 2011). Besides their role in lipid and calcium transfer, MERCS appear to be functional platforms where the coordinated cooperation of cytoskeleton, ER and the mitochondria drives Drp1 assembly and mitochondrial fission (Hatch et al., 2014; Ji et al., 2015). The activity of Drp1 is highly regulated by post-translational modifications, for example, phosphorylation (Taguchi et al., 2007), sumoylation/ubiquitination (Figueroa-Romero et al., 2009), S-nitrosilation (Cho et al., 2009) and O-GlcNacylationation (Gawlowski et al., 2012) and mitochondrial lipids (Bustillo-Zabalbeitia et al., 2014).

On the other hand, mitochondrial fusion involves mixing of matrix contents, implying that both the MOM and MIM must coordinately fuse.

These two processes are tightly harmonized, although they are exclusive and mechanistically distinct. MOM fusion is regulated by two membrane-anchored proteins termed mitofusins (Mfn1 and Mfn2) (Chen et al., 2003). MIM fusion seems to be mediated by another dynamin family member, OPA1 (Frezza et al., 2006; Olichon et al., 2002, 2003). Another protein involved in mitochondrial dynamics is endophilin B1 (or Bif-1), which regulates membrane curvature (Peter et al., 2004). Bif-1 membrane remodeling function appears to be modulated by the intrinsic curvature of different mitochondrial lipids (Etxebarria et al., 2009). Importantly, mitochondrial dynamics are regulated by two mitochondrial signature phospholipids, the phosphatidic acid (PA) and CL. These two lipids interact with the core components of mitochondrial fusion/fission machinery, including Drp1, OPA1, and mitofusins and provide suitable building blocks for the highly curved structures formed during mitochondrial division and elongation (Kameoka et al., 2018).

Finally, mitochondrial cristae remodeling is a fundamental step for the differential release of apoptotic factors from the mitochondria to the cytosol (Fig. 1C) (McArthur et al., 2018; Riley et al., 2018; Scorrano et al., 2002). OPA1, apart from its role in mitochondrial fusion, also regulates apoptotic cristae remodeling (Cipolat et al., 2006). Again, the CL-tBID tandem was proposed to play a key role in OPA1 regulation and therefore in cristae remodeling. While tBID widens the cristae junctions and disrupts OPA1 oligomers, the lack of CL association appears to be related with deficient OPA1 activity and DOA (dominant optic atrophy) (Ban et al., 2010; Frezza et al., 2006).

3. Role of lipids in nonapoptotic regulated cell death

In recent years, a number of genetically encoded machineries that induce cell death different to apoptosis have emerged, including necroptosis, pyroptosis and ferroptosis (Linkermann et al., 2013; Vince and Silke, 2016; Wallach et al., 2016). Despite the clear biochemical difference among these forms of nonapoptotic cell death, they all end up with the rupture of the plasma membrane and the release of intracellular contents. The PM represents the main target in these processes and the increase in its permeability together with other membrane alterations are key steps at the core of their molecular mechanism (Espiritu et al., 2019; Galluzzi et al., 2014a).

The PM is formed by a phospholipid bilayer in which proteins are embedded and functions as a barrier that delimitates the extracellular and intracellular aqueous compartments. The average idealized mammalian plasma membrane contains sterols, PC, sphingomyelin (SM), and gangliosides predominantly in the outer leaflet and PE, PS, and other charged lipids such as PI in the inner leaflet. Proteins and lipids embedded within the phospholipid bilayer carry out specific functions including selective transport of molecules and ions, signaling and cell–cell recognition. Maintaining PM integrity is essential for ion homeostasis and protecting the cells from the environment (Bilgin et al., 2011; Ingolfsson et al., 2014; van Meer and de Kroon, 2011).

Emerging evidence suggests that lipids and lipid metabolism play an important role in different forms of nonapoptotic cell death. The contribution of lipids in these cases can also be associated with their function as triggers, executors or modulators of PM components that act as platforms for cell death execution (Lizardo et al., 2018; Magtanong et al., 2016; Parisi et al., 2018). In this section, we focus on the essential interplay existing between lipids, membranes and the delicate protein machinery that mediates necroptosis and ferroptosis, the two forms of regulated necrosis for which the contribution of lipids is best understood. We use them to highlight how the relevance of lipid molecules and membrane structures in regulated cell death goes beyond classic apoptosis.

3.1 Lipid-protein reshaping during necroptosis

Necroptosis is an inflammatory type of cell death that is linked to cellular demise in several pathophysiological processes such as bacterial (Bleriot and Lecuit, 2016) and viral (Upton et al., 2012; Wang et al., 2014b) infections, autoimmunity (Alvarez-Diaz et al., 2016), chronic inflammation (Taraborrelli et al., 2018) and cancer (Fulda, 2014; Galluzzi et al., 2017; Su et al., 2016). Mechanistically, it is a caspase-independent form of regulated cell death that critically depends on the pseudokinase Mixed Lineage Kinase domain-Like (MLKL) (Fig. 3A) (Galluzzi et al., 2018). The protein machinery leading to necroptosis, can be triggered through the activation of death receptors (e.g., tumor necrosis factor receptor 1) or pathogen recognition receptors (e.g., Toll-like receptor 3/4), under conditions of caspase-8 inhibition. All these alternative pathways converge in the activation of MLKL in the necrosome, a cytosolic protein complex containing the kinases RIP1 and RIP3 in its canonical form.

MLKL is produced as a cytosolic protein and requires large conformational changes for membrane insertion. RIP3-mediated phosphorylation

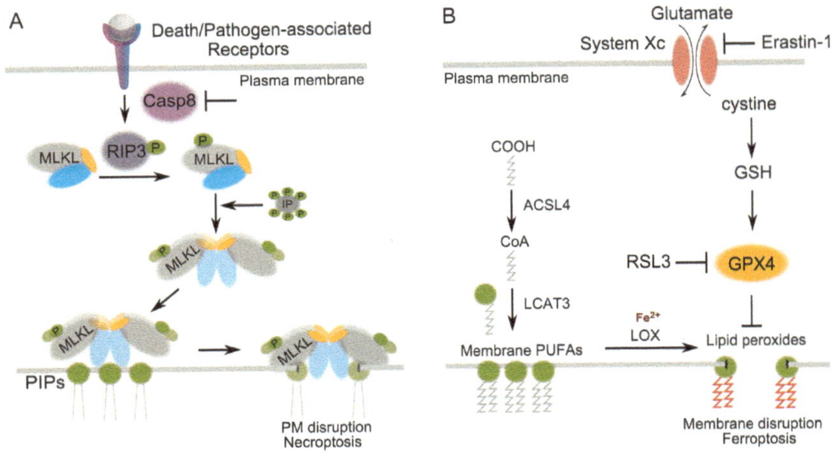

Fig. 3 Role of lipids in necroptosis and ferroptosis. (A) Necroptosis is initiated by various stimuli that all result in the formation of the necrosome, phosphorylation of MLKL and its subsequent translocation to the plasma membrane. MLKL oligomerization is also modulated by its interaction with highly phosphorylated IP, which are soluble products of lipid metabolism. MLKL activation facilitates its weak interaction with PIPs at the membrane. Membrane binding promotes the exposure of new high-affinity sites and strong membrane interaction. (B) Ferroptosis is a caspase-independent cell death mechanism that involves the production of lipid peroxides in cellular membranes. Loss of GPX4 activity is the main trigger of ferroptosis. RSL3, which directly inhibits GPX4, and erastin-1, which affects the catalytic cycle of GPX4 by perturbing the levels of GSH, both trigger ferroptosis. ACSL4 and LPCAT3 mediate the incorporation of PUFAs into membrane. LOX catalyzes the incorporation of oxygen into the long chain PUFAs, to produce lipid hydroperoxides, which are accumulated at the membrane in the absence of GPX4 activity.

induces a conformational change that activates MLKL and drives its oligomerization and translocation to the plasma membrane (Fig. 3A) (Galluzzi et al., 2014a,b). A current model of MLKL activation assumes that its interaction with soluble inositol phosphates and membrane lipids is also important for the molecular switch required for MLKL to exert its membrane destabilizing function (Dovey et al., 2018; McNamara et al., 2018; Quarato et al., 2016). MLKL-mediated plasma membrane disruption represents the key step for the execution of necroptotic cell death (de Almagro and Vucic, 2015; Petrie et al., 2017).

The way necroptosis is orchestrated indicates that lipids and MLKL are two essential pieces of the necroptosis-inducing machinery that reshape each other along the way to kill the cell. On the one hand, the membrane structure is modified upon MLKL insertion. On the other hand, MLKL interaction with lipids and/or membranes contributes to trigger the conformational changes required for its evolution from the soluble and monomeric inactive state to the oligomeric and active membrane-inserted form. In the next sections, we will analyze the function of lipids as regulators of MLKL and necroptosis and discuss the implications of MLKL-induced membrane perturbations.

3.1.1 Lipids as regulators of necroptosis

Specific membrane lipids are needed for MLKL engagement at the plasma membrane. In this regard, phosphatidyl inositol phosphates (PIPs) including phosphatidylinositol-5-phosphate (PIP5) and phosphatidylinositol-4,5-bisphosphate (PIP4,5) act as lipid receptors of MLKL in the inner leaflet of the plasma membrane (Fig. 3A). MLKL is recruited to the membrane through the interaction of a cluster of positively charged amino acids located on the surface of its four helical bundle (4HB) domain with these negatively-charged phospholipids (Dondelinger et al., 2014). MLKL is proposed to insert into the membrane in a multistep process. Upon activation by phosphorylation in the cytosol, MLKL undergoes a conformational change that would initially facilitate a weak interaction of the 4HB with PIPs. Subsequently, lipid binding would promote the exposure of new high-affinity sites within the same domain and the establishment of stronger interactions with the membrane (Quarato et al., 2016).

A recent study has shown that the soluble intermediates of lipid metabolism inositol phosphates (IPs), can act as molecular co-factors of MLKL (Fig. 3A). Specifically, the binding of highly phosphorylated metabolites (i.e., IP6 > IP5 > IP4) to specific regions of MLKL in the 4HB domain promotes a conformational change that releases its auto-inhibition by the displacement of the adjacent brace region. This switch takes places downstream of RIP3-mediated MLKL phosphorylation and is essential for its further oligomerization and plasma membrane recruitment. Interestingly, neither low phosphorylated IPs (which are the soluble proxy for the polar head group of the PIPs) nor PIP4,5, the lipid receptor of MLKL at the PM, are sufficient to trigger a similar switch in MLKL, which supports the specificity of this conformational change prior to membrane binding (Dovey et al., 2018; McNamara et al., 2018).

In addition to PIPs, other lipids are starting to be recognized as regulators of necroptosis. It was found that very long chain fatty acids (VLCFAs) significantly change in the lipidome of necroptotic cells probably due to upregulation of fatty acid biosynthesis and elongation. Accumulation of

VLCFAs seems to be required for membrane permeabilization and necroptotic cell death. Apparently, VLCFAs are not related to the machinery of necroptosis induction but are rather required for the membrane alterations that facilitate cell death by changing the organization of lipids and/or proteins in the membrane (Parisi et al., 2017, 2018).

Recent studies have shown that MLKL interaction with sulfatides plays a role in its membrane destabilizing activity during sciatic nerve injury. In this biological context, the ability of MLKL to perturb the PM does not kill the cell but promotes myelin sheath breakdown. Most likely, this alternative lipid/MLKL interaction results as a consequence of its ability to adopt a different conformation stabilized by the RIP3-independent phosphorylation of S441 (Ying et al., 2018).

3.1.2 Plasma membrane permeabilization as the key execution step in necroptosis

Despite the central role of MLKL in necroptosis, the molecular mechanism of PM permeabilization remains highly controversial. Competing models propose that plasma membrane rupture and cell death are indirectly triggered by MLKL through the activation of endogenous ion channels (Cai et al., 2014; Chen et al., 2014) or, alternatively, that it is directly mediated by MLKL oligomerization at the plasma membrane. The direct mode of membrane perturbation could be the result of its partial insertion into the plasma membrane and/or of its ability to form selective ion channels (Fang et al., 2016), pores (Dondelinger et al., 2014; Wang et al., 2014a) or amyloid-like fibers (Liu et al., 2017) in the PM.

One of the most immediate consequences of MLKL translocation to the PM is the perturbation of cellular ion homeostasis. Upon disruption of the PM barrier function, water uptake is driven by osmotic pressure, resulting in cell swelling and ultimately membrane rupture. Different ion fluxes (e.g., Ca^{2+}, Mg^{2+}, Na^{+}, and K^{+}) have appeared as potential mediators of necroptotic cell death (Cai et al., 2014; Chen et al., 2014; Fang et al., 2016; Gong et al., 2017a,b; Ousingsawat et al., 2017; Ros et al., 2017; Zhang et al., 2016). As calcium is the most versatile secondary intracellular messenger (Clapham, 2007; Zhivotovsky and Orrenius, 2011), early studies on necroptosis focused on this ion. There is mounting evidence showing that cytosolic calcium concentration increases upon necroptosis induction dependent on MLKL. It remains to be seen how exactly the alteration of PM permeability is connected with sustained high calcium fluxes and if this ion plays an additional role as a mediator of

MLKL activation or other intracellular alterations including membrane remodeling in necroptosis.

Similar to apoptosis, necroptotic cells also expose PS on the outer leaflet of the PM that acts as an "eat-me" signal and mediates the recognition of the dying cell by macrophages. However, the mechanistic details of PS externalization during necroptosis are much less understood compared to apoptosis. In fact, the molecular actors implicated in apoptosis appear dispensable in necroptosis (Espiritu et al., 2019). PS exposure in necroptosis occurs much earlier than the breakdown of the PM (Gong et al., 2017a,b; Zargarian et al., 2017) and almost simultaneously to the rise of cytosolic calcium (Espiritu et al., 2019). Yet, the identification of the proteins responsible for PS externalization in necroptosis is a matter of ongoing research. Interestingly, PS exposure follows activation and membrane translocation of MLKL, which has led to the hypothesis about a possible direct role of MLKL in lipid scrambling (Zargarian et al., 2017). Alternatively, PS externalization could be the result of MLKL forming toroidal or lipid/protein pores. In this type of pores, which are similar to those formed by BAX during apoptosis, the generation of high membrane curvature and the continuity between the inner and outer leaflets of the PM at the pore edge would allow lipid scrambling to the surface (Pokorny and Almeida, 2004).

Remarkably, MLKL targeting of the PM does not represent a point of no return in necroptosis progression. Phosphorylated MLKL can be removed from the membrane through different membrane-repair mechanisms including shedding of damaged membrane regions through the action of the endosomal sorting complexes required for transport machinery (ESCRT-III) (Gong et al., 2017a,b) or flotillin-mediated endocytosis followed by lysosomal degradation and ALIX-syntenin-1 mediated exocytosis (Fan et al., 2019). It is argued that ESCRT-mediated membrane repair could delay necroptosis to allow cells to produce and release critical cytokines that are required for specific inflammatory or immune modulatory responses in necroptosis (Gong et al., 2017a,b). Moreover, these new findings highlighted the importance of membrane resealing as a dynamic mechanism that permits restoring cellular homeostasis when the necroptotic stimulus is removed (Gong et al., 2017a,b).

3.2 Central role of lipid peroxidation in ferroptosis

3.2.1 Antioxidant and lipid metabolism in ferroptosis

Ferroptosis is a type of regulated cell death of necrotic nature that is linked to different pathologies including ischemia/reperfusion injury (Li et al., 2019a;

Tuo et al., 2017), neurodegenerative diseases and cancer (Stockwell et al., 2017). By definition, ferroptosis is caspase-independent, under constitutive control of glutathione peroxidase 4 (GPX4) and specifically initiated by the generation of lipid peroxides (Fig. 3B) (Dixon et al., 2012; Galluzzi et al., 2018). Cellular iron is also recognized as a key factor in ferroptosis. Excess of active iron generation facilitates ROS production based on the Fenton reaction, which promotes lipid peroxidation and ferroptosis (Dixon et al., 2012). In contrast to apoptosis and necroptosis, which possess a specific protein machinery to mediate the execution step of cell death, there are no proteins directly involved in the execution of ferroptosis identified so far. Instead, it seems that ferroptosis is induced by a failure in the antioxidant defenses of the cell. How execution of ferroptotic cell death is mediated at the molecular level remains unknown.

The metabolism of polyunsaturated fatty acid (PUFAs) and the generation of specific lipid peroxides through iron-dependent enzymatic reactions are key processes in ferroptosis (Agmon and Stockwell, 2017; Yang et al., 2016). Membrane phospholipids containing the long chain PUFAs such as arachidonic acid ($20{:}4n-6$) are specifically peroxidized during this type of cell death (Dixon, 2017; Dixon et al., 2012; Forcina and Dixon, 2019). In contrast, exogenous monounsaturated fatty acids can suppress ferroptosis by replacing PUFAs from the plasma membrane (Magtanong et al., 2019). Among the different membrane phospholipids, specific oxidation of phosphatidylethanolamine-containing PUFAs has been recently suggested to mediate ferroptosis (Wenzel et al., 2017). However, this model is still under debate.

GPX4 activity reduces toxic peroxides of phospholipids in membranes to nontoxic alcohols. Therefore, depletion or inhibition of this unique enzyme is considered the main trigger of ferroptosis (Seiler et al., 2008; Yang et al., 2016). Erastin-1 promotes ferroptosis via a mechanism that involves the inhibition of the cystine/glutamate antiporter system Xc, and the consequent decrease of intracellular cysteine and GSH (which is the co-factor of GPX4). Moreover, the Ras-selective lethal small molecule 3 (RSL3) induces ferroptosis by directly inhibiting GPX4. As PUFAs are highly susceptible to peroxidation, loss of activity of GPX4 results in the accumulation of lipid hydroperoxides. Together with these lipids, lysophospholipids also accumulate in ferroptosis, probably resulting from the cleavage of peroxidized PUFAs from the glycerophospholipid backbones (Forcina and Dixon, 2019; Proneth and Conrad, 2019).

Classic enzymes of lipid metabolism have been linked to ferroptosis. In line with the essential role of PUFAs, acyl-CoA synthetase long chain family

4 (ACSL4) and lysophosphatidylcholine acyltransferase 3 (LPCAT3) play a role by mediating the incorporation of PUFAs into membrane phospholipids (Doll et al., 2017; Kagan et al., 2017; Proneth and Conrad, 2019). In fact, genetic deletion of these enzymes causes resistance to ferroptosis. Multiple lipoxygenases (i.e., ALOX12 and 15) have also been directly associated with ferroptosis signaling (Shintoku et al., 2017; Wang et al., 2017; Yang et al., 2016). These are iron-dependent dioxygenases that catalyze the incorporation of oxygen into the long chain PUFAs, to produce lipid hydroperoxides intermediates of leukotriene metabolic pathways (Wenzel et al., 2017). As a result, the balance between the production of lipid hydroperoxides by LOX and its reduction to the nontoxic alcohol by GPX4 might serve as a pivotal point between cell survival and ferroptotic cell death. However, the role of LOX enzymes in ferroptosis has been recently challenged. A recent hypothesis defends that although these proteins kick off lipid peroxidation, oxidation of lipids in ferroptosis is derived from nonenzymatic autoxidation in a chain reaction (Proneth and Conrad, 2019). Beyond the function of LOX in ferroptosis, these enzymes and the prostaglandin-endoperoxide synthase (PTGS) are controlled by the levels of cellular peroxides, which could have implications for the inflammatory effects of ferroptosis by modulating the levels of prostaglandins and leukotrienes produced during cell death (Proneth and Conrad, 2019).

3.2.2 Cellular membranes as targets of lipid oxidation

Ferroptosis is a form of regulated cell death that intrinsically involves the alteration of the properties of the lipids in membranes. We are just starting to grasp some of the membrane remodeling effects potentiated in this form of cell death. Oxidable phospholipids are spread throughout cellular membranes including the PM, mitochondria, ER and lysosomes, which could act as sites of lipid ROS accumulation. However, the specific membrane target of lipid oxidation during ferroptosis remains unclear. Both plasma and mitochondrial membranes have been identified as possible main sites of lipid oxidation in ferroptotic cells. However, the role of mitochondria in ferroptosis remains controversial and seems to be context-dependent. Up to date, there is no evidence whether ferroptosis involves the formation of protein-based or lipid-based pores and research is still ongoing in this direction. One hypothesis proposes that oxidative destruction of the cell can result from the formation of small gaps or discontinuities in the PM that impair ion homeostasis. In this regard, it is not clear whether peroxidation of PUFA lipids at the PM during ferroptosis would be sufficient to cause its permeabilization or, alternatively, change the membrane properties in a way that alters the function of key proteins required for cell viability (Agmon and Stockwell, 2017). It seems that peroxidized lipids produced during ferroptosis can relocate into the aqueous phase resulting in membrane thinning (Wong-Ekkabut et al., 2007). Furthermore, a recent molecular dynamics simulation study suggests that the accumulation of oxidized-PUFA-containing phospholipids in the bilayer during ferroptosis would induce membrane thinning together with an increase in membrane curvature, which would enhance the accessibility of the membrane to additional oxidants and promote micellization with the consequent damage to membrane integrity (Agmon and Stockwell, 2017). However, all these hypotheses require further experimental validation in the coming years.

4. Concluding remarks

Membranes are more than mere scaffolds where proteins associate to rule cellular processes. They provide compartmentalization, thereby enabling gradients, which are fundamental for cellular homeostasis. There are many pathways by which cells can mediate their own dismissal. Each of them is characterized by a unique membrane alteration pattern, but they all share the loosening of the membrane integrity. For the execution of cell death, membrane lipid composition determines specific protein-lipid interaction profiles and signaling events required for the activation of the deadly machinery. Lipids play a fundamental role in these processes as components of the bilayer structure of the cellular membranes, as direct triggers or as modulators of the protein machinery required to kill the cells.

Mitochondrial lipids regulate intrinsic apoptosis, both indirectly by changing the mechanical proteins of the membrane, or directly by specifically modulating protein targeting, structure and function. In this chapter, we discussed the role of different mitochondrial lipids that affect cell fate, with particular interest in CL, a mitochondrial specific lipid. CL has emerged as one of the most studied regulators of apoptotic cell death, due to its link to ROS production, mitochondrial potential loss and reshaping processes and its influence in BCL2 interaction network. Importantly, alterations in the abundance and molecular forms of CL are associated with several diseases, such as Barth syndrome, heart failure, aging, and cancer, among others. Despite the complexity of establishing an exclusive role for the CL or other mitochondrial lipids in those human pathologies, the potential role of lipid targeting in therapeutics cannot be diminished.

On the other hand, PM constitutes the locus for multiple cellular processes, including those that affect cellular integrity like necroptosis and ferroptosis. Lipids are emerging as key components of the mechanism of action of these poorly studied types of cell death. In necroptosis, its ultimate executor MLKL is regulated by lipids: its membrane targeting requires PIPs and a direct interaction with specific membrane lipids seems necessary for its membrane permeabilizing activity. Ferroptosis is probably the kind of cell death where lipids and lipid metabolism have an obvious impact on cell death. In this type of nonapoptotic regulated cell death, the oxidative modification of different membrane lipids seems to affect both the integrity of the PM and lipid interactions with protein components.

Altogether, it is becoming clear that lipids mediate in important cellular processes not only during healthy conditions, but also upon cell death regulation. In RCD, lipids can directly induce membrane rupture by modulating membrane properties or by interacting with concrete proteins (e.g., RCD activators and/or last executors). Moreover, lipid metabolism and membrane dynamics constitute essential pieces in the regulation of the cellular fate. This knowledge must boost the development of therapies that target lipids and lipid metabolic processes, as dysregulation of the cellular lipidome has undeniably a clear impact on multiple human pathologies linked to RCD.

Acknowledgments

We thank Dr. John S. Danial and Malina John for carefully reading our manuscript and for their comments. Work in our lab is supported by the European Research Council and by the Deutsche Forschungsgemeinschaft. A.J.G.S. is a Young Investigator from EMBO.

References

Abate, M., Festa, A., Falco, M., Lombardi, A., Luce, A., Grimaldi, A., Zappavigna, S., Sperlongano, P., Irace, C., Caraglia, M., Misso, G., 2019. Mitochondria as playmakers of apoptosis, autophagy and senescence. Semin. Cell Dev. Biol. https://doi.org/10.1016/j.semcdb.2019.05.022.

Agmon, E., Stockwell, B.R., 2017. Lipid homeostasis and regulated cell death. Curr. Opin. Chem. Biol. 39, 83–89.

Alvarez-Diaz, S., Dillon, C.P., Lalaoui, N., Tanzer, M.C., Rodriguez, D.A., Lin, A., Lebois, M., Hakem, R., Josefsson, E.C., O'reilly, L.A., Silke, J., Alexander, W.S., Green, D.R., Strasser, A., 2016. The pseudokinase MLKL and the kinase RIPK3 have distinct roles in autoimmune disease caused by loss of death-receptor-induced apoptosis. Immunity 45, 513–526.

Aouacheria, A., Rech De Laval, V., Combet, C., Hardwick, J.M., 2013. Evolution of Bcl-2 homology motifs: homology versus homoplasy. Trends Cell Biol. 23, 103–111.

Aouacheria, A., Baghdiguian, S., Lamb, H.M., Huska, J.D., Pineda, F.J., Hardwick, J.M., 2017. Connecting mitochondrial dynamics and life-or-death events via Bcl-2 family proteins. Neurochem. Int. 109, 141–161.

Arana, L., Gangoiti, P., Ouro, A., Trueba, M., Gomez-Munoz, A., 2010. Ceramide and ceramide 1-phosphate in health and disease. Lipids Health Dis. 9, 15.

Ardail, D., Privat, J.P., Egret-Charlier, M., Levrat, C., Lerme, F., Louisot, P., 1990. Mitochondrial contact sites. Lipid composition and dynamics. J. Biol. Chem. 265, 18797–18802.

Baggetto, L.G., Clottes, E., Vial, C., 1992. Low mitochondrial proton leak due to high membrane cholesterol content and cytosolic creatine kinase as two features of the deviant bioenergetics of Ehrlich and AS30-D tumor cells. Cancer Res. 52, 4935–4941.

Ban, T., Heymann, J.A., Song, Z., Hinshaw, J.E., Chan, D.C., 2010. OPA1 disease alleles causing dominant optic atrophy have defects in cardiolipin-stimulated GTP hydrolysis and membrane tubulation. Hum. Mol. Genet. 19, 2113–2122.

Basanez, G., Nechushtan, A., Drozhinin, O., Chanturiya, A., Choe, E., Tutt, S., Wood, K.A., Hsu, Y., Zimmerberg, J., Youle, R.J., 1999. Bax, but not Bcl-xL, decreases the lifetime of planar phospholipid bilayer membranes at subnanomolar concentrations. Proc. Natl. Acad. Sci. U. S. A. 96, 5492–5497.

Basanez, G., Zhang, J., Chau, B.N., Maksaev, G.I., Frolov, V.A., Brandt, T.A., Burch, J., Hardwick, J.M., Zimmerberg, J., 2001. Pro-apoptotic cleavage products of Bcl-xL form cytochrome c-conducting pores in pure lipid membranes. J. Biol. Chem. 276, 31083–31091.

Basanez, G., Sharpe, J.C., Galanis, J., Brandt, T.B., Hardwick, J.M., Zimmerberg, J., 2002. Bax-type apoptotic proteins porate pure lipid bilayers through a mechanism sensitive to intrinsic monolayer curvature. J. Biol. Chem. 277, 49360–49365.

Basanez, G., Soane, L., Hardwick, J.M., 2012. A new view of the lethal apoptotic pore. PLoS Biol. 10, e1001399.

Beverly, L.J., Howell, L.A., Hernandez-Corbacho, M., Casson, L., Chipuk, J.E., Siskind, L.J., 2013. BAK activation is necessary and sufficient to drive ceramide synthase-dependent ceramide accumulation following inhibition of BCL2-like proteins. Biochem. J. 452, 111–119.

Bilgin, M., Markgraf, D.F., Duchoslav, E., Knudsen, J., Jensen, O.N., De Kroon, A.I., Ejsing, C.S., 2011. Quantitative profiling of PE, MMPE, DMPE, and PC lipid species by multiple precursor ion scanning: a tool for monitoring PE metabolism. Biochim. Biophys. Acta 1811, 1081–1089.

Billen, L.P., Kokoski, C.L., Lovell, J.F., Leber, B., Andrews, D.W., 2008. Bcl-XL inhibits membrane permeabilization by competing with Bax. PLoS Biol. 6, e147.

Birbes, H., El Bawab, S., Hannun, Y.A., Obeid, L.M., 2001. Selective hydrolysis of a mitochondrial pool of sphingomyelin induces apoptosis. FASEB J. 15, 2669–2679.

Birbes, H., Luberto, C., Hsu, Y.T., El Bawab, S., Hannun, Y.A., Obeid, L.M., 2005. A mitochondrial pool of sphingomyelin is involved in TNFalpha-induced Bax translocation to mitochondria. Biochem. J. 386, 445–451.

Birkinshaw, R.W., Czabotar, P.E., 2017. The BCL-2 family of proteins and mitochondrial outer membrane permeabilisation. Semin. Cell Dev. Biol. 72, 152–162.

Bleicken, S., Landeta, O., Landajuela, A., Basanez, G., Garcia-Saez, A.J., 2013a. Proapoptotic Bax and Bak proteins form stable protein-permeable pores of tunable size. J. Biol. Chem. 288, 33241–33252.

Bleicken, S., Wagner, C., Garcia-Saez, A.J., 2013b. Mechanistic differences in the membrane activity of Bax and Bcl-xL correlate with their opposing roles in apoptosis. Biophys. J. 104, 421–431.

Bleicken, S., Jeschke, G., Stegmueller, C., Salvador-Gallego, R., Garcia-Saez, A.J., Bordignon, E., 2014. Structural model of active Bax at the membrane. Mol. Cell 56, 496–505.

Bleicken, S., Hantusch, A., Das, K.K., Frickey, T., Garcia-Saez, A.J., 2017. Quantitative interactome of a membrane Bcl-2 network identifies a hierarchy of complexes for apoptosis regulation. Nat. Commun. 8, 73.

Bleriot, C., Lecuit, M., 2016. The interplay between regulated necrosis and bacterial infection. Cell. Mol. Life Sci. 73, 2369–2378.

Boldogh, I.R., Pon, L.A., 2006. Interactions of mitochondria with the actin cytoskeleton. Biochim. Biophys. Acta 1763, 450–462.

Bordt, E.A., Clerc, P., Roelofs, B.A., Saladino, A.J., Tretter, L., Adam-Vizi, V., Cherok, E., Khalil, A., Yadava, N., Ge, S.X., Francis, T.C., Kennedy, N.W., Picton, L.K., Kumar, T., Uppuluri, S., Miller, A.M., Itoh, K., Karbowski, M., Sesaki, H., Hill, R.B., Polster, B.M., 2017. The putative Drp1 inhibitor mdivi-1 is a reversible mitochondrial complex I inhibitor that modulates reactive oxygen species. Dev. Cell 40, 583–594 e6.

Bustillo-Zabalbeitia, I., Montessuit, S., Raemy, E., Basanez, G., Terrones, O., Martinou, J.C., 2014. Specific interaction with cardiolipin triggers functional activation of Dynamin-Related Protein 1. PLoS One 9, e102738.

Cai, Z., Jitkaew, S., Zhao, J., Chiang, H.C., Choksi, S., Liu, J., Ward, Y., Wu, L.G., Liu, Z.G., 2014. Plasma membrane translocation of trimerized MLKL protein is required for TNF-induced necroptosis. Nat. Cell Biol. 16, 55–65.

Campbell, K.J., Tait, S.W.G., 2018. Targeting BCL-2 regulated apoptosis in cancer. Open Biol. 8, 180002. https://doi.org/10.1098/rsob.180002.

Canals, D., Hannun, Y.A., 2013. Novel chemotherapeutic drugs in sphingolipid cancer research. Handb. Exp. Pharmacol. 215, 211–238.

Cartron, P.F., Arokium, H., Oliver, L., Meflah, K., Manon, S., Vallette, F.M., 2005. Distinct domains control the addressing and the insertion of Bax into mitochondria. J. Biol. Chem. 280, 10587–10598.

Chen, H., Detmer, S.A., Ewald, A.J., Griffin, E.E., Fraser, S.E., Chan, D.C., 2003. Mitofusins Mfn1 and Mfn2 coordinately regulate mitochondrial fusion and are essential for embryonic development. J. Cell Biol. 160, 189–200.

Chen, L., Willis, S.N., Wei, A., Smith, B.J., Fletcher, J.I., Hinds, M.G., Colman, P.M., Day, C.L., Adams, J.M., Huang, D.C., 2005. Differential targeting of prosurvival Bcl-2 proteins by their BH3-only ligands allows complementary apoptotic function. Mol. Cell 17, 393–403.

Chen, X., Li, W., Ren, J., Huang, D., He, W.T., Song, Y., Yang, C., Li, W., Zheng, X., Chen, P., Han, J., 2014. Translocation of mixed lineage kinase domain-like protein to plasma membrane leads to necrotic cell death. Cell Res. 24, 105–121.

Chen, H.C., Kanai, M., Inoue-Yamauchi, A., Tu, H.C., Huang, Y., Ren, D., Kim, H., Takeda, S., Reyna, D.E., Chan, P.M., Ganesan, Y.T., Liao, C.P., Gavathiotis, E., Hsieh, J.J., Cheng, E.H., 2015. An interconnected hierarchical model of cell death regulation by the BCL-2 family. Nat. Cell Biol. 17, 1270–1281.

Cheng, E.H., Kirsch, D.G., Clem, R.J., Ravi, R., Kastan, M.B., Bedi, A., Ueno, K., Hardwick, J.M., 1997. Conversion of Bcl-2 to a Bax-like death effector by caspases. Science 278, 1966–1968.

Chipuk, J.E., Mcstay, G.P., Bharti, A., Kuwana, T., Clarke, C.J., Siskind, L.J., Obeid, L.M., Green, D.R., 2012. Sphingolipid metabolism cooperates with BAK and BAX to promote the mitochondrial pathway of apoptosis. Cell 148, 988–1000.

Cho, D.H., Nakamura, T., Fang, J., Cieplak, P., Godzik, A., Gu, Z., Lipton, S.A., 2009. S-nitrosylation of Drp1 mediates beta-amyloid-related mitochondrial fission and neuronal injury. Science 324, 102–105.

Cipolat, S., Rudka, T., Hartmann, D., Costa, V., Serneels, L., Craessaerts, K., Metzger, K., Frezza, C., Annaert, W., D'adamio, L., Derks, C., Dejaegere, T., Pellegrini, L., D'hooge, R., Scorrano, L., De Strooper, B., 2006. Mitochondrial rhomboid PARL regulates cytochrome c release during apoptosis via OPA1-dependent cristae remodeling. Cell 126, 163–175.

Clapham, D.E., 2007. Calcium signaling. Cell 131, 1047–1058.
Claypool, S.M., Koehler, C.M., 2012. The complexity of cardiolipin in health and disease. Trends Biochem. Sci. 37, 32–41.
Clem, R.J., Cheng, E.H., Karp, C.L., Kirsch, D.G., Ueno, K., Takahashi, A., Kastan, M.B., Griffin, D.E., Earnshaw, W.C., Veliuona, M.A., Hardwick, J.M., 1998. Modulation of cell death by Bcl-XL through caspase interaction. Proc. Natl. Acad. Sci. U. S. A. 95, 554–559.
Colbeau, A., Nachbaur, J., Vignais, P.M., 1971. Enzymic characterization and lipid composition of rat liver subcellular membranes. Biochim. Biophys. Acta 249, 462–492.
Colombini, M., 2010. Ceramide channels and their role in mitochondria-mediated apoptosis. Biochim. Biophys. Acta 1797, 1239–1244.
Colombini, M., 2017. Ceramide channels and mitochondrial outer membrane permeability. J. Bioenerg. Biomembr. 49, 57–64.
Cosentino, K., Garcia-Saez, A.J., 2014. Mitochondrial alterations in apoptosis. Chem. Phys. Lipids 181, 62–75.
Cosentino, K., Garcia-Saez, A.J., 2017. Bax and Bak pores: are we closing the circle? Trends Cell Biol. 27, 266–275.
Crain, R.C., Clark, R.W., Harvey, B.E., 1983. Role of lipid transfer proteins in the abnormal lipid content of Morris hepatoma mitochondria and microsomes. Cancer Res. 43, 3197–3202.
Cremesti, A., Paris, F., Grassme, H., Holler, N., Tschopp, J., Fuks, Z., Gulbins, E., Kolesnick, R., 2001. Ceramide enables fas to cap and kill. J. Biol. Chem. 276, 23954–23961.
Csordas, G., Renken, C., Varnai, P., Walter, L., Weaver, D., Buttle, K.F., Balla, T., Mannella, C.A., Hajnoczky, G., 2006. Structural and functional features and significance of the physical linkage between ER and mitochondria. J. Cell Biol. 174, 915–921.
Csordas, G., Varnai, P., Golenar, T., Roy, S., Purkins, G., Schneider, T.G., Balla, T., Hajnoczky, G., 2010. Imaging interorganelle contacts and local calcium dynamics at the ER-mitochondrial interface. Mol. Cell 39, 121–132.
Csordas, G., Weaver, D., Hajnoczky, G., 2018. Endoplasmic reticular-mitochondrial contactology: structure and signaling functions. Trends Cell Biol. 28, 523–540.
Czabotar, P.E., Westphal, D., Dewson, G., Ma, S., Hockings, C., Fairlie, W.D., Lee, E.F., Yao, S., Robin, A.Y., Smith, B.J., Huang, D.C., Kluck, R.M., Adams, J.M., Colman, P.M., 2013. Bax crystal structures reveal how BH3 domains activate Bax and nucleate its oligomerization to induce apoptosis. Cell 152, 519–531.
Czabotar, P.E., Lessene, G., Strasser, A., Adams, J.M., 2014. Control of apoptosis by the BCL-2 protein family: implications for physiology and therapy. Nat. Rev. Mol. Cell Biol. 15, 49–63.
Czubowicz, K., Jesko, H., Wencel, P., Lukiw, W.J., Strosznajder, R.P., 2019. The role of ceramide and sphingosine-1-phosphate in Alzheimer's disease and other neurodegenerative disorders. Mol. Neurobiol. 56, 5436–5455.
Dadsena, S., Bockelmann, S., Mina, J.G.M., Hassan, D.G., Korneev, S., Razzera, G., Jahn, H., Niekamp, P., Muller, D., Schneider, M., Tafesse, F.G., Marrink, S.J., Melo, M.N., Holthuis, J.C.M., 2019. Ceramides bind VDAC2 to trigger mitochondrial apoptosis. Nat. Commun. 10, 1832.
Daum, G., 1985. Lipids of mitochondria. Biochim. Biophys. Acta 822, 1–42.
De Almagro, M.C., Vucic, D., 2015. Necroptosis: pathway diversity and characteristics. Semin. Cell Dev. Biol. 39, 56–62.
De Kroon, A.I., Dolis, D., Mayer, A., Lill, R., De Kruijff, B., 1997. Phospholipid composition of highly purified mitochondrial outer membranes of rat liver and Neurospora crassa. Is cardiolipin present in the mitochondrial outer membrane? Biochim. Biophys. Acta 1325, 108–116.

Dewson, G., Kratina, T., Sim, H.W., Puthalakath, H., Adams, J.M., Colman, P.M., Kluck, R.M., 2008. To trigger apoptosis, Bak exposes its BH3 domain and homodimerizes via BH3: groove interactions. Mol. Cell 30, 369–380.

Dixon, S.J., 2017. Ferroptosis: bug or feature? Immunol. Rev. 277, 150–157.

Dixon, S.J., Lemberg, K.M., Lamprecht, M.R., Skouta, R., Zaitsev, E.M., Gleason, C.E., Patel, D.N., Bauer, A.J., Cantley, A.M., Yang, W.S., Morrison, B., Stockwell, B.R., 2012. Ferroptosis: an iron-dependent form of nonapoptotic cell death. Cell 149, 1060–1072.

Dlugosz, P.J., Billen, L.P., Annis, M.G., Zhu, W., Zhang, Z., Lin, J., Leber, B., Andrews, D.W., 2006. Bcl-2 changes conformation to inhibit Bax oligomerization. EMBO J. 25, 2287–2296.

Doll, S., Proneth, B., Tyurina, Y.Y., Panzilius, E., Kobayashi, S., Ingold, I., Irmler, M., Beckers, J., Aichler, M., Walch, A., Prokisch, H., Trumbach, D., Mao, G., Qu, F., Bayir, H., Fullekrug, J., Scheel, C.H., Wurst, W., Schick, J.A., Kagan, V.E., Angeli, J.P., Conrad, M., 2017. ACSL4 dictates ferroptosis sensitivity by shaping cellular lipid composition. Nat. Chem. Biol. 13, 91–98.

Dondelinger, Y., Declercq, W., Montessuit, S., Roelandt, R., Goncalves, A., Bruggeman, I., Hulpiau, P., Weber, K., Sehon, C.A., Marquis, R.W., Bertin, J., Gough, P.J., Savvides, S., Martinou, J.C., Bertrand, M.J., Vandenabeele, P., 2014. MLKL compromises plasma membrane integrity by binding to phosphatidylinositol phosphates. Cell Rep. 7, 971–981.

Dovey, C.M., Diep, J., Clarke, B.P., Hale, A.T., McNamara, D.E., Guo, H., Brown Jr., N.W., Cao, J.Y., Grace, C.R., Gough, P.J., Bertin, J., Dixon, S.J., Fiedler, D., Mocarski, E.S., Kaiser, W.J., Moldoveanu, T., York, J.D., Carette, J.E., 2018. MLKL requires the inositol phosphate code to execute necroptosis. Mol. Cell 70, 936–948 e7.

Edlich, F., Banerjee, S., Suzuki, M., Cleland, M.M., Arnoult, D., Wang, C., Neutzner, A., Tjandra, N., Youle, R.J., 2011. Bcl-x(L) retrotranslocates Bax from the mitochondria into the cytosol. Cell 145, 104–116.

Epand, R.F., Martinou, J.C., Montessuit, S., Epand, R.M., 2003. Transbilayer lipid diffusion promoted by Bax: implications for apoptosis. Biochemistry 42, 14576–14582.

Epand, R.F., Schlattner, U., Wallimann, T., Lacombe, M.L., Epand, R.M., 2007. Novel lipid transfer property of two mitochondrial proteins that bridge the inner and outer membranes. Biophys. J. 92, 126–137.

Espiritu, R.A., Pedrera, L., Ros, U., 2019. Tuning the Way to Die: Implications of Membrane Perturbations in Necroptosis. vol. 29. Elsevier, pp. 201–247.

Esposti, M.D., Erler, J.T., Hickman, J.A., Dive, C., 2001. Bid, a widely expressed proapoptotic protein of the Bcl-2 family, displays lipid transfer activity. Mol. Cell. Biol. 21, 7268–7276.

Etxebarria, A., Terrones, O., Yamaguchi, H., Landajuela, A., Landeta, O., Antonsson, B., Wang, H.G., Basanez, G., 2009. Endophilin B1/Bif-1 stimulates BAX activation independently from its capacity to produce large scale membrane morphological rearrangements. J. Biol. Chem. 284, 4200–4212.

Fan, W., Guo, J., Gao, B., Zhang, W., Ling, L., Xu, T., Pan, C., Li, L., Chen, S., Wang, H., Zhang, J., Wang, X., 2019. Flotillin-mediated endocytosis and ALIX-syntenin-1-mediated exocytosis protect the cell membrane from damage caused by necroptosis. Sci. Signal. 12, eaaw3423.

Figueroa-Romero, C., Iniguez-Lluhi, J.A., Stadler, J., Chang, C.R., Arnoult, D., Keller, P.J., Hong, Y., Blackstone, C., Feldman, E.L., 2009. SUMOylation of the mitochondrial fission protein Drp1 occurs at multiple nonconsensus sites within the B domain and is linked to its activity cycle. FASEB J. 23, 3917–3927.

Flannery, P.J., Trushina, E., 2019. Mitochondrial dynamics and transport in Alzheimer's disease. Mol. Cell. Neurosci. 98, 109–120.

Fleischer, S., Rouser, G., Fleischer, B., Casu, A., Kritchevsky, G., 1967. Lipid composition of mitochondria from bovine heart, liver and kidney. J. Lipid. Res. 8, 170–180.
Flores-Romero, H., Garcia-Saez, A.J., 2019a. The incomplete puzzle of the BCL2 proteins. Cells 8, E1176. https://doi.org/10.3390/cells8101176.
Flores-Romero, H., Garcia-Saez, A.J., 2019b. MAVS-induced mitochondrial membrane remodeling. FEBS J. 286, 1540–1542.
Flores-Romero, H., Garcia-Porras, M., Basanez, G., 2017. Membrane insertion of the BAX core, but not latch domain, drives apoptotic pore formation. Sci. Rep. 7, 16259.
Flores-Romero, H., Landeta, O., Ugarte-Uribe, B., Cosentino, K., Garcia-Porras, M., Garcia-Saez, A.J., Basanez, G., 2018. BFL1 modulates apoptosis at the membrane level through a bifunctional and multimodal mechanism showing key differences with BCLXL. Cell Death Differ. 26, 1880–1894. https://doi.org/10.1038/s41418-018-0258-5.
Fonseca, T.B., Sanchez-Guerrero, A., Milosevic, I., Raimundo, N., 2019. Mitochondrial fission requires DRP1 but not dynamins. Nature 570, E34–E42.
Forcina, G.C., Dixon, S.J., 2019. GPX4 at the crossroads of lipid homeostasis and ferroptosis. Proteomics 19, e1800311.
Frank, S., Gaume, B., Bergmann-Leitner, E.S., Leitner, W.W., Robert, E.G., Catez, F., Smith, C.L., Youle, R.J., 2001. The role of dynamin-related protein 1, a mediator of mitochondrial fission, in apoptosis. Dev. Cell 1, 515–525.
Frezza, C., Cipolat, S., Martins De Brito, O., Micaroni, M., Beznoussenko, G.V., Rudka, T., Bartoli, D., Polishuck, R.S., Danial, N.N., De Strooper, B., Scorrano, L., 2006. OPA1 controls apoptotic cristae remodeling independently from mitochondrial fusion. Cell 126, 177–189.
Friedman, J.R., Lackner, L.L., West, M., Dibenedetto, J.R., Nunnari, J., Voeltz, G.K., 2011. ER tubules mark sites of mitochondrial division. Science 334, 358–362.
Friedmann Angeli, J.P., Krysko, D.V., Conrad, M., 2019. Ferroptosis at the crossroads of cancer-acquired drug resistance and immune evasion. Nat. Rev. Cancer 19, 405–414.
Fuertes, G., 2011. Baxα5 at Lipid Membranes: Structure, Assembly and Pore Formation. Ph.D thesis, Universitat de València.
Fuertes, G., Garcia-Saez, A.J., Esteban-Martin, S., Gimenez, D., Sanchez-Munoz, O.L., Schwille, P., Salgado, J., 2010. Pores formed by Baxalpha5 relax to a smaller size and keep at equilibrium. Biophys. J. 99, 2917–2925.
Fulda, S., 2014. Therapeutic exploitation of necroptosis for cancer therapy. Semin. Cell Dev. Biol. 35, 51–56.
Galluzzi, L., Kepp, O., Krautwald, S., Kroemer, G., Linkermann, A., 2014a. Molecular mechanisms of regulated necrosis. Semin. Cell Dev. Biol. 35, 24–32.
Galluzzi, L., Kepp, O., Kroemer, G., 2014b. MLKL regulates necrotic plasma membrane permeabilization. Cell Res. 24, 139–140.
Galluzzi, L., Kepp, O., Chan, F.K., Kroemer, G., 2017. Necroptosis: mechanisms and relevance to disease. Annu. Rev. Pathol. 12, 103–130.
Galluzzi, L., Vitale, I., Aaronson, S.A., Abrams, J.M., Adam, D., Agostinis, P., Alnemri, E.S., Altucci, L., Amelio, I., Andrews, D.W., Annicchiarico-Petruzzelli, M., Antonov, A.V., Arama, E., Baehrecke, E.H., Barlev, N.A., Bazan, N.G., Bernassola, F., Bertrand, M.J.M., Bianchi, K., Blagosklonny, M.V., Blomgren, K., Borner, C., Boya, P., Brenner, C., Campanella, M., Candi, E., Carmona-Gutierrez, D., Cecconi, F., Chan, F.K., Chandel, N.S., Cheng, E.H., Chipuk, J.E., Cidlowski, J.A., Ciechanover, A., Cohen, G.M., Conrad, M., Cubillos-Ruiz, J.R., Czabotar, P.E., D'angiolella, V., Dawson, T.M., Dawson, V.L., De Laurenzi, V., De Maria, R., Debatin, K.M., Deberardinis, R.J., Deshmukh, M., Di Daniele, N., Di Virgilio, F., Dixit, V.M., Dixon, S.J., Duckett, C.S., Dynlacht, B.D., El-Deiry, W.S., Elrod, J.W., Fimia, G.M., Fulda, S., Garcia-Saez, A.J., Garg, A.D., Garrido, C.,

Gavathiotis, E., Golstein, P., Gottlieb, E., Green, D.R., Greene, L.A., Gronemeyer, H., Gross, A., Hajnoczky, G., Hardwick, J.M., Harris, I.S., Hengartner, M.O., Hetz, C., Ichijo, H., Jaattela, M., Joseph, B., Jost, P.J., Juin, P.P., Kaiser, W.J., Karin, M., Kaufmann, T., Kepp, O., Kimchi, A., Kitsis, R.N., Klionsky, D.J., Knight, R.A., Kumar, S., Lee, S.W., Lemasters, J.J., Levine, B., Linkermann, A., Lipton, S.A., Lockshin, R.A., Lopez-Otin, C., Lowe, S.W., Luedde, T., Lugli, E., Macfarlane, M., Madeo, F., Malewicz, M., Malorni, W., Manic, G., et al., 2018. Molecular mechanisms of cell death: recommendations of the Nomenclature Committee on Cell Death 2018. Cell Death Differ. 25, 486–541.

Gandre-Babbe, S., Van Der Bliek, A.M., 2008. The novel tail-anchored membrane protein Mff controls mitochondrial and peroxisomal fission in mammalian cells. Mol. Biol. Cell 19, 2402–2412.

Ganesan, V., Perera, M.N., Colombini, D., Datskovskiy, D., Chadha, K., Colombini, M., 2010. Ceramide and activated Bax act synergistically to permeabilize the mitochondrial outer membrane. Apoptosis 15, 553–562.

Garcia-Saez, A.J., Mingarro, I., Perez-Paya, E., Salgado, J., 2004. Membrane-insertion fragments of Bcl-xL, Bax, and Bid. Biochemistry 43, 10930–10943.

Garcia-Saez, A.J., Coraiola, M., Dalla Serra, M., Mingarro, I., Menestrina, G., Salgado, J., 2005. Peptides derived from apoptotic Bax and Bid reproduce the poration activity of the parent full-length proteins. Biophys. J. 88, 3976–3990.

Garcia-Saez, A.J., Coraiola, M., Serra, M.D., Mingarro, I., Muller, P., Salgado, J., 2006. Peptides corresponding to helices 5 and 6 of Bax can independently form large lipid pores. FEBS J. 273, 971–981.

Garcia-Saez, A.J., Ries, J., Orzaez, M., Perez-Paya, E., Schwille, P., 2009. Membrane promotes tBID interaction with BCL(XL). Nat. Struct. Mol. Biol. 16, 1178–1185.

Garofalo, T., Misasi, R., Mattei, V., Giammarioli, A.M., Malorni, W., Pontieri, G.M., Pavan, A., Sorice, M., 2003. Association of the death-inducing signaling complex with microdomains after triggering through CD95/Fas. Evidence for caspase-8-ganglioside interaction in T cells. J. Biol. Chem. 278, 8309–8315.

Garofalo, T., Giammarioli, A.M., Misasi, R., Tinari, A., Manganelli, V., Gambardella, L., Pavan, A., Malorni, W., Sorice, M., 2005. Lipid microdomains contribute to apoptosis-associated modifications of mitochondria in T cells. Cell Death Differ. 12, 1378–1389.

Gawlowski, T., Suarez, J., Scott, B., Torres-Gonzalez, M., Wang, H., Schwappacher, R., Han, X., Yates 3rd, J.R., Hoshijima, M., Dillmann, W., 2012. Modulation of dynamin-related protein 1 (DRP1) function by increased O-linked-beta-N-acetylglucosamine modification (O-GlcNAc) in cardiac myocytes. J. Biol. Chem. 287, 30024–30034.

Gebert, N., Joshi, A.S., Kutik, S., Becker, T., McKenzie, M., Guan, X.L., Mooga, V.P., Stroud, D.A., Kulkarni, G., Wenk, M.R., Rehling, P., Meisinger, C., Ryan, M.T., Wiedemann, N., Greenberg, M.L., Pfanner, N., 2009. Mitochondrial cardiolipin involved in outer-membrane protein biogenesis: implications for Barth syndrome. Curr. Biol. 19, 2133–2139.

Giorgi, C., De Stefani, D., Bononi, A., Rizzuto, R., Pinton, P., 2009. Structural and functional link between the mitochondrial network and the endoplasmic reticulum. Int. J. Biochem. Cell Biol. 41, 1817–1827.

Gomez-Munoz, A., 2006. Ceramide 1-phosphate/ceramide, a switch between life and death. Biochim. Biophys. Acta 1758, 2049–2056.

Gong, Y.N., Guy, C., Crawford, J.C., Green, D.R., 2017a. Biological events and molecular signaling following MLKL activation during necroptosis. Cell Cycle 16, 1748–1760.

Gong, Y.N., Guy, C., Olauson, H., Becker, J.U., Yang, M., Fitzgerald, P., Linkermann, A., Green, D.R., 2017b. ESCRT-III acts downstream of MLKL to regulate necroptotic cell death and its consequences. Cell 169, 286–300.e16.

Gonzalvez, F., Gottlieb, E., 2007. Cardiolipin: setting the beat of apoptosis. Apoptosis 12, 877–885.
Gonzalvez, F., Pariselli, F., Dupaigne, P., Budihardjo, I., Lutter, M., Antonsson, B., Diolez, P., Manon, S., Martinou, J.C., Goubern, M., Wang, X., Bernard, S., Petit, P.X., 2005. tBid interaction with cardiolipin primarily orchestrates mitochondrial dysfunctions and subsequently activates Bax and Bak. Cell Death Differ. 12, 614–626.
Gonzalvez, F., Schug, Z.T., Houtkooper, R.H., Mackenzie, E.D., Brooks, D.G., Wanders, R.J., Petit, P.X., Vaz, F.M., Gottlieb, E., 2008. Cardiolipin provides an essential activating platform for caspase-8 on mitochondria. J. Cell Biol. 183, 681–696.
Gonzalvez, F., Pariselli, F., Jalmar, O., Dupaigne, P., Sureau, F., Dellinger, M., Hendrickson, E.A., Bernard, S., Petit, P.X., 2010. Mechanistic issues of the interaction of the hairpin-forming domain of tBid with mitochondrial cardiolipin. PLoS One 5, e9342.
Graham, J.M., Green, C., 1970. The properties of mitochondria enriched in vitro with cholesterol. Eur. J. Biochem. 12, 58–66.
Green, D.R., Kroemer, G., 2004. The pathophysiology of mitochondrial cell death. Science 305, 626–629.
Grijalba, M.T., Vercesi, A.E., Schreier, S., 1999. Ca^{2+}-induced increased lipid packing and domain formation in submitochondrial particles. A possible early step in the mechanism of Ca^{2+}-stimulated generation of reactive oxygen species by the respiratory chain. Biochemistry 38, 13279–13287.
Grosse, L., Ambree, O., Jorgens, S., Jawahar, M.C., Singhal, G., Stacey, D., Arolt, V., Baune, B.T., 2016. Cytokine levels in major depression are related to childhood trauma but not to recent stressors. Psychoneuroendocrinology 73, 24–31.
Grunewald, A., Kumar, K.R., Sue, C.M., 2019. New insights into the complex role of mitochondria in Parkinson's disease. Prog. Neurobiol. 177, 73–93.
Hajnoczky, G., Csordas, G., 2010. Calcium signalling: fishing out molecules of mitochondrial calcium transport. Curr. Biol. 20, R888–R891.
Hajnoczky, G., Csordas, G., Das, S., Garcia-Perez, C., Saotome, M., Sinha Roy, S., Yi, M., 2006. Mitochondrial calcium signalling and cell death: approaches for assessing the role of mitochondrial Ca^{2+} uptake in apoptosis. Cell Calcium 40, 553–560.
Harner, M., Korner, C., Walther, D., Mokranjac, D., Kaesmacher, J., Welsch, U., Griffith, J., Mann, M., Reggiori, F., Neupert, W., 2011. The mitochondrial contact site complex, a determinant of mitochondrial architecture. EMBO J. 30, 4356–4370.
Hatch, A.L., Gurel, P.S., Higgs, H.N., 2014. Novel roles for actin in mitochondrial fission. J. Cell Sci. 127, 4549–4560.
Heit, B., Yeung, T., Grinstein, S., 2011. Changes in mitochondrial surface charge mediate recruitment of signaling molecules during apoptosis. Am. J. Physiol. Cell Physiol. 300, C33–C41.
Hirabayashi, Y., Kwon, S.K., Paek, H., Pernice, W.M., Paul, M.A., Lee, J., Erfani, P., Raczkowski, A., Petrey, D.S., Pon, L.A., Polleux, F., 2017. ER-mitochondria tethering by PDZD8 regulates Ca^{2+} dynamics in mammalian neurons. Science 358, 623–630.
Hoppins, S., Nunnari, J., 2012. Cell Biology. Mitochondrial dynamics and apoptosis—the ER connection. Science 337, 1052–1054.
Horvath, S.E., Daum, G., 2013. Lipids of mitochondria. Prog. Lipid Res. 52, 590–614.
Hwang, M.S., Boulanger, J., Howe, J.D., Albecka, A., Pasche, M., Muresan, L., Modis, Y., 2019. MAVS polymers smaller than 80 nm induce mitochondrial membrane remodeling and interferon signaling. FEBS J. 286, 1543–1560.
Ingolfsson, H.I., Melo, M.N., Van Eerden, F.J., Arnarez, C., Lopez, C.A., Wassenaar, T.A., Periole, X., De Vries, A.H., Tieleman, D.P., Marrink, S.J., 2014. Lipid organization of the plasma membrane. J. Am. Chem. Soc. 136, 14554–14559.

Jalmar, O., Francois-Moutal, L., Garcia-Saez, A.J., Perry, M., Granjon, T., Gonzalvez, F., Gottlieb, E., Ayala-Sanmartin, J., Klosgen, B., Schwille, P., Petit, P.X., 2013. Caspase-8 binding to cardiolipin in giant unilamellar vesicles provides a functional docking platform for bid. PLoS One 8, e55250.

Ji, W.K., Hatch, A.L., Merrill, R.A., Strack, S., Higgs, H.N., 2015. Actin filaments target the oligomeric maturation of the dynamin GTPase Drp1 to mitochondrial fission sites. Elife 4, e11553.

Kagan, V.E., Borisenko, G.G., Tyurina, Y.Y., Tyurin, V.A., Jiang, J., Potapovich, A.I., Kini, V., Amoscato, A.A., Fujii, Y., 2004. Oxidative lipidomics of apoptosis: redox catalytic interactions of cytochrome c with cardiolipin and phosphatidylserine. Free Radic. Biol. Med. 37, 1963–1985.

Kagan, V.E., Tyurin, V.A., Jiang, J., Tyurina, Y.Y., Ritov, V.B., Amoscato, A.A., Osipov, A.N., Belikova, N.A., Kapralov, A.A., Kini, V., Vlasova, I., Zhao, Q., Zou, M., Di, P., Svistunenko, D.A., Kurnikov, I.V., Borisenko, G.G., 2005. Cytochrome c acts as a cardiolipin oxygenase required for release of proapoptotic factors. Nat. Chem. Biol. 1, 223–232.

Kagan, V.E., Mao, G., Qu, F., Angeli, J.P., Doll, S., Croix, C.S., Dar, H.H., Liu, B., Tyurin, V.A., Ritov, V.B., Kapralov, A.A., Amoscato, A.A., Jiang, J., Anthonymuthu, T., Mohammadyani, D., Yang, Q., Proneth, B., Klein-Seetharaman, J., Watkins, S., Bahar, I., Greenberger, J., Mallampalli, R.K., Stockwell, B.R., Tyurina, Y.Y., Conrad, M., Bayir, H., 2017. Oxidized arachidonic and adrenic PEs navigate cells to ferroptosis. Nat. Chem. Biol. 13, 81–90.

Kale, J., Osterlund, E.J., Andrews, D.W., 2018. BCL-2 family proteins: changing partners in the dance towards death. Cell Death Differ. 25, 65–80.

Kameoka, S., Adachi, Y., Okamoto, K., Iijima, M., Sesaki, H., 2018. Phosphatidic acid and cardiolipin coordinate mitochondrial dynamics. Trends Cell Biol. 28, 67–76.

Kawai, F., Shoda, M., Harashima, R., Sadaie, Y., Hara, H., Matsumoto, K., 2004. Cardiolipin domains in Bacillus subtilis marburg membranes. J. Bacteriol. 186, 1475–1483.

Kim, H., Tu, H.C., Ren, D., Takeuchi, O., Jeffers, J.R., Zambetti, G.P., Hsieh, J.J., Cheng, E.H., 2009. Stepwise activation of BAX and BAK by tBID, BIM, and PUMA initiates mitochondrial apoptosis. Mol. Cell 36, 487–499.

Kroemer, G., Galluzzi, L., Brenner, C., 2007. Mitochondrial membrane permeabilization in cell death. Physiol. Rev. 87, 99–163.

Kuwana, T., Mackey, M.R., Perkins, G., Ellisman, M.H., Latterich, M., Schneiter, R., Green, D.R., Newmeyer, D.D., 2002. Bid, Bax, and lipids cooperate to form supramolecular openings in the outer mitochondrial membrane. Cell 111, 331–342.

Lai, Y.C., Li, C.C., Sung, T.C., Chang, C.W., Lan, Y.J., Chiang, Y.W., 2019. The role of cardiolipin in promoting the membrane pore-forming activity of BAX oligomers. Biochim. Biophys. Acta. Biomembr. 1861, 268–280.

Landeta, O., Garcia Valero, J., Flores-Romero, H., Bustillo-Zabalbeitia, I., Landajuela, A., Garcia-Porras, M., Terrones, O., Basanez, G., 2014. Lipid-dependent bimodal MCL1 membrane activity. ACS Chem. Biol. 9, 2852–2863.

Leber, B., Lin, J., Andrews, D.W., 2007. Embedded together: the life and death consequences of interaction of the Bcl-2 family with membranes. Apoptosis 12, 897–911.

Li, Y., Feng, D., Wang, Z., et al., 2019a. Ischemia-induced ACSL4 activation contributes to ferroptosis-mediated tissue injury in intestinal ischemia/reperfusion. Cell Death Differ. 26, 2284–2299. https://doi.org/10.1038/s41418-019-0299-4.

Li, M., Mandal, A., Tyurin, V.A., Delucia, M., Ahn, J., Kagan, V.E., Van Der Wel, P.C.A., 2019b. Surface-binding to cardiolipin nanodomains triggers cytochrome c pro-apoptotic peroxidase activity via localized dynamics. Structure 27, 806–815.e4.

Lin, B., Kolluri, S.K., Lin, F., Liu, W., Han, Y.H., Cao, X., Dawson, M.I., Reed, J.C., Zhang, X.K., 2004. Conversion of Bcl-2 from protector to killer by interaction with nuclear orphan receptor Nur77/TR3. Cell 116, 527–540.

Lindsay, J., Esposti, M.D., Gilmore, A.P., 2011. Bcl-2 proteins and mitochondria—specificity in membrane targeting for death. Biochim. Biophys. Acta 1813, 532–539.
Linkermann, A., Brasen, J.H., Darding, M., Jin, M.K., Sanz, A.B., Heller, J.O., De Zen, F., Weinlich, R., Ortiz, A., Walczak, H., Weinberg, J.M., Green, D.R., Kunzendorf, U., Krautwald, S., 2013. Two independent pathways of regulated necrosis mediate ischemia-reperfusion injury. Proc. Natl. Acad. Sci. U. S. A. 110, 12024–12029.
Liu, J., Chen, J., Dai, Q., Lee, R.M., 2003a. Phospholipid scramblase 3 is the mitochondrial target of protein kinase C delta-induced apoptosis. Cancer Res. 63, 1153–1156.
Liu, X., Dai, S., Zhu, Y., Marrack, P., Kappler, J.W., 2003b. The structure of a Bcl-xL/Bim fragment complex: implications for Bim function. Immunity 19, 341–352.
Liu, S., Liu, H., Johnston, A., Hanna-Addams, S., Reynoso, E., Xiang, Y., Wang, Z., 2017. MLKL forms disulfide bond-dependent amyloid-like polymers to induce necroptosis. Proc. Natl. Acad. Sci. U. S. A. 114, E7450–E7459.
Lizardo, D.Y., Parisi, L.R., Li, N., Atilla-Gokcumen, G.E., 2018. Noncanonical roles of lipids in different cellular fates. Biochemistry 57, 22–29.
Llambi, F., Moldoveanu, T., Tait, S.W., Bouchier-Hayes, L., Temirov, J., McCormick, L.L., Dillon, C.P., Green, D.R., 2011. A unified model of mammalian BCL-2 protein family interactions at the mitochondria. Mol. Cell 44, 517–531.
Lovell, J.F., Billen, L.P., Bindner, S., Shamas-Din, A., Fradin, C., Leber, B., Andrews, D.W., 2008. Membrane binding by tBid initiates an ordered series of events culminating in membrane permeabilization by Bax. Cell 135, 1074–1084.
Lucken-Ardjomande, S., Montessuit, S., Martinou, J.C., 2008. Bax activation and stress-induced apoptosis delayed by the accumulation of cholesterol in mitochondrial membranes. Cell Death Differ. 15, 484–493.
Magtanong, L., Ko, P.J., Dixon, S.J., 2016. Emerging roles for lipids in non-apoptotic cell death. Cell Death Differ. 23, 1099–1109.
Magtanong, L., Ko, P.J., To, M., Cao, J.Y., Forcina, G.C., Tarangelo, A., Ward, C.C., Cho, K., Patti, G.J., Nomura, D.K., Olzmann, J.A., Dixon, S.J., 2019. Exogenous monounsaturated fatty acids promote a ferroptosis-resistant cell state. Cell. Chem. Biol. 26, 420–432.e9.
Malorni, W., Giammarioli, A.M., Garofalo, T., Sorice, M., 2007. Dynamics of lipid raft components during lymphocyte apoptosis: the paradigmatic role of GD3. Apoptosis 12, 941–949.
Mannella, C.A., 2006. Structure and dynamics of the mitochondrial inner membrane cristae. Biochim. Biophys. Acta 1763, 542–548.
Martinou, J.C., Youle, R.J., 2006. Which came first, the cytochrome c release or the mitochondrial fission? Cell Death Differ. 13, 1291–1295.
Martinou, J.C., Youle, R.J., 2011. Mitochondria in apoptosis: Bcl-2 family members and mitochondrial dynamics. Dev. Cell 21, 92–101.
McArthur, K., Whitehead, L.W., Heddleston, J.M., Li, L., Padman, B.S., Oorschot, V., Geoghegan, N.D., Chappaz, S., Davidson, S., San Chin, H., Lane, R.M., Dramicanin, M., Saunders, T.L., Sugiana, C., Lessene, R., Osellame, L.D., Chew, T.L., Dewson, G., Lazarou, M., Ramm, G., Lessene, G., Ryan, M.T., Rogers, K.L., Van Delft, M.F., Kile, B.T., 2018. BAK/BAX macropores facilitate mitochondrial herniation and mtDNA efflux during apoptosis. Science 359, eaao6047.
McNamara, D.E., Quarato, G., Guy, C.S., Green, D.R., Moldoveanu, T., 2018. Characterization of MLKL-mediated plasma membrane rupture in necroptosis. J. Vis. Exp. (138). https://doi.org/10.3791/58088.
Menoret, E., Gomez-Bougie, P., Surget, S., Trichet, V., Oliver, L., Pellat-Deceunynck, C., Amiot, M., 2010. Mcl-1(128-350) fragment induces apoptosis through direct interaction with Bax. FEBS Lett. 584, 487–492.

Michels, J., O'neill, J.W., Dallman, C.L., Mouzakiti, A., Habens, F., Brimmell, M., Zhang, K.Y., Craig, R.W., Marcusson, E.G., Johnson, P.W., Packham, G., 2004. Mcl-1 is required for Akata6 B-lymphoma cell survival and is converted to a cell death molecule by efficient caspase-mediated cleavage. Oncogene 23, 4818–4827.

Moldoveanu, T., Liu, Q., Tocilj, A., Watson, M., Shore, G., Gehring, K., 2006. The X-ray structure of a BAK homodimer reveals an inhibitory zinc binding site. Mol. Cell 24, 677–688.

Moldoveanu, T., Grace, C.R., Llambi, F., Nourse, A., Fitzgerald, P., Gehring, K., Kriwacki, R.W., Green, D.R., 2013. BID-induced structural changes in BAK promote apoptosis. Nat. Struct. Mol. Biol. 20, 589–597.

Montero, J., Morales, A., Llacuna, L., Lluis, J.M., Terrones, O., Basanez, G., Antonsson, B., Prieto, J., Garcia-Ruiz, C., Colell, A., Fernandez-Checa, J.C., 2008. Mitochondrial cholesterol contributes to chemotherapy resistance in hepatocellular carcinoma. Cancer Res. 68, 5246–5256.

Moro, K., Nagahashi, M., Gabriel, E., Takabe, K., Wakai, T., 2019. Clinical application of ceramide in cancer treatment. Breast Cancer 26, 407–415.

Muchmore, S.W., Sattler, M., Liang, H., Meadows, R.P., Harlan, J.E., Yoon, H.S., Nettesheim, D., Chang, B.S., Thompson, C.B., Wong, S.L., Ng, S.L., Fesik, S.W., 1996. X-ray and NMR structure of human Bcl-xL, an inhibitor of programmed cell death. Nature 381, 335–341.

Nagata, S., Tanaka, M., 2017. Programmed cell death and the immune system. Nat. Rev. Immunol. 17, 333–340.

Nechushtan, A., Smith, C.L., Hsu, Y.T., Youle, R.J., 1999. Conformation of the Bax C-terminus regulates subcellular location and cell death. EMBO J. 18, 2330–2341.

Obeid, L.M., Linardic, C.M., Karolak, L.A., Hannun, Y.A., 1993. Programmed cell death induced by ceramide. Science 259, 1769–1771.

Olichon, A., Emorine, L.J., Descoins, E., Pelloquin, L., Brichese, L., Gas, N., Guillou, E., Delettre, C., Valette, A., Hamel, C.P., Ducommun, B., Lenaers, G., Belenguer, P., 2002. The human dynamin-related protein OPA1 is anchored to the mitochondrial inner membrane facing the inter-membrane space. FEBS Lett. 523, 171–176.

Olichon, A., Baricault, L., Gas, N., Guillou, E., Valette, A., Belenguer, P., Lenaers, G., 2003. Loss of OPA1 perturbates the mitochondrial inner membrane structure and integrity, leading to cytochrome c release and apoptosis. J. Biol. Chem. 278, 7743–7746.

Ong, S.B., Hausenloy, D.J., 2017. Mitochondrial dynamics as a therapeutic target for treating cardiac diseases. Handb. Exp. Pharmacol. 240, 251–279.

Ortiz, A., Killian, J.A., Verkleij, A.J., Wilschut, J., 1999. Membrane fusion and the lamellar-to-inverted-hexagonal phase transition in cardiolipin vesicle systems induced by divalent cations. Biophys. J. 77, 2003–2014.

Osman, C., Voelker, D.R., Langer, T., 2011. Making heads or tails of phospholipids in mitochondria. J. Cell Biol. 192, 7–16.

Otera, H., Wang, C., Cleland, M.M., Setoguchi, K., Yokota, S., Youle, R.J., Mihara, K., 2010. Mff is an essential factor for mitochondrial recruitment of Drp1 during mitochondrial fission in mammalian cells. J. Cell Biol. 191, 1141–1158.

Ott, M., Robertson, J.D., Gogvadze, V., Zhivotovsky, B., Orrenius, S., 2002. Cytochrome c release from mitochondria proceeds by a two-step process. Proc. Natl. Acad. Sci. U. S. A. 99, 1259–1263.

Ousingsawat, J., Cabrita, I., Wanitchakool, P., Sirianant, L., Krautwald, S., Linkermann, A., Schreiber, R., Kunzelmann, K., 2017. Ca(2+) signals, cell membrane disintegration, and activation of TMEM16F during necroptosis. Cell. Mol. Life Sci. 74, 173–181.

Ow, Y.P., Green, D.R., Hao, Z., Mak, T.W., 2008. Cytochrome c: functions beyond respiration. Nat. Rev. Mol. Cell Biol. 9, 532–542.

Palmer, C.S., Osellame, L.D., Laine, D., Koutsopoulos, O.S., Frazier, A.E., Ryan, M.T., 2011. MiD49 and MiD51, new components of the mitochondrial fission machinery. EMBO Rep. 12, 565–573.

Parisi, L.R., Li, N., Atilla-Gokcumen, G.E., 2017. Very long chain fatty acids are functionally involved in necroptosis. Cell. Chem. Biol. 24, 1445–1454.e8.

Parisi, L.R., Morrow, L.M., Visser, M.B., Atilla-Gokcumen, G.E., 2018. Turning the spotlight on lipids in non-apoptotic cell death. ACS Chem. Biol. 13, 506–515.

Patwardhan, G.A., Beverly, L.J., Siskind, L.J., 2016. Sphingolipids and mitochondrial apoptosis. J. Bioenerg. Biomembr. 48, 153–168.

Pecot, J., Maillet, L., Le Pen, J., Vuillier, C., Trecesson, S.C., Fetiveau, A., Sarosiek, K.A., Bock, F.J., Braun, F., Letai, A., Tait, S.W.G., Gautier, F., Juin, P.P., 2016. Tight sequestration of BH3 proteins by BCL-xL at subcellular membranes contributes to apoptotic resistance. Cell Rep. 17, 3347–3358.

Peter, B.J., Kent, H.M., Mills, I.G., Vallis, Y., Butler, P.J., Evans, P.R., McMahon, H.T., 2004. BAR domains as sensors of membrane curvature: the amphiphysin BAR structure. Science 303, 495–499.

Petit, P.X., Dupaigne, P., Pariselli, F., Gonzalvez, F., Etienne, F., Rameau, C., Bernard, S., 2009. Interaction of the alpha-helical H6 peptide from the pro-apoptotic protein tBid with cardiolipin. FEBS J. 276, 6338–6354.

Petrie, E.J., Hildebrand, J.M., Murphy, J.M., 2017. Insane in the membrane: a structural perspective of MLKL function in necroptosis. Immunol. Cell Biol. 95, 152–159.

Petros, A.M., Nettesheim, D.G., Wang, Y., Olejniczak, E.T., Meadows, R.P., Mack, J., Swift, K., Matayoshi, E.D., Zhang, H., Thompson, C.B., Fesik, S.W., 2000. Rationale for Bcl-xL/Bad peptide complex formation from structure, mutagenesis, and biophysical studies. Protein Sci. 9, 2528–2534.

Phan, T.K., Williams, S.A., Bindra, G.K., Lay, F.T., Poon, I.K.H., Hulett, M.D., 2019. Phosphoinositides: multipurpose cellular lipids with emerging roles in cell death. Cell Death Differ. 26, 781–793.

Plati, J., Bucur, O., Khosravi-Far, R., 2011. Apoptotic cell signaling in cancer progression and therapy. Integr. Biol. (Camb) 3, 279–296.

Pokorny, A., Almeida, P.F., 2004. Kinetics of dye efflux and lipid flip-flop induced by delta-lysin in phosphatidylcholine vesicles and the mechanism of graded release by amphipathic, alpha-helical peptides. Biochemistry 43, 8846–8857.

Proneth, B., Conrad, M., 2019. Ferroptosis and necroinflammation, a yet poorly explored link. Cell Death Differ. 26, 14–24.

Qian, S., Wang, W., Yang, L., Huang, H.W., 2008. Structure of transmembrane pore induced by Bax-derived peptide: evidence for lipidic pores. Proc. Natl. Acad. Sci. U. S. A. 105, 17379–17383.

Quarato, G., Guy, C.S., Grace, C.R., Llambi, F., Nourse, A., Rodriguez, D.A., Wakefield, R., Frase, S., Moldoveanu, T., Green, D.R., 2016. Sequential engagement of distinct MLKL phosphatidylinositol-binding sites executes necroptosis. Mol. Cell 61, 589–601.

Raemy, E., Montessuit, S., Pierredon, S., Van Kampen, A.H., Vaz, F.M., Martinou, J.C., 2016. Cardiolipin or MTCH2 can serve as tBID receptors during apoptosis. Cell Death Differ. 23, 1165–1174.

Reichert, A.S., Neupert, W., 2002. Contact sites between the outer and inner membrane of mitochondria-role in protein transport. Biochim. Biophys. Acta 1592, 41–49.

Riley, J.S., Quarato, G., Cloix, C., Lopez, J., O'prey, J., Pearson, M., Chapman, J., Sesaki, H., Carlin, L.M., Passos, J.F., Wheeler, A.P., Oberst, A., Ryan, K.M., Tait, S.W., 2018. Mitochondrial inner membrane permeabilisation enables mtDNA release during apoptosis. EMBO J. 37, e99238.

Rippo, M.R., Malisan, F., Ravagnan, L., Tomassini, B., Condo, I., Costantini, P., Susin, S.A., Rufini, A., Todaro, M., Kroemer, G., Testi, R., 2000. GD3 ganglioside directly targets mitochondria in a bcl-2-controlled fashion. FASEB J. 14, 2047–2054.

Ros, U., Pena-Blanco, A., Hanggi, K., Kunzendorf, U., Krautwald, S., Wong, W.W., Garcia-Saez, A.J., 2017. Necroptosis execution is mediated by plasma membrane nanopores independent of calcium. Cell Rep. 19, 175–187.

Salvador-Gallego, R., Mund, M., Cosentino, K., Schneider, J., Unsay, J., Schraermeyer, U., Engelhardt, J., Ries, J., Garcia-Saez, A.J., 2016. Bax assembly into rings and arcs in apoptotic mitochondria is linked to membrane pores. EMBO J. 35, 389–401.

Salvador-Gil, D., Ortego, L., Herrera, R.P., Marzo, I., Gimeno, M.C., 2017. Highly active group 11 metal complexes with alpha-hydrazidophosphonate ligands. Dalton Trans. 46, 13745–13755.

Schellenberg, B., Wang, P., Keeble, J.A., Rodriguez-Enriquez, R., Walker, S., Owens, T.W., Foster, F., Tanianis-Hughes, J., Brennan, K., Streuli, C.H., Gilmore, A.P., 2013. Bax exists in a dynamic equilibrium between the cytosol and mitochondria to control apoptotic priming. Mol. Cell 49, 959–971.

Schlame, M., Ren, M., 2009. The role of cardiolipin in the structural organization of mitochondrial membranes. Biochim. Biophys. Acta 1788, 2080–2083.

Schlattner, U., Tokarska-Schlattner, M., Ramirez, S., Tyurina, Y.Y., Amoscato, A.A., Mohammadyani, D., Huang, Z., Jiang, J., Yanamala, N., Seffouh, A., Boissan, M., Epand, R.F., Epand, R.M., Klein-Seetharaman, J., Lacombe, M.L., Kagan, V.E., 2013. Dual function of mitochondrial Nm23-H4 protein in phosphotransfer and intermembrane lipid transfer: a cardiolipin-dependent switch. J. Biol. Chem. 288, 111–121.

Schug, Z.T., Gottlieb, E., 2009. Cardiolipin acts as a mitochondrial signalling platform to launch apoptosis. Biochim. Biophys. Acta 1788, 2022–2031.

Scorrano, L., Ashiya, M., Buttle, K., Weiler, S., Oakes, S.A., Mannella, C.A., Korsmeyer, S.J., 2002. A distinct pathway remodels mitochondrial cristae and mobilizes cytochrome c during apoptosis. Dev. Cell 2, 55–67.

Seiler, A., Schneider, M., Förster, H., Roth, S., Wirth, E.K., Culmsee, C., Plesnila, N., Kremmer, E., Rådmark, O., Wurst, W., Bornkamm, G.W., Schweizer, U., Conrad, M., 2008. Glutathione peroxidase 4 senses and translates oxidative stress into 12/15-lipoxygenase dependent- and AIF-mediated cell death. Cell Metab. 8 (3), 237–248. https://doi.org/10.1016/j.cmet.2008.07.005. 18762024.

Selzner, M., Bielawska, A., Morse, M.A., Rudiger, H.A., Sindram, D., Hannun, Y.A., Clavien, P.A., 2001. Induction of apoptotic cell death and prevention of tumor growth by ceramide analogues in metastatic human colon cancer. Cancer Res. 61, 1233–1240.

Shah, S.Z.A., Zhao, D., Hussain, T., Yang, L., 2017. Role of the AMPK pathway in promoting autophagic flux via modulating mitochondrial dynamics in neurodegenerative diseases: insight into prion diseases. Ageing Res. Rev. 40, 51–63.

Shamas-Din, A., Kale, J., Leber, B., Andrews, D.W., 2013. Mechanisms of action of Bcl-2 family proteins. Cold Spring Harb. Perspect. Biol. 5, a008714.

Shintoku, R., Takigawa, Y., Yamada, K., Kubota, C., Yoshimoto, Y., Takeuchi, T., Koshiishi, I., Torii, S., 2017. Lipoxygenase-mediated generation of lipid peroxides enhances ferroptosis induced by erastin and RSL3. Cancer Sci. 108, 2187–2194.

Simbeni, R., Pon, L., Zinser, E., Paltauf, F., Daum, G., 1991. Mitochondrial membrane contact sites of yeast. Characterization of lipid components and possible involvement in intramitochondrial translocation of phospholipids. J. Biol. Chem. 266, 10047–10049.

Singh, A., Kukreti, R., Saso, L., Kukreti, S., 2019. Oxidative stress: a key modulator in neurodegenerative diseases. Molecules 24, E1583. https://doi.org/10.3390/molecules24081583.

Siskind, L.J., 2005. Mitochondrial ceramide and the induction of apoptosis. J. Bioenerg. Biomembr. 37, 143–153.

Siskind, L.J., Mullen, T.D., Romero Rosales, K., Clarke, C.J., Hernandez-Corbacho, M.J., Edinger, A.L., Obeid, L.M., 2010. The BCL-2 protein Bak is required for long-chain ceramide generation during apoptosis. J. Biol. Chem. 285, 11818–11826.

Sorice, M., Manganelli, V., Matarrese, P., Tinari, A., Misasi, R., Malorni, W., Garofalo, T., 2009. Cardiolipin-enriched raft-like microdomains are essential activating platforms for apoptotic signals on mitochondria. FEBS Lett. 583, 2447–2450.

Stockwell, B.R., Angeli, J.P.F., Bayir, H., Bush, A.I., Conrad, M., Dixon, S.J., Fulda, S., Gascon, S., Hatzios, S.K., Kagan, V.E., Noel, K., Jiang, X.J., Linkermann, A., Murphy, M.E., Overholtzer, M., Oyagi, A., Pagnussat, G.C., Park, J., Ran, Q., Rosenfeld, C.S., Salnikow, K., Tang, D.L., Torti, F.M., Torti, S.V., Toyokuni, S., Woerpel, K.A., Zhang, D.D., 2017. Ferroptosis: a regulated cell death nexus linking metabolism, redox biology, and disease. Cell 171, 273–285.

Strack, S., Cribbs, J.T., 2012. Allosteric modulation of Drp1 mechanoenzyme assembly and mitochondrial fission by the variable domain. J. Biol. Chem. 287, 10990–11001.

Su, Z., Yang, Z., Xie, L., Dewitt, J.P., Chen, Y., 2016. Cancer therapy in the necroptosis era. Cell Death Differ. 23, 748–756.

Suarez-Rivero, J.M., Villanueva-Paz, M., De La Cruz-Ojeda, P., De La Mata, M., Cotan, D., Oropesa-Avila, M., De Lavera, I., Alvarez-Cordoba, M., Luzon-Hidalgo, R., Sanchez-Alcazar, J.A., 2016. Mitochondrial dynamics in mitochondrial diseases. Diseases 5, E1. https://doi.org/10.3390/diseases5010001.

Subburaj, Y., Cosentino, K., Axmann, M., Pedrueza-Villalmanzo, E., Hermann, E., Bleicken, S., Spatz, J., Garcia-Saez, A.J., 2015. Bax monomers form dimer units in the membrane that further self-assemble into multiple oligomeric species. Nat. Commun. 6, 8042.

Suen, D.F., Norris, K.L., Youle, R.J., 2008. Mitochondrial dynamics and apoptosis. Genes Dev. 22, 1577–1590.

Suzuki, M., Youle, R.J., Tjandra, N., 2000. Structure of Bax: coregulation of dimer formation and intracellular localization. Cell 103, 645–654.

Taguchi, N., Ishihara, N., Jofuku, A., Oka, T., Mihara, K., 2007. Mitotic phosphorylation of dynamin-related GTPase Drp1 participates in mitochondrial fission. J. Biol. Chem. 282, 11521–11529.

Tait, S.W., Parsons, M.J., Llambi, F., Bouchier-Hayes, L., Connell, S., Munoz-Pinedo, C., Green, D.R., 2010. Resistance to caspase-independent cell death requires persistence of intact mitochondria. Dev. Cell 18, 802–813.

Tamura, Y., Endo, T., Iijima, M., Sesaki, H., 2009. Ups1p and Ups2p antagonistically regulate cardiolipin metabolism in mitochondria. J. Cell Biol. 185, 1029–1045.

Taraborrelli, L., Peltzer, N., Montinaro, A., Kupka, S., Rieser, E., Hartwig, T., Sarr, A., Darding, M., Draber, P., Haas, T.L., Akarca, A., Marafioti, T., Pasparakis, M., Bertin, J., Gough, P.J., Bouillet, P., Strasser, A., Leverkus, M., Silke, J., Walczak, H., 2018. LUBAC prevents lethal dermatitis by inhibiting cell death induced by TNF, TRAIL and CD95L. Nat. Commun. 9, 3910.

Terrones, O., Antonsson, B., Yamaguchi, H., Wang, H.G., Liu, J., Lee, R.M., Herrmann, A., Basanez, G., 2004. Lipidic pore formation by the concerted action of proapoptotic BAX and tBID. J. Biol. Chem. 279, 30081–30091.

Todt, F., Cakir, Z., Reichenbach, F., Emschermann, F., Lauterwasser, J., Kaiser, A., Ichim, G., Tait, S.W., Frank, S., Langer, H.F., Edlich, F., 2015. Differential retrotranslocation of mitochondrial Bax and Bak. EMBO J. 34, 67–80.

Tonnus, W., Meyer, C., Paliege, A., Belavgeni, A., Von Massenhausen, A., Bornstein, S.R., Hugo, C., Becker, J.U., Linkermann, A., 2019. The pathological features of regulated necrosis. J. Pathol. 247, 697–707.

Tuo, Q., Lei, P., Jackman, K., et al., 2017. Tau-mediated iron export prevents ferroptotic damage after ischemic stroke. Mol. Psychiatry 22, 1520–1530. https://doi.org/10.1038/mp.2017.171.

Ueda, N., 2015. Ceramide-induced apoptosis in renal tubular cells: a role of mitochondria and sphingosine-1-phosphate. Int. J. Mol. Sci. 16, 5076–5124.
Ugarte-Uribe, B., Garcia-Saez, A.J., 2017. Apoptotic foci at mitochondria: in and around Bax pores. Philos. Trans. R. Soc. Lond. B Biol. Sci. 372(1726), 20160217. https://doi.org/10.1098/rstb.2016.0217.
Unsay, J.D., Cosentino, K., Subburaj, Y., Garcia-Saez, A.J., 2013. Cardiolipin effects on membrane structure and dynamics. Langmuir 29, 15878–15887.
Upton, J.W., Kaiser, W.J., Mocarski, E.S., 2012. DAI/ZBP1/DLM-1 complexes with RIP3 to mediate virus-induced programmed necrosis that is targeted by murine cytomegalovirus vIRA. Cell Host Microbe 11, 290–297.
Valera-Alberni, M., Canto, C., 2018. Mitochondrial stress management: a dynamic journey. Cell Stress 2, 253–274.
Van Brocklyn, J.R., Williams, J.B., 2012. The control of the balance between ceramide and sphingosine-1-phosphate by sphingosine kinase: oxidative stress and the seesaw of cell survival and death. Comp. Biochem. Physiol. B Biochem. Mol. Biol. 163, 26–36.
Van Meer, G., De Kroon, A.I., 2011. Lipid map of the mammalian cell. J. Cell Sci. 124, 5–8.
Van Meer, G., Voelker, D.R., Feigenson, G.W., 2008. Membrane lipids: where they are and how they behave. Nat. Rev. Mol. Cell Biol. 9, 112–124.
Vance, J.E., 1990. Phospholipid synthesis in a membrane fraction associated with mitochondria. J. Biol. Chem. 265, 7248–7256.
Vince, J.E., Silke, J., 2016. The intersection of cell death and inflammasome activation. Cell. Mol. Life Sci. 73, 2349–2367.
Vyas, S., Zaganjor, E., Haigis, M.C., 2016. Mitochondria and cancer. Cell 166, 555–566.
Wallach, D., Kang, T.B., Dillon, C.P., Green, D.R., 2016. Programmed necrosis in inflammation: toward identification of the effector molecules. Science 352, aaf2154.
Wang, H., Sun, L., Su, L., Rizo, J., Liu, L., Wang, L.F., Wang, F.S., Wang, X., 2014a. Mixed lineage kinase domain-like protein MLKL causes necrotic membrane disruption upon phosphorylation by RIP3. Mol. Cell 54, 133–146.
Wang, X., Li, Y., Liu, S., Yu, X., Li, L., Shi, C., He, W., Li, J., Xu, L., Hu, Z., Yu, L., Yang, Z., Chen, Q., Ge, L., Zhang, Z., Zhou, B., Jiang, X., Chen, S., He, S., 2014b. Direct activation of RIP3/MLKL-dependent necrosis by herpes simplex virus 1 (HSV-1) protein ICP6 triggers host antiviral defense. Proc. Natl. Acad. Sci. U. S. A. 111, 15438–15443.
Wang, Y., Gao, W., Shi, X., Ding, J., Liu, W., He, H., Wang, K., Shao, F., 2017. Chemotherapy drugs induce pyroptosis through caspase-3 cleavage of a gasdermin. Nature 547, 99–103.
Wenzel, S.E., Tyurina, Y.Y., Zhao, J.M., Croix, C.M.S., Dar, H.H., Mao, G.W., Tyurin, V.A., Anthonymuthu, T.S., Kapralov, A.A., Amoscato, A.A., Mikulska-Ruminska, K., Shrivastava, I.H., Kenny, E.M., Yang, Q., Rosenbaum, J.C., Sparvero, L.J., Emlet, D.R., Wen, X.Y., Minami, Y., Qu, F., Watkins, S.C., Holman, T.R., Vandemark, A.P., Kellum, J.A., Bahar, I., Bayir, H., Kagan, V.E., 2017. PEBP1 wardens ferroptosis by enabling lipoxygenase generation of lipid death signals. Cell 171, 628.
Westphal, D., Dewson, G., Menard, M., Frederick, P., Iyer, S., Bartolo, R., Gibson, L., Czabotar, P.E., Smith, B.J., Adams, J.M., Kluck, R.M., 2014. Apoptotic pore formation is associated with in-plane insertion of Bak or Bax central helices into the mitochondrial outer membrane. Proc. Natl. Acad. Sci. U. S. A. 111, E4076–E4085.
Wong-Ekkabut, J., Xu, Z., Triampo, W., Tang, I.M., Tieleman, D.P., Monticelli, L., 2007. Effect of lipid peroxidation on the properties of lipid bilayers: a molecular dynamics study. Biophys. J. 93, 4225–4236.
Xia, B., Fang, S., Chen, X., Hu, H., Chen, P., Wang, H., Gao, Z., 2016. MLKL forms cation channels. Cell Res. 26, 517–528.

Yang, W.S., Kim, K.J., Gaschler, M.M., Patel, M., Shchepinov, M.S., Stockwell, B.R., 2016. Peroxidation of polyunsaturated fatty acids by lipoxygenases drives ferroptosis. Proc. Natl. Acad. Sci. U. S. A. 113, E4966–E4975.

Yethon, J.A., Epand, R.F., Leber, B., Epand, R.M., Andrews, D.W., 2003. Interaction with a membrane surface triggers a reversible conformational change in Bax normally associated with induction of apoptosis. J. Biol. Chem. 278, 48935–48941.

Yeung, T., Terebiznik, M., Yu, L., Silvius, J., Abidi, W.M., Philips, M., Levine, T., Kapus, A., Grinstein, S., 2006. Receptor activation alters inner surface potential during phagocytosis. Science 313, 347–351.

Ying, Z., Pan, C., Shao, T., Liu, L., Li, L., Guo, D., Zhang, S., Yuan, T., Cao, R., Jiang, Z., Chen, S., Wang, F., Wang, X., 2018. Mixed lineage kinase domain-like protein MLKL breaks down myelin following nerve injury. Mol. Cell 72, 457–468 e5.

Youle, R.J., Strasser, A., 2008. The BCL-2 protein family: opposing activities that mediate cell death. Nat. Rev. Mol. Cell Biol. 9, 47–59.

Zamzami, N., Marchetti, P., Castedo, M., Decaudin, D., Macho, A., Hirsch, T., Susin, S.A., Petit, P.X., Mignotte, B., Kroemer, G., 1995. Sequential reduction of mitochondrial transmembrane potential and generation of reactive oxygen species in early programmed cell death. J. Exp. Med. 182, 367–377.

Zargarian, S., Shlomovitz, I., Erlich, Z., Hourizadeh, A., Ofir-Birin, Y., Croker, B.A., Regev-Rudzki, N., Edry-Botzer, L., Gerlic, M., 2017. Phosphatidylserine externalization, "necroptotic bodies" release, and phagocytosis during necroptosis. PLoS Biol. 15, e2002711.

Zhang, K., Wang, H., 2015. Cancer Genome Atlas Pan-cancer analysis project. Zhongguo Fei Ai Za Zhi 18, 219–223.

Zhang, Z., Lapolla, S.M., Annis, M.G., Truscott, M., Roberts, G.J., Miao, Y., Shao, Y., Tan, C., Peng, J., Johnson, A.E., Zhang, X.C., Andrews, D.W., Lin, J., 2004. Bcl-2 homodimerization involves two distinct binding surfaces, a topographic arrangement that provides an effective mechanism for Bcl-2 to capture activated Bax. J. Biol. Chem. 279, 43920–43928.

Zhang, T., Zhang, Y., Cui, M., Jin, L., Wang, Y., Lv, F., Liu, Y., Zheng, W., Shang, H., Zhang, J., Zhang, M., Wu, H., Guo, J., Zhang, X., Hu, X., Cao, C.M., Xiao, R.P., 2016. CaMKII is a RIP3 substrate mediating ischemia- and oxidative stress-induced myocardial necroptosis. Nat. Med. 22, 175–182.

Zhang, Y., Chen, X., Gueydan, C., Han, J., 2018. Plasma membrane changes during programmed cell deaths. Cell Res. 28, 9–21.

Zhivotovsky, B., Orrenius, S., 2011. Calcium and cell death mechanisms: a perspective from the cell death community. Cell Calcium 50, 211–221.

FSC
www.fsc.org
MIX
Papier | Fördert
gute Waldnutzung
FSC® C083411

Kolibri 360 GmbH
Ferdinand-Jühlke-Straße 7
99095 Erfurt, Deutschland
produktsicherheit@kolibri360.de